Fundamental Theories of Physics

Volume 167

For further volumes:
http://www.springer.com/series/6001

Preface

In 1908 Hermann Minkowski gave the four-dimensional (spacetime) formulation of special relativity [1]. In fact, Henri Poincaré [2] first noticed in 1906 that the Lorentz transformations had a geometric interpretation as rotations in a four-dimensional space with time as the fourth dimension. However it was Minkowski, who successfully decoded the profound message about the dimensionality of the world hidden in the relativity postulate, which reflects the experimental fact that natural laws are the same in all inertial reference frames. Unlike Poincaré, Minkowski did not regard spacetime – the unification of space and time – as a convenient *mathematical* space, but insisted that this absolute four-dimensional world, as Minkowski called it, represents physical phenomena and the world more adequately than the relativity postulate: "the word *relativity-postulate*... seems to me very feeble. Since the postulate comes to mean that only the four-dimensional world in space and time is given by the phenomena... I prefer to call it the *postulate of the absolute world*" [3].

The impact of Minkowski's ideas on the twentieth century physics has been so immense that one cannot imagine modern physics without the notion of spacetime. It would hardly be an exaggeration to say that spacetime has been the greatest discovery in physics of all times. The only other discovery that comes close to spacetime is Einstein's general relativity, which revealed that gravity is a manifestation of the curvature of spacetime. But it was the discovery of spacetime, which paved the way for this deep understanding of what gravity really is. Einstein saw the link between the geometry of spacetime and gravitation only after he overcame his initial hostile attitude toward the notion of spacetime.

The implications of Minkowski's revolutionary ideas of space and time for philosophy, and especially for the philosophy of space and time, have also been enormous. I think just one example will suffice to demonstrate the extent of those implications. The views of time flow, becoming, and ultimately of what exists are all defined in terms of *simultaneity*. For instance, the present – the three-dimensional world at the present moment – is defined as everything that exists *simultaneously* at the moment 'now'. When Einstein published his special relativity in 1905 the implications of one of its major consequences – *relativity of simultaneity* – for the view of reality had not been immediately realized. But it is now evident that the widely held presentist view of the world contradicts relativity of simultaneity – on the presentist view it is only the present (the class of *absolutely simultaneous* events at the present moment) that exists, whereas according to special relativity two observers in relative motion have different classes of simultaneous events. The temptation to interpret the fact that every observer has a different class of simultaneous events in a sense that what exists is *relative* to an observer could hardly be defended, especially in view of the understanding, stressed by Minkowski, that the very appearance of relative quantities in a theory is a manifestation of the existence of an absolute underlying reality. And indeed, the *different* presents (the different classes of simultaneous events) of two observers in relative motion are merely different three-dimensional 'cross-sections' of spacetime and the truly challenging question then is – Is reality an absolute four-dimensional world?

In 2008 the one hundredth anniversary of Minkowski's talk "Space and Time" given on September 21, 1908 in Cologne provided an excellent opportunity to commemorate his major contribution to physics and its profound implications for physics, philosophy, and our entire worldview. There were several events which marked this anniversary. The Third International Conference on the Nature and Ontology of Spacetime (http://www.spacetimesociety.org/conferences/2008/) held at Concordia University, Montreal, on June 13–15, 2008 was dedicated to the centennial anniversary of Minkowski's talk. On September 7–12, 2008 the 414th WE-Heraeus-Seminar (http://www.uni-koeln.de/minkowski/) also commemorated Minkowski's famous lecture at its meeting "Space and Time 100 Years after Minkowski" in the Physikzentrum Bad Honnef, Germany. The September–October 2008 special issue of *Annalen der Physik* "The Minkowski spacetime of special relativity theory – 100 years after its discovery" [5] was dedicated to the memory of Minkowski. The volume *Minkowski Spacetime: A Hundred Years Later* [4] published in the Springer series *Fundamental Theories of Physics* contains a new translation of Minkowski's talk "Raum und Zeit", accompanied by the original German version, and papers by physicists specifically written on the occasion of Minkowski's anniversary.

This volume is part of the celebration of the centennial anniversary of spacetime and compliments [4] by exploring the implications of Minkowski's discovery for issues in physics not covered in [4] and most importantly for the physical foundations and the philosophy of space and time, which alone warrants the publication of a separate collection of papers. The volume contains selected papers by physicists and philosophers, most of which were presented at the Third International Conference on the Nature and Ontology of Spacetime. One of the selection criteria was to have examples of the influence of Minkowski's ideas on different issues in physics, philosophy, and other disciplines. The first six papers, comprising Part I of the book, provide examples of the impact of Minkowski's spacetime representation of special relativity on the twentieth century physics. Part II also contains six papers which deal with implications of Minkowski's ideas for the philosophy of space and time. The last part is represented by two papers which explore the influence of Minkowski's ideas beyond the philosophy of space and time.

Montreal
June 2010

Vesselin Petkov

References

1. H. Minkowski, "Raum und Zeit", *Physikalische Zeitschrift* **10** (1909) pp 104–111; *Jahresbericht der Deutschen Mathematiker-Vereinigung* **18** (1909) pp 75–88
2. H. Poincaré, "Sur la dynamique de l'électron", *Rendiconti del Circolo matematico Rendiconti del Circolo di Palermo* **21** (1906) pp 129–176
3. H. Minkowski, "Space and Time." New translation in [4] (pp xiv–xlii) p xxv
4. V. Petkov (ed.), *Minkowski Spacetime: A Hundred Years Later* (Springer, Heidelberg 2010)
5. *Ann. Phys.* (Berlin) **17**, No. 9–10, pp 613–851 (2008)

Contents

Contributors

Richard T.W. Arthur Department of Philosophy, McMaster University, Hamilton, ON, Canada L8S 4K1, rarthur@mcmaster.ca

Jonathan Bain Department of Humanities and Social Sciences, Polytechnic Institute of New York University, Brooklyn, NY 11201, USA, jbain@duke.poly.edu

Ronny Desmet Center for Logic and Philosophy of Science, Vrije Universiteit Brussel, Brussels, Belgium, Ronald.Desmet@vub.ac.be

Mauro Dorato Department of Philosophy, University of Rome 3, Rome, Italy, dorato@uniroma3.it

Laura Felline Department of Philosophy, University of Cagliari, Cagliari, Italy, felline@uniroma3.it

Eduardo V. Flores Department of Physics and Astronomy, Rowan University, Glassboro, NJ 08028, USA, flores@rowan.edu

Hans Herlof Grelland Department of Engineering Science, University of Agder, Grimstad, Norway, hans.grelland@uia.no

Stephen N. Lyle Freelance science editor, Alzen 09240, France, steven.lyle@wanadoo.fr

Graham Nerlich Department of Philosophy, University of Adelaide, Adelaide, Australia, graham.nerlich@adelaide.edu.au

Paul O'Hara Department of Mathematics, Northeastern Illinois University, Chicago, IL, USA, pohara@neiu.edu

James Overduin Gravity Probe B Theory Group, Hansen Experimental Physics Laboratory, Stanford University, Stanford, CA 94305, USA, joverduin@towson.edu
and
Department of Physics, Astronomy and Geosciences, Towson University, Towson, MD 21252, USA

Daniel Peterson University of Michigan Philosophy Department, 2215 Angell Hall, 435 South State Street, Ann Arbor, MI 48109-1003, petersod@umich.edu

Michael Silberstein Department of Philosophy, Elizabethtown College, Elizabethtown, PA 17022, USA
and
Department of Philosophy, University of Maryland, College Park, MD 20742, USA, silbermd@etown.edu

Edward Slowik Department of Philosophy, Winona State University, Winona, MN, USA, ESlowik@winona.edu

Friedel Weinert SSH, University of Bradford (UK), f.weinert@brad.ac.uk

Christian Wüthrich Department of Philosophy, University of California, San Diego, CA, USA, wuthrich@ucsd.edu

Part I
Minkowski's Representation of Special Relativity – Examples of Its Impact on the Twentieth Century Physics

The Minkowskian Background of Whitehead's Theory of Gravitation

Ronny Desmet

Abstract Whitehead's 1922 theory of gravity, an alternative to Einstein's 1916 theory of general relativity, cannot be understood well apart from the context of the British reception of both Einstein's special and his general relativity, and more specifically, apart from the Minkowskian nature of that reception. The aim of this essay is to emphasize the latter: the (indirect) influence of Minkowski on Whitehead was a major one.

1 Introduction

Alfred North Whitehead (1861–1947) is known by many for his *Principia Mathematica* collaboration with Bertrand Russell, and by some for his later philosophical works. However, in order to discover Whitehead's Minkowskian background, we must not primarily focus on the mathematics of his Cambridge period (1880–1910), nor on the metaphysics of his Harvard period (1924–1947), but on his involvement with relativity during the London period of his professional career (1910–1924). This involvement culminated in an alternative rendering of Albert Einstein's general theory of relativity, outlined in a number of publications, most notably in his 1922 book, *The Principle of Relativity with applications to Physical Science*.

Whitehead's alternative theory of gravitation is a Minkowski background-dependent theory of gravity, both in the historical sense of being rooted in a Minkowskian context, and in the technical sense of describing the gravitational field against a Minkowskian space-time background. In 1920 Whitehead wrote: "A tribute should be paid to the genius of Minkowski. It was he who stated in its full generality the conception of a four-dimensional world embracing space and time [...] He built on Einstein's foundations and his work forms an essential factor in the evolution of relativistic theory." (*ESP* 334–335) The aim of this paper is to show that the latter sentence is not only true in general, but also holds for Whitehead in particular – Minkowski's work forms an essential factor in the genesis of Whitehead's relativistic theory.

V. Petkov (ed.), *Space, Time, and Spacetime*, Fundamental Theories of Physics 167,
DOI 10.1007/978-3-642-13538-5_1, © Springer-Verlag Berlin Heidelberg 2010

2 Why Whitehead Was Interested in Special Relativity

The following publications are representative for the British knowledge about Einstein's special theory of relativity in the decade after its conception[1]:

- Edwin Bidwell Wilson and Gilbert Newton Lewis's 1912 memoir "The Space-Time Manifold of Relativity"
- Ludwik Silberstein's 1914 monograph, *The Theory of Relativity*
- Ebenezer Cunningham's 1914 and 1915 books, *The Principle of Relativity* and *Relativity and the Electron Theory*

Linking Whitehead with the Wilson and Lewis 1912 memoir, and with Silberstein's 1914 monograph, is straightforward. Indeed, in 1919, Whitehead wrote: "In connection with the theory of relativity I have received suggestive stimulus from Dr L. Silberstein's *Theory of Relativity*, and from an important Memoir ('The Space-Time Manifold of Relativity,' *Proc. of the Amer. Acad. of Arts and Sciences*, Vol. XLVIII, 1912) by Profs. E. B. Wilson and G. N. Lewis." (*PNK* vii) However, in order to understand *why* the co-author of *Principia Mathematica* took interest in these writings, as well as in Cunningham's writings, it is important to stress three biographical facts on Whitehead.

First, we should avoid the mistake of reducing Whitehead to a pure mathematician, but take into account Russell's remark that "Clerk Maxwell's great book on electricity and magnetism [was] the subject of Whitehead's Fellowship dissertation," and that "on this ground, Whitehead was always regarded at Cambridge as an applied, rather than a pure, mathematician." (Russell 1959: 33) In fact, looking at Whitehead's Cambridge training, we can notice a remarkable similarity with his near contemporaries J. H. Poynting, J. J. Thomson, and Joseph Larmor – three major proponents of the second generation of British Maxwellians. Poynting, Thomson, Larmor, and Whitehead can be qualified as similar Cambridge products.[2] Poynting did his Cambridge Mathematical Tripos exam in 1876, Thomson and Larmor in 1880, and Whitehead in 1883; all four were coached by Edward Routh, who excelled during the Tripos examination of 1854, and beat Maxwell into second place; and all four attended the intercollegiate courses on Maxwell's 1873 *Treatise on Electricity and Magnetism*, given by Maxwell's friend W. D. Niven. Being slightly younger, Whitehead also attended Thomson's lectures on electromagnetism. As is manifest in his writings, Whitehead developed a life-long interest in Maxwell's theory of electromagnetism, Poynting's theorem on the energy flow of an the electromagnetic field, and Thomson and Larmor's electronic theory of matter. In line with Hermann Minkowski's electromagnetic worldview, Wilson and Lewis in 1912, Silberstein in 1914, and Cunningham in 1914 and 1915, presented Einstein's special theory of relativity primarily as a contribution to electromagnetism, and more specifically, as

[1] Cf. Eddington (1918: vi–vii) and Henry Brose's Translator's Note in Freundlich (1920: vii).

[2] Cf. Warwick (2003: 333–398) and Lowe (1985: 92–109).

a contribution to the electronic theory of matter. This constitutes a first explanation of why Whitehead took interest in them.

Secondly, we should avoid the mistake of reducing Whitehead's mathematical research to the *Principia Mathematica* project, and his philosophy of mathematics to Russell's logicism. Prior to his collaboration with Russell, Whitehead's mathematical research had already given birth to his 1898 *Treatise on Universal Algebra with Applications* – a publication that led to Whitehead's election as a Fellow of the Royal Society.[3] In Book VII of his *Universal Algebra*, Whitehead forged a vector calculus from Hermann Grassmann's algebra of extensions, applicable in various branches of physics, especially hydrodynamics and electrodynamics. This was an important first step in Whitehead's career to make the philosophical dream of applied mathematics come true, "that in the future these applications will unify themselves into a mathematical theory of a hypothetical substructure of the universe, uniform under all the diverse phenomena." (*ESP* 285) Whitehead was well aware of the similar approach by Josiah Willard Gibbs at Yale University.[4] Gibbs forged a three-dimensional vector calculus from William Rowan Hamilton's algebra of quaternions, and one might say that it belonged to Whitehead's core business to pay attention to the further development of both Grassmannian and Hamiltonian vector calculus. In line with Minkowski's formal developments, Wilson and Lewis in 1912, Silberstein in 1914, and Cunningham in 1914 and 1915, each presented a tailor-made four-dimensional vector calculus to deal with special relativity. This constitutes a second explanation of Whitehead's interest in them.

Thirdly, we should avoid the mistake of reducing the *Principia Mathematica* project to the three volumes that have been published. Early on in their collaboration, Russell and Whitehead decided that the latter would write a fourth volume in which all of geometry was going to be based on the symbolic logic of relations.[5] This was an obvious decision, given the prominence of Euclidean and non-Euclidean, projective and descriptive geometry in Whitehead's earlier *Universal Algebra* research. However, following the lure of the applied mathematician, Whitehead's attention shifted from the logical reformulation of all known pure geometries to the search for an answer, in terms of the symbolic logic of relations, to a question that had long occupied him: How is the geometry of physics rooted in experience?[6] Not only did this question lead Whitehead into an area of research that had been dominated by men like Hermann von Helmholtz, Henri Poincaré, and Ernst Mach, hence necessitating Whitehead to position himself with respect to these giants, it also made him hypersensitive to the impact of special relativity, for this theory required the replacement of Euclidean space as the object of physical geometry by Minkowskian space-time. Of course, the Minkowskian unification of space and

[3] Cf. Lowe (1966: 137). That Whitehead was an FRS explains his presence at the famous meeting of the Royal Society and the Royal Astronomical Society on November 6th, 1919, a meeting that will be delt with in the remainder of this paper.

[4] Cf. *UA* 573.

[5] Cf. Lowe (1990: 12, 14–15, 92–95, 273).

[6] Cf. *PNK* v.

time was the point of departure of Wilson and Lewis in 1912, Silberstein in 1914, and Cunningham in 1914 and 1915, and hence, constitutes a third explanation of Whitehead's interest in them.

To summarize: For Whitehead, the special relativistic writings of Wilson and Lewis, Silberstein, and Cunningham, represented a threefold attraction. This attraction can safely be called 'Minkowskian,' for it is associated with the imperative unification of space and time, with the mathematics developed to formulate physical laws against the background of this unified space-time, and with the thus reformulated electromagnetic worldview.

3 Cunningham

According to Whitehead's biographer,[7] in June 1911, Karl Pearson vacated the Goldschmidt chair of Applied Mathematics and Mechanics at University College, London, and Ebenezer Cunningham – by then Pearson's assistant – was asked to continue Pearson's teachings prior to naming a final successor. In July 1911, however, Cunningham was already released to accept a lectureship at Cambridge, and Whitehead – who had moved from Cambridge to London in 1910, and was in search for a job – gladly accepted to replace Cunningham during the interregnum year 1911–1912. Whitehead hoped to be the final successor of Pearson, but mid March 1912, his hopes were destroyed when he learned of the appointment of another applied mathematician (L. N. G. Filon). Yet, Whitehead stayed at University College during the years 1912–1913 and 1913–1914, occupying a chair in pure mathematics, prior to leaving it for the Imperial College of Science and Technology, where he was able to secure a professorship in applied mathematics.

Anyway, the fact that Whitehead succeeded Cunningham in 1911 is one of the factors to conclude that, most likely, he was familiar with Cunningham's work on special relativity. Other factors are: their similar training and teaching curricula; their common interest in Thomson and Larmor's electronic theory of matter; the text-book status of Cunningham's *The Principle of Relativity*. Moreover, Whitehead and Cunningham met at least once in the context of the annual meetings of the British Association for the Advancement of Science, namely, in 1916, when Whitehead presided over Section A (mathematics and physics), and when Cunningham and Eddington introduced Einstein's general theory of relativity to the British scientific community.[8]

---

[7] Cf. Lowe (1990: 6–14).

[8] Cf. *Nature*, Vol. 98, Nr. 2450 (October 12, 1916), p. 120, and Sanchez-Ron (1992: 60 and 76).

4 Silberstein and the Aristotelian Society

According to Ludwik Silberstein's biographers,[9] this physicist from Polish origin, German student of, e.g., Helmholtz and Max Planck, and Italian lecturer in mathematical physics, moved from Italy to London in 1912, where he obtained a lectureship at University College, London. Consequently, Whitehead and Silberstein became University College colleagues that year. Moreover, it is Silberstein's University College course of lectures on the special theory of relativity, delivered during the academic year 1912–1913, which developed into his 1914 monograph, *The Theory of Relativity*.[10] Hence, Whitehead most likely knew of Silberstein's work prior to Silberstein's 1914 publication. Next to the fact that during a brief period Whitehead and Silberstein were colleagues, there are quite a number of other facts that imply a more personal relationship. These facts are related with Whitehead being elected a member of the London Aristotelian Society in 1915 – in Whitehead's words: "a pleasant center of discussion," where "close friendships were formed." (*ESP* 14)

As from 1912, the Aristotelian Society now and again welcomed Silberstein to take part in the discussions.[11] And when Whitehead joined it in 1915, Wildon Carr was its president, Samuel Alexander and Lord Haldane were among its vice-presidents, Percy Nunn was its treasurer, and Charles Dunbar Broad was one of its younger members.[12] That these Aristotelian Society members became Whitehead's friends, even close friends in the case of Haldane and Nunn,[13] is not the only reason for mentioning them here. All these men were deeply engaged in the philosophical issues with regard to relativity. This involvement with relativity – most likely one of the major reason for the mutual attraction between Carr, Alexander, Haldane, Nunn, Broad, Silberstein, and Whitehead – culminated in Carr's *The General Principle of Relativity, in Its Philosophical and Historical Aspect* (1920), Alexander's *Space, Time and Deity* (1920), Haldane's *The Reign of Relativity* (1921), Nunn's *Relativity and Gravitation* (1923), and Broad's *Scientific Thought* (1923). Whitehead figures in the latter four books, especially in Haldane's, and Silberstein figures in Nunn's book. In fact, Nunn and Silberstein were close. In 1914 Nunn already read the proofs of Silberstein's *Theory of Relativity*,[14] and in 1922 Nunn recalled that he "mixed a

[9] Cf. Duerbeck and Flin (2006: 1087–1089).

[10] Cf. Silberstein (1914: v).

[11] For instance, on November 4th, 1912, when Russell gave a lecture on "The Notion of Cause," and on January 5th, 1914, when a paper was read on "Philosophy as the Co-ordination of Science." Cf. *Proceedings of the Aristotelian Society*, New Series, Vol. 13, p. 362 and Vol. 14, p. 425. Notice that even though Silberstein participated in the discussions prior to Whitehead, Whitehead became a member prior to Silberstein, for Whitehead was elected in 1915, and Silberstein in 1919, only a year before he left London for New York. Cf. *Proceedings of the Aristotelian Society*, New Series, Vol. 19, p. 310.

[12] Cf. *Proceedings of the Aristotelian Society*, New Series, Vol. 15, p. 437.

[13] Cf. Lowe (1990).

[14] Cf. Silberstein (1914: v).

good deal with men like Silberstein, who are keen followers and even developers of the theory of relativity when it first came among us."[15]

So: Whitehead and Silberstein were colleagues during the academic year 1912–1913; as from 1915, they were both active in the Aristotelian Society; and they had a close friend in common (Nunn). All this leads to the conjecture that Whitehead and Silberstein knew each other well, and frequently met. The latter is confirmed by the minutes of the meetings of the Aristotelian Society. On January 3rd, 1916, when Whitehead read some explanatory notes on his first relativity paper, "Space, Time, and Relativity,"[16] Silberstein was present, and took part in the subsequent discussion. And on December 18th, 1916, when Whitehead read "The Organization of Thought,"[17] Silberstein was again present, and again joined the discussion.[18] Of course, the hypothesis that Whitehead and Silberstein frequently met is also supported by their joint presences at other Aristotelian Society meetings, e.g., on January 6th, 1919, and on March 3rd, 1919, when both took part in the discussion.[19] And finally, one should not forget that they were both present at the famous joint meeting of the Royal Society and the Royal Astronomical Society on November 6th, 1919, when Eddington presented the observational data gathered during the May 1919 solar eclipse, and when Silberstein, contrary to Eddington, pointed out that they were insufficient to confirm Einstein's general theory of relativity.[20]

5 Minkowski's 1908 Papers

Whitehead's acquaintance with the work of Wilson and Lewis, Silberstein, and Cunningham – and hence, with Einstein and Minkowski's unification of space and time, with the Minkowskian mathematics to formulate physical laws against the background of Minkowski's unified space-time, and with the Minkowskian reformulation of electromagnetism, naturally led him to the work of Minkowski himself. In May 1941, Whitehead told his biographer, Victor Lowe: "Minkowski's paper was published in 1908, but its influence on me was postponed approximately ten years." (Lowe 1990: 15) Given the fact that Whitehead got to know Minkowski's work via Wilson and Lewis, Silberstein, and Cunningham, accounts for a retardation of the direct influence of Minkowski's 1908 paper on Whitehead, although, as Lowe adds:

---

[15] Letter of Nunn to Haldane, dated July 8th, 1922. Cf. National Library of Scotland, Haldane Papers, Manuscript 5915, Folio 192.

[16] This paper of Whitehead was first read to Section A (mathematics and physics) at the Manchester Meeting of the British Association for the Advancement of Science in 1915. Cf. *OT* 191–228.

[17] This paper of Whitehead was his Presidential Address to Section A at the Newcastle Meeting of the British Association for the Advancement of Science in September 1916. Cf. *OT* 105–133.

[18] Cf. *Proceedings of the Aristotelian Society*, New Series, Vol. 16, p. 364 and Vol. 17, p. 481.

[19] Cf. *Proceedings of the Aristotelian Society*, New Series, Vol. 19, p. 293 and p. 294.

[20] Cf. *SMW* 10 for Whitehead's presence, and Duerbeck and Flin (2005: 191 and 200–203) for Silberstein's presence and intervention.

"*Ten* may be an overstatement by one to three years." Also, it is not immediately clear whether Whitehead pointed at Minkowski's 1908 paper "Die Grundgleichungen für die elektromagnetischen Vorgänge in bewegten Körpern," or to his famous 1908 Cologne lecture "Raum und Zeit" – the two texts of Minkowski to which Wilson and Lewis, as well as Silberstein, frequently refer.[21]

It was in the *Grundgleichungen* that Minkowski first employed the term "space-time" (Walter 2007: 219), but it was in his famous "Space and Time" lecture that he said: "Henceforth space by itself, and time by itself, are doomed to fade away into mere shadows, and only a kind of union of the two will preserve an independent reality." (Minkowski 1952: 75) The *Grundgleichungen*, with its treatment of space-time vectors of the first and the second kind, and of the matrix-method to operate with these vectors,[22] is more mathematical than "Space and Time," but the latter, with its vision of the whole universe being resolved into world-lines, and of a world-line as "the everlasting career of the substantial point" (Minkowski 1952: 76), is more likely to be remembered by an 80 year old philosopher – Whitehead in 1941 – whose notion of 'historical routes' in *The Principle of Relativity* is a slightly more concrete version of Minkowski's abstract notion of 'world-lines.'[23] Moreover, in "Space and Time" Minkowski holds that "physical laws might find their most perfect expression as reciprocal relations between those world-lines," and after describing the electrodynamical relations between the world-lines of point-charges in terms of the Maxwell-Lorentz electron theory and the Liénard-Wiechert retarded potentials, Minkowski expresses his belief that the resolution of the universe in world-lines of point-charges can be seen as "the true nucleus of an electromagnetic image of the world." (Minkowski 1952: 91) Well, with some exaggeration, one might say that Whitehead's relativity of historical routes of events is the true nucleus of the philosophical image of the world as presented in his later works, again implying that most likely Whitehead pointed at Minkowski's famous 1908 Cologne lecture, "Space and Time," when telling Lowe about Minkowski's influence on him.

6 The Search for a Relativistic Theory of Gravitation

Next to space-time unification, mathematical formalism, and electromagnetic world-view, there is another important aspect in both Minkowski's appendix to the *Grundgleichungen*[24] and his "Space and Time" lecture, as well as in the Wilson and Lewis

[21] E.g., Wilson and Lewis (1912: 391 and 495), and Silberstein (1914:127, 129–130, 143, 266, 282). Remarkably, even though Cunningham devotes a whole part of his 1914 book to Minkowski's work (Part II Minkowski's Four-Dimensional World, pp. 85–134), he does not explicitly refer to any of Minkowski's papers or lectures.

[22] Cf. Minkowski (1910: 483–486 and 495–503).

[23] Cf. *R* 30.

[24] This appendix is titled "Mechanics and the Relativity Postulate." For an English translation, see Minkowski (2007).

memoir, and the Silberstein and Cunningham textbooks. Inspired by Poincaré – "Poincaré's scientific output fascinated Göttingen scientists in general, and Minkowski in particular" (Walter 2007: 214) – Minkowski sought to bring gravitation within the purview of Einstein's principle of relativity.

In the appendix to the *Grundgleichungen* Minkowski wrote that "it would be highly unsatisfactory" if Einstein's principle of relativity "could be accepted as valid for only a subfield of physics" (Minkowski 2007: 274), and he proposed a first relativistic law of gravitation, formulated in terms of a four-scalar gravitational potential, and inspired by his own reformulation of Maxwell's equations in terms of a four-vector electromagnetic potential.[25]

In "Space and Time" Minkowski expressed the belief that the gravitational relations between the world-lines of point-masses should be treated just like the electromagnetic relations in the case of point-charges, and he accordingly proposed a second relativistic law of gravitation, expressing the driving gravitational force in terms of a four-vector gravitational potential.[26]

However, the challenge to solve the problem of the incorporation of gravitation in a relativistic image of the word remained, because, as Scott Walter poignantly puts it: "By proposing two laws instead of one, Minkowski tacitly acknowledged defeat," and "could hardly claim to have solved unambiguously the problem of gravitation." (Walter 2007: 234)

At the end of their 1912 memoir, Wilson and Lewis echo Minkowski's vision that the searched for formulae expressing the gravitational force and potential must be "completely analogous" to the new formulae expressing the electromagnetic force and potential, and suggest – by analogy – the use of the term "gravito-magnetic" instead of gravitational. (Wilson and Lewis 1912: 496) Silberstein – in his 1914 book – mentions Poincaré's 1906 attempt to use the general form of the Lorentz transformations for the treatment of both the dynamics of the electron and universal gravitation, and notices the advantage Minkowski's approach seems to offer for a relativistic theory of gravity.[27] The most elaborate treatment of the search for a relativistic theory of gravity, however, is given in Cunningham's *The Principle of Relativity*.

While dealing with the electron theory in Minkowskian format, and, more specifically, with the Lorentz covariant four-vector expression of the Liénard-Wiechert potentials for the field due to the motion of a single point-charge, Cunningham refers to "the work founded on that of Poincaré for modifying the law of gravitation to conform to the Principle of Relativity." (Cunningham 1914: 109) Contrary to Wilson and Lewis, and to Silberstein, Cunningham does not leave it at a simple suggestion of the gravitation-electrodynamics analogy. An important part of his treatment of the dynamics of a particle is devoted to the search for a relativistic

[25] Cf. Walter (2007: 224).

[26] Cf. Walter (2007: 234).

[27] Cf. Silberstein (1914: 87 and 241).

theory of gravitation.[28] Moreover, Cunningham does not only refer to the 1906 paper of Poincaré – "Sur la dynamique de l'électron" – but also treats the 1911 paper of the Dutch astronomer Willem de Sitter – "On the Bearing of the Principle of Relativity on Gravitational Astronomy" – which has been called "the most authoritative account in English of the astronomical importance of the principle of relativity [...] before the appearance of Einstein's general theory." (Warwick 2003: 453)

None of the relativistic theories of gravitation Whitehead encountered in the writings on relativity at his disposal prior to 1916 was satisfactory. None of the theories which can be found in Poincaré's 1906 paper, Minkowski's 1908 papers, de Sitter's 1911 paper, and Cunningham's 1914 book, were in accordance with the astronomical observations of the secular motion of the perihelion of Mercury, as de Sitter and Cunningham clearly highlight.[29]

It seems to have been Minkowski's opinion that the incorporation of gravitation into relativistic thinking was not a major problem, and – as Minkowski died on January 12, 1909, at the age of 44 of a sudden and violent attack of appendicitis – he never lived to see how elusive and difficult the task would turn out to be.[30] By 1913, Einstein characterized the attempt to find a relativistic generalization of Newton's law of gravitation as "a hopeless undertaking," at least, in the absence of some good physical guiding principles, such as the "laws of energy and momentum conservation," and the "equality of the *inertial* and the *gravitational* mass," and Einstein adds:

> To see this clearly, one need only imagine being in the following analogous situation: suppose that of all electromagnetic phenomena, only those of electrostatics are known experimentally. Yet one knows that electrical effects cannot propagate with superluminal velocity. Who would have been able to develop Maxwell's theory of electromagnetic processes on the basis of these data? Our knowledge of gravitation corresponds precisely to this hypothetical case: we only know the interaction between masses at rest, and probably only in the first approximation. (Einstein 2007: 544)

Contrary to Minkowski, but in line with Einstein, when Whitehead read his first relativity paper, "Space, Time, and Relativity," before the members of the Aristotelian Society, he added the following comment: "We have begun to expect that all physical influences require time for their propagation in space. This generalization is a long way from being proved. Gravitation stands like a lion in the path." (*OT* 225) However, in September 1916, 8 months after making this comment, Whitehead first learned about Einstein's general theory of relativity – a theory that claimed to have defeated the lion that blocked the road to an empirically adequate relativistic treatment of gravitation.

[28] Cf. Cunningham (1914: 171–180).

[29] Cf. Cunningham (1914: 180).

[30] Cf. Corry (2004: 192 and 227).

7 Eddington and de Sitter

Linking Whitehead and Eddington is easy, because in 1902, Eddington – thanks to his outstanding ability in mathematics and physics – was granted a natural science scholarship to Trinity College, Cambridge, where he was coached by Robert Herman,[31] and where "among the formal lectures which Eddington and most of his group attended were those of [. . .] A. N. Whitehead." (Douglas 1957: 10) So when Whitehead presided over Section A (mathematics and physics) at the 86th meeting of the British Association for the Advancement of Science in Newcastle-On-Tyne in September 1916,[32] he already knew Eddington personally, at the very least as his former student. And that the two men met at this section, and, moreover, that their meeting was related to Einstein's general theory of relativity, is made clear by the following account of it in the October 12th, 1916, issue of *Nature*:

> The first of the two organized discussions arranged for this section was on "Gravitation." The discussion followed immediately after Prof. Whitehead's presidential address,[33] and it happened that the arrangement was appropriate, for the president's exposition of the logical texture of geometry had carried us far from the ordinary conceptions of space, and paved the way for the revolutionary ideas associated with the space-time world of Einstein and Minkowski. Mr. E. Cunningham, who opened the discussion, and Prof. A. S. Eddington, who followed, dealt with Einstein's recent work, which brings gravitation within the scope of the principle of relativity. (*Nature*, Vol. 98, Nr. 2450, p. 120)

This *Nature* quote is in harmony with two of this paper's claims: Whitehead's *Principia Mathematica Volume 4* research on the logical texture of geometry formed his pathway to relativity; and, in 1916, Cunningham was one of the most prominent British mathematicians engaged in the quest of bringing gravitation within the scope of the principle of relativity. At the same time, the quote also links Whitehead with Eddington's research on general relativity. It must be said, however, that in September 1916, Cunningham and Eddington's research on general relativity was still premature, and according to Andrew Warwick: "It is a measure of Cunningham and Eddington's ignorance of Einstein's work at this time that the official account of the session (published in 1917) made no mention of their presentations, but referred the reader directly to de Sitter's first two papers in the *Monthly Notices*." (Warwick 2003: 462–463)

This Warwick quote does not only point at the British ignorance of Einstein's general theory of relativity in 1916. It also points at the fact that the official account of Section A, presided by Whitehead, referred to two of de Sitter's general relativity papers, which immediately establishes a link between Whitehead and de Sitter's general relativity output. Actually, both elements – the British ignorance with regard to general relativity in 1916, and the fact that the ignorant British readers, including

[31] Cf. Douglas (1957: 5 and 9–10) and Warwick (2003: 449–451).

[32] Cf. Sanchez-Ron (1992: 76).

[33] "The Organization of Thought," to which I already referred, because Whitehead also read it at the Aristotelian Society in December 1916.

Whitehead, were referred to De Sitter's papers instead of Einstein's papers – are closely related. A common cause was World War I.

Germany and Britain being at war, the German publications of Einstein did not easily reach the British, partially explaining their ignorance.[34] However, the Netherlands was neutral, and the news of Einstein's completion of the general theory of relativity in November 1915 reached Britain in the form of a letter from the Netherlands.[35] Indeed, on the one hand, the Dutch astronomer de Sitter was one of the three Leiden University physicists who acted as Einstein's sounding board during the development of his general theory of relativity – the other two were Hendrik Antoon Lorentz and Paul Ehrenfest. So, de Sitter was well informed on Einstein's struggle with, and completion of, a new theory of gravitation. On the other hand, when Einstein published a general summary of his new theory in May 1916, including some discussion of its cosmological consequences,[36] de Sitter realized its importance for astronomers in the English-speaking world, while at the same time realizing that one could not very well reprint the work of a German in a British journal during the wartime. So, in June 1916, de Sitter wrote a letter to inform Eddington, then Secretary of the Royal Astronomical Society, and he offered to submit a paper of his own on the subject. In his reply to de Sitter, Eddington confirmed that he was immensely interested, and he encouraged de Sitter to submit the promised paper. In the event, de Sitter published a paper in *The Observatory* magazine, and a series of three papers in the *Monthly Notices of the Royal Astronomical Society*.

In a letter of July 4th, 1916, Eddington informed de Sitter:

> We are having a discussion at the British Association on Gravitation – at Newcastle, Dec. 5–8. I wish we could have invited you to come over to take part; but we are not inviting any foreign guests this year because Newcastle is a "restricted area" and aliens are not allowed in it. [. . .] I feel sure you will allow me to make use of the papers you send, in making my contribution to the discussion. So far as I can make out, no one in England has yet been able to see Einstein's paper and many are very curious to know the new theory. So I propose to give an account of it at the Meeting.[37]

---

[34] Einstein's struggle to formulate a relativistic theory of gravitation started not long after the 1905 publication of his special theory of relativity, and hence, prior to World War I. However, his pre-war attempts, and his corresponding publications, were undervalued in Britain. This undervaluation has to be included in order to fully explain the British ignorance in 1916, and is exemplified by both Cunningham and Eddington. In 1914, Cunningham wrote: "No attempt has been made to present the highly speculative attempt of Einstein at a generalization of the principle [of relativity] in connection with a physical theory of gravitation." (Cunningham 1914: vi) And whereas Cunningham dismissed Einstein's pre-war attempts as too speculative, Eddington – who knew Einstein's 1911 paper "On the Influence of Gravitation on the Propagation of Light" – mainly focused on Einstein's empirical predictions, and undervalued the fact that these predictions were based upon a new hypothesis concerning the physical nature of gravity (Einstein's equivalence principle). Cf. Warwick (2003: 455–457).

[35] For more complete accounts than the one I can give here, cf. Stachel (2002: 455–456), Warwick (2003: 457–462), and Crelinsten (2006: 94–98).

[36] Cf. Einstein (1916).

[37] This quote is taken from Stachel (2002: 456). The British Association meeting was held in September.

Whitehead listened at Eddington's account at Newcastle, but this account was not included in the official report. However, a good idea of what Whitehead heard can be formed by reading Eddington's first published paper devoted to the general theory of relativity, "Gravitation and the Principle of Relativity," published in the December 28th, 1916, issue of *Nature*. The reason is offered by John Stachel's remark that "it is presumably based on his talk on the same subject to the British Association for the Advancement of Science." (Stachel 2002: 457) Eddington refers his readers – and hence, presumably referred his audience – to the following three paper on the subject: Einstein's May 1916 paper "Die Grundlage der algemeinen Relativitätstheorie," de Sitter's October 1916 *Observatory* paper, and de Sitter's first 1916 *Monthly Notices* paper. This means that, most likely, Whitehead was referred to de Sitter's writings on general relativity prior to the appearance of the official report of the Newcastle meeting in 1917.[38]

Of course, *The Observatory*, the *Monthly Notices*, and *Nature*, were readily available to Whitehead, but we do not know whether Eddington offered his former Cambridge lecturer the opportunity to read the paper that led to all the excitement in the first place – Einstein's summary paper, which Eddington got from de Sitter. Likewise, we do not know whether, half a year later, Silberstein offered his former University College colleague a reprint of Einstein's summary paper. Silberstein, by then, had also received reprints via a neutral country, namely via Michele Besso in Switzerland. In fact, on May 7, 1917, Einstein wrote to his close friend: "Lieber Michele! Ich sende Dir einige Abhandlungen mit der Bitte, Sie an Herrn Dr. L. Silberstein, 4 Anson Road Cricklewood London N.W.2. weiterzusenden, der mich darum gebeten hat."[39] But what we do know is that Whitehead had both the appropriate personal contacts, and the references to all 1916–1917 English articles on the topic, to get acquainted with Einstein's general theory of relativity.

Despite his remarkable speed to master new mathematical theories, and despite his prior knowledge on differential geometry, acquired a decade earlier in the lectures of his coach, Herman, at Trinity College, it took Eddington almost 2 years to master Einstein's general theory of relativity, and even then, upon completion of his 1918 official *Report on the Relativity Theory of Gravitation* for the Physical Society of London, he asked for de Sitter's "general criticism and detection of blunders."[40] Nonetheless, as Andrew Warwick puts it, "for many British mathematicians and physicists the *Report* represented the definitive English-language account

[38] Further research might provide an answer to the following question: Was Whitehead himself, having been the Section A president in 1916, responsible for the official report on that section or not?

[39] "Dear Michele! I'm sending you some reprints, asking you to forward them to Dr. L. Silberstein, 4 Anson Road Cricklewood London N.W.2., who has requested them." The German quote in the main text is a quote from Document 335 in The Collected Papers of Albert Einstein, Volume 8, Part A, p. 446. For the English translation, and more details on the Silberstein-Einstein correspondence, cf. Duerbeck and Flin (2006: 1089).

[40] Letter of Eddington to de Sitter, dated August 16th, 1918. Cf. Warwick (2003: 467–468).

of general relativity and further established Eddington's emergent reputation as the theory's master and champion in Britain." (Warwick 2003: 468)

One cannot imagine that Whitehead – whose research dealt with the question of how to derive, by means of the *Principia Mathematica* logic, the space-time geometry of physics from the spatio-temporal texture of our experience; who repeatedly discussed relativity with his Aristotelian Society friends, e.g., with Alexander on July 5th, 1918, following Alexander's address on "Space-Time"[41].; whose lecture courses on applied mathematics at the Imperial College of Science and Technology culminated in his postgraduate lecture course on "Relativity and the Nature of Space"[42]; and who started Herbert Dingle's lifelong interest in the theory of relativity, and encouraged him to write *Relativity for All* [43] – ignored the definitive English-language account of general relativity, written by his former student, and meanwhile famous astronomer, Eddington.

8 Whitehead's Enquiry

After all that has been said and done in this paper to link Whitehead with Eddington and de Sitter, a surprise awaits the reader when turning to Whitehead's first book on relativity, his 1919 *Enquiry Concerning the Principles of Natural Knowledge*. The book shows no trace of any Eddington or de Sitter impact! However, the explanation is straightforward.

Whitehead's *Enquiry* is the apex of his research to find – in terms of the logic of relations – an answer to the question: "How is space rooted in experience?" The first output of this research, written in 1905, and published in 1906, was Whitehead's Royal Society memoir "On Mathematical Concepts of the Material World," in which 'space' still meant 'Euclidean space,' and in which 'points' were logically defined by Whitehead in terms of 'linear objective reals' – entities closely resembling Faraday and Maxwell's spatial lines of force. However, Einstein and Minkowski's unification of space and time prompted Whitehead to replace 'space' with 'space-time', 'substantial points' with 'event-particles', the electromagnetic lines of force with Minkowski's world lines, and so on. In other words, special relativity caused an update of Whitehead's research question into: "How can Minkowski's space-time geometry be logically abstracted from our experience of spatio-temporal events?"

By the time he wrote his *Enquiry*, Whitehead had developed a method – the method of extensive abstraction – to do just that; a method which harmonized the world of physics with the world of everyday experience – Whitehead's main philosophical motivation; and hence, a method he was not willing to put aside because Einstein's general relativity invited us to give up Minkowski's non-curved space-

---

[41] Cf. *Proceedings of the Aristotelian Society*, New Series, Vol. 18, p. 640, and *CN* viii

[42] Cf. Lowe (1990: 64–65).

[43] In the July 1921 Preface of *Relativity for All* Dingle wrote: "The author is glad to acknowledge his deep indebtedness to Professor Whitehead for invaluable help and unwearying kindness in unveiling the mysteries of a difficult subject." (Dingle 1922: vi) Cf. also Lowe (1990: 65).

time geometry in favor of a variably curved space-time geometry. In the April 20th, 1919, Preface of his *Enquiry*, Whitehead expresses this concern as follows:

> The whole investigation is based on the principle that the scientific concepts of space and time are the first outcome of the simplest generalizations from experience, and that they are not to be looked for at the tail end of a welter of differential equations. This position does not mean that Einstein's recent theory of general relativity and of gravitation is to be rejected. The divergence is purely a matter of interpretation. [. . .] It has certainly resulted from Einstein's investigations that a modification of the gravitational law [. . .] will account for the more striking outstanding difficulties otherwise unexplained by the law of gravitation. This is a remarkable discovery for which the utmost credit is due to the author. Now that the fact is known, it is easy to see that it is the sort of modification which on the simple electromagnetic theory of relativity is likely to be required for this law. I have however been anxious to disentangle the considerations of the main positions in this enquiry from theories designed to explain special laws of nature. Also at the date of writing the evidence for some of the consequences of Einstein's theory is ambiguous and even adverse. In connection with the theory of relativity I have received suggestive stimulus from Dr L Silberstein's *Theory of Relativity* [. . .]. (*PNK* vi–vii)

The first sentence of this quote confirms Whitehead's main philosophical challenge – to avoid the bifurcation of nature into the mathematical world discovered by Einstein (and his predecessors) "at the tail end of a welter of differential equations," and the common word of our day-to-day experience – and it confirms Whitehead's main answer at the time, both to this challenge, and to his research question: a method of "the simplest generalizations" – the method of extensive abstraction.

Clearly, Whitehead was well aware of the general theory of relativity when writing his Preface, and he did not reject its new law of gravitation. On the contrary, he credited Einstein for solving the outstanding difficulties of Newton's law of gravitation, such as the difficulty of accounting for the observed precession of the perihelion of Mercury, a difficulty Poincaré, Minkowski, de Sitter, and Cunningham, were unable to solve. At the same time, Whitehead distanced himself from Einstein's general relativistic interpretation of his new law of gravitation, and already gave two hints on how to reinterpret it: (1) by learning from how Einstein modified Newton's law, and by performing a similar modification while adhering more closely to the special theory of relativity, that is, while respecting the main position of his research, that the Minkowskian space-time structure was at one with the spatio-temporal texture of our experience; and (2) by disentangling the problem of discovering the general structure of space-time from the problem of discovering the particular character of physical laws, or, in other words, by separating again what Einstein had unified, space-time geometry and physics.

To summarize, the Preface of *An Enquiry Concerning the Principles of Natural Knowledge* clearly confirms our claim that Whitehead was familiar with Einstein's general theory of relativity on April 20, 1919, while at the same time providing an explanation of why no trace leading to Eddington or de Sitter can be found in the book. Whitehead's research aimed at thinking together the geometrical world of Minkowski, and the spatio-temporal world of the events we experience. And with his *Enquiry*, Whitehead wanted to communicate how he had managed to do so, without explicitly addressing and answering the next question on his research agenda:

"How to interpret Einstein's new law of gravitation in terms of a gravitational field against the background of Minkowski's space-time, instead of accepting Einstein's interpretation of his new law in terms of the identification of the field of gravitation with a variably curved space-time?"

9 Silberstein's 1918 Paper

Early 1919, Whitehead had a good reason for not yet addressing his new research question. Its relevance was dependent on "the evidence for some of the consequences of Einstein's theory," and when writing the Preface of his *Enquiry*, this evidence was still "ambiguous and even adverse," even though a month later it was going to be strengthened thanks to some relevant observations by British astronomers, including Eddington, at the occasion of the May 29th, 1919, solar eclipse.

It is no coincidence that Whitehead indicated to have received "suggestive stimulus from Dr L Silberstein's *Theory of Relativity*." I claim that Whitehead's referral to Silberstein in the context of reinterpreting Einstein's new law of gravitation, disentangling space-time geometry and physics, and highlighting adverse evidence with regard to empirical consequences, points to the fact that Silberstein was a source of inspiration to help Whitehead answer his new, but still private, research question of reinterpreting general relativity. I will now first add an element to substantiate the latter claim, and then return to the solar eclipse.

In 1923, George Temple – a mathematician who had taken his first degree as an evening student, and at the time was working as a research assistant at Birkbeck College, London – gave a lecture on "A Generalization of Professor Whitehead's Theory of Relativity" at the Physical Society of London. The importance of mentioning Temple's lecture at this point is formed by the following facts. Temple treated Silberstein's 1918 paper, "General Relativity without the Equivalence Hypothesis," as a precursor of Whitehead's alternative theory of gravitation. Whitehead was present, and responded to Temple's paper. His extensive response is registered in the Proceedings of the Physical Society of London, and shows that he was very pleased with this paper "from the pen of a young scientist whose work augurs a very distinguished career."[44] At no point did Whitehead object to treating Silberstein's 1918 paper as a precursor of his own 1920–1922 theory. The opposite is true. Whitehead emphasized that at the heart of his alternative theory of gravitation lies the distinction "between space-time relations as universally valid and physical relations as contingent."[45] In other words, Whitehead stressed the importance of separating the general space-time structure from the more particular physical structures, a separation that is central in Silberstein's 1918 paper, in which Einstein's equivalence principle – Einstein's identification of inertial and gravitational descriptions

[44] Temple (1923: 192).

[45] Temple (1923: 193).

to the point of identifying space-time geometry and gravitational physics – was rejected. This adds to what has been said before on the Whitehead-Silberstein link, and thus helps to substantiate the claim that Silberstein was a source of inspiration for Whitehead.

Silberstein, who had a kind of love-hate relationship with Einstein's general theory of relativity, willing to accept it wholeheartedly, and yet, relentlessly criticizing it, has been called Einstein's *"advocatus diaboli"* (Pais 1983: 305), as well as "Einstein's antagonist" (Duerbeck and Flin 2005: 186). This might suggest that Whitehead mainly derived his critical attitude towards Einstein from Silberstein. However, in order not to overestimate the influence of Silberstein on Whitehead, an important comment is due. Whitehead's critique of Einstein's approach has many more sources, too many to list here. Some date from his Cambridge period, others from his London period. Some are to be found in the domain of mathematical physics, others in the domain of philosophy.

Also, one must not forget that from its introduction in Britain, Einstein's general theory of relativity was exposed to critique. Most importantly, the 1916–1917 *Monthly Notices* papers of de Sitter reflect an at that time ongoing Einstein-de Sitter debate on Einstein's Machian explanation of inertia, and on the priority of matter over space-time.[46] So there never was a "pure" or "uncritical" transmission of Einstein's general theory of relativity from the Continent to Britain to start with. When Eddington, Silberstein, and Whitehead learned about it, Einstein's theory was already wrapped in de Sitter's anti-Machian critique, elements of which became part of their critiques. I deliberately include Eddington, because in his 1918 *Report*, he clearly sided with de Sitter in the debate with Einstein on the various cosmological hypotheses at the time.

No wonder that Whitehead, inspired by de Sitter, Eddington, and Silberstein, repeatedly attacks the Machian explanation of inertia in his 1920–1922 writings on relativity, and that he replaced Einstein's theory, in which matter has priority over space-time, and in which space-time is constantly curved at the local and at the cosmological scale (respectively zero curved and non-zero curved), while being variably curved at the intermediate scale (e.g., of the solar system), with an alternative in which space-time has priority over matter, and in which the universe is Minkowskian (and hence at one with our common experience) at all scales.

10 The 1919 Solar Eclipse

One of the most important consequences of the general theory of relativity was Einstein's prediction concerning the deflection of rays of starlight passing near the limb of the sun. If the starry sky is photographed twice, once by night, and once

[46] For an account of the 1916–1917 Einstein-De Sitter dialogue, see Janssen (1998: 351–357) (can also be found on http://www.tc.umn.edu/~janss011/), Crelinsten (2006: 103–108) and Matteo Realdi's (2007) lecture "The Universe of Willem de Sitter" (can be found on http://www.phil-inst.hu/~szekely/PIRT_BUDAPEST/).

during a solar eclipse, all other things being equal, then, upon comparison of the two pictures, we will observe exactly calculated deflections of rays of starlight (that is, shifts of starlight spots on the pictures) near the solar corona. The pictures taken by English astronomers during the solar eclipse on May 29th, 1919, seemed to confirm Einstein's prediction, and when Eddington made this confirmation public on November 6th, 1919, at a joint meeting of the Royal Society and the Royal Astronomical Society, it immediately launched Einstein's career to superstar-heights, despite Silberstein's unease with Eddington's way of handling the solar eclipse data, and his warning to await confirmation of Einstein's red shift prediction. As said before, Whitehead was present at this meeting, and in an account published years later in *Science and the Modern World*, Whitehead wrote:

> The whole atmosphere of tense interest was exactly that of the Greek drama: we were the chorus commenting on the decree of destiny as disclosed in the development of a supreme incident. There was dramatic quality in the very staging: – the traditional ceremonial, and in the background the picture of Newton to remind us that the greatest of scientific generalizations was now, after more than two centuries, to receive its first modification. Nor was the personal interest wanting: a great adventure of thought had at length come safe to shore. (*SMW* 10)

Not only was Whitehead present at this memorable meeting, by November 1919, the British considered Whitehead as an authority on the subject of general relativity. On November 15th, 1919, his first article on the momentous confirmation of Einstein's revolutionary theory appeared in *The Nation* under the title "A Revolution in Science." Also, Wildon Carr (already mentioned in this paper as one of Whitehead's Aristotelian Society friends), Frederick Lindemann (a famous Oxford physicist), and Whitehead, were asked to write a contribution on "Einstein's Theory" for the readers of the Educational Supplement of *The Times*. Carr's article was published on January 22nd, 1920, Lindemann's on January 29th, and Whitehead's on February 12th.[47]

The opening of Whitehead's 1920 article reads: "The articles on this subject, which appeared on January 22 and 29, summarized the general philosophical theory of relativity and the physical ideas involved in Einstein's researches. The purpose of the present article is in some respects critical, with the object of suggesting an alternative explanation of Einstein's great achievement." (*ESP* 332) Whereas Whitehead's 1919 article, "A Revolution in Science," does not give away Whitehead's critical attitude, and is as orthodox as Lindemann's *Times* article, his 1920 article, "Einstein's theory," not only reveals Whitehead's critique of Einstein, but also gives a first outline of his alternative theory of gravitation.

In it, Whitehead starts with the analysis of Einstein's work in three factors: "a principle, a procedure, and an explanation." (*ESP* 332) According to Whitehead, Einstein's principle is the unification of space and time into space-time, and he writes: "What I call Einstein's principle is the connexion between time and space." (*ESP* 332) Whitehead does not give similarly clear definitions of Einstein's

---

[47] Whitehead's "Einstein's Theory" is reprinted in *ESP* (pp. 332–342) and in *IS* (pp. 125–135).

procedure and Einstein's explanation, but their meaning is rendered clear by the continuation of his article. According to Whitehead, Einstein's procedure is the procedure to formulate invariant tensor laws to describe physical phenomena in general, and gravitational phenomena in particular, and Einstein's explanation is the explanation in terms of Mach's principle (inertia is determined by matter), and the closely related equivalence principle (inertia and gravity are identical). So, when Whitehead writes: "Einstein's [...] discovery of the principle and the procedure constitute an epoch in science. I venture, however, to think that the explanation is faulty" (*ESP* 332), this 1920 way of expressing himself is completely in line with his 1922 way of putting things: "My whole course of thought presupposes the magnificent stroke of genius by which Einstein and Minkowski assimilated time and space. It also presupposes the general method of seeking tensor or invariant relations as general expressions for the laws of the physical field, a method due to Einstein. But the worst homage we can pay to genius is to accept uncritically formulations of truths which we owe to it." (*R* 88)

By rejecting Einstein's explanation, Whitehead rejects the principles that were already criticized by de Sitter in 1916 and 1917, and by Eddington and Silberstein in 1918. Whitehead was fully aware that he thus dropped two of the major principles that actually guided Einstein's search for a new law of gravitation. Indeed, Mach's principle and the principle of equivalence "formed the clue by which Einstein guided himself along the path from his principle to his procedure." However, as Whitehead immediately adds: "It is no novelty to the history of science that factors of thought which guided genius to its goal should be subsequently discarded. The names of Kepler and Maupertuis at once occur in illustration." (*ESP* 332) Of course, the rejection of Einstein's explanation implies the challenge to offer an alternative explanation. No wonder Whitehead ends his 1920 article with a brief outline of such an alternative.

11 Whitehead's Alternative Theory of Gravitation

Whitehead writes that his alternative theory of gravitation starts from the general theory of time and space which is explained in his *Enquiry*, in other words, from Minkowski's space-time, and that it also starts "from Einstein's great discovery that the physical field in the neighborhood of an event-particle should be defined in terms of ten elements" (*ESP* 342), meaning that the gravitational field should be defined as a symmetrical second rank tensor.

Einstein's gravitational field tensor is called the fundamental tensor, and does not define a gravitational field apart from space-time, but space-time as being equal to the gravitational field. Contrary to Einstein, Whitehead calls his gravitational field tensor the impetus tensor, and he uses it to define the gravitational field against the background of Minkowski's space-time. This implies that Whithead keeps field physics and space-time geometry apart, as Silberstein did in 1918. Consequently, he writes: "According to Einstein such elements [the ten elements of the gravitational field tensor] merely define the properties of space and time in the neighborhood.

I interpret them as defining in Euclidean space [or better: in Minkowskian space-time] a definite physical property of the field which I call the 'impetus."' (*ESP* 342)

Einstein's law of gravitation equates two tensors: the Einstein tensor, which results from second order differential operations on the elements of his fundamental tensor; and the energy-momentum-stress tensor, which represents the source of gravitation. However, Whitehead's treatment of gravitation focuses on point-masses (event-particles) as the sources of gravitation, and it does not take into account the more general case of a continuous mass-energy distribution. Hence, when comparing his law of gravitation with Einstein's, Whitehead is only taking into account the case in which the energy-momentum-stress tensor is the zero tensor – the case in which Einstein's law equates the Einstein tensor with the zero tensor, hence expressing the vanishing of this invariant tensor.

Consequently, differentiating his law from Einstein's law, Whitehead writes that "the essence of the divergence of the two methods lies in the fact that my law of gravitation is not expressed as the vanishing of an invariant expression, but in the more familiar way by the expression of the ten elements in terms of [...] what I call the 'associate potential."' (*ESP* 342) Whitehead means that the elements of Einstein's fundamental tensor are determined by solving the equation 'Einstein tensor = 0,' whereas the elements of his impetus tensor are determined by some kind of potential.

The gravitational potential Whitehead refers to, is a scalar potential satisfying the wave equation. In fact, it is the scalar and gravitational equivalent of the retarded Liénard-Wiechert four-vector potential of electrodynamics, as expressed by Cunningham in 1914. So, whereas in Einstein's procedure the ten elements of the fundamental tensor are solutions of a tensor equation, Whitehead's alternative defines the ten elements of the gravitational field tensor in terms of a single scalar – a Liénard-Wiechert-like retarded potential satisfying the familiar wave equation. In other words, Whitehead fulfils the hope that Minkowski expressed in "Space and Time," and describes the gravitodynamic relation between point-masses in terms of an electrodynamic-like retarded potential.

Enough has been said about Whitehead's 1920 article to claim that Whitehead's alternative theory of gravity was a Minkowkian theory, and that it was largely developed by February 1920. In his *Enquiry*, Whitehead had not explicitly addressed, let alone answered, the question: "How to reinterpret Einstein's new law of gravitation in terms of a gravitational field against the background of Minkowski's space-time?" However, the April Preface of his 1919 book *did* make it clear that this was the next question on his agenda. Hence we are led to the conclusion that Whitehead developed his alternative theory of gravitation during the academic year 1919–1920, most likely in conjunction with the development of his postgraduate lecture course, "Relativity and the Nature of Space."[48]

Apart from his 1920 article, "Einstein's Theory," Whitehead also gave an outline of his Minkowskian theory of gravitation in his 1920 lecture for the students of

[48] Lowe (1990: 65).

the Chemical Society of the Imperial College of Science and Technology (Chapter VIII of *CN*). However, his most elaborated account was offered in 1922, in *The Principle of Relativity*. For further details on its content, I must refer the reader to the book itself, for the aim of my paper was not to present a fully-fledged account of Whitehead's theory of gravitation. My only aim was to reveal the Minkowskian background of Whitehead's theory.

References

Alexander, Samuel. *Space, Time and Deity*. 1927 Impression of the 1920 First Edition. Kessinger Publishing's Rare Reprints.

Broad, Charles Dunbar. *Scientific Thought*. London: Kegan Paul, Trench, Trubner & Co., 1923.

Carr, Herbert Wildon. "Einstein's Theory." *The Times, Educational Supplement* of January 22, 1920: 47.

Carr, Herbert Wildon. *The General Principle of Relativity, In Its Philosophical and Historical Aspect*. London: MacMillan, 1920.

Corry, Leo. *David Hilbert and the Axiomatization of Physics (1898–1918)*. Dordrecht: Kluwer Academic, 2004.

Crelinsten, Jeffrey. *Einstein's Jury: The Race to Test Relativity*. Princeton: Princeton Universtiy Press, 2006.

Cunningham, Ebenezer. *The Principle of Relativity*. Cambridge: Cambridge University Press, 1914.

Cunningham, Ebenezer. *Relativity and the Electron Theory*. Monographs on Physics. London: Longmans, Green & Co., 1915.

De Sitter, Willem. "Space, Time, and Gravitation." 1916a. *The Observatory* 505 (October 1916): 412–419.

De Sitter, Willem. "On Einstein's Theory of Gravitation, and Its Astronomical Consequences." 1916b. *Monthly Notices of the Royal Astronomical Society* LXXVI (Supplementary Number 1916): 699–728.

De Sitter, Willem. "On Einstein's Theory of Gravitation, and Its Astronomical Consequences." 1916c. *Monthly Notices of the Royal Astronomical Society* LXXVII (December 1916): 155–184.

De Sitter, Willem. "On Einstein's Theory of Gravitation, and Its Astronomical Consequences." *Monthly Notices of the Royal Astronomical Society* LXXVIII. I, (November 1917): 3–28.

Dingle, Herbert. *Relativity for All*. London: Methuen & Co., 1922.

Douglas, Alice Vibert. *Arthur Stanley Eddington*. Toronto and New York: Thomas Nelson and Sons, 1957.

Duerbeck, Hilmar W. & Flin, Piotr. "Ludwik Silberstein – Einsteins Antagonist." *Einsteins Kosmos*. Eds. H.W. Duerbeck & W.R. Dick, Frankfurt am Main: Verlag Harri Deutch, 2005: 186–209.

Duerbeck, Hilmar W. & Flin, Piotr. "Silberstein, General Relativity and Cosmology." *Albert Einstein Century International Conference*. Eds. J.-M. Alimi & A. Füzfa. New York: American Institute of Physics, 2006: 1087–1094.

Eddington, Arthur. "Gravitation and The Principle of Relativity." *Nature* Vol. 98 (2461) (December 28, 1916): 328–330.

Eddington, Arthur. *Report on the Relativity Theory of Gravitation*. 1918. New York: Dover Phoenix Editions, 2006.

Einstein, Albert. "The Foundations of the General Theory of Relativity." 1916. *The Principle of Relativity*. Ed. A. Sommerfeld. New York: Dover, 1952: 109–164.

Einstein, Albert. *The Collected Papers of Albert Einstein. Volume 8. The Berlin Years: Correspondence, 1914–1918, Part A: 1914–1917*. Eds. R. Schulmann, A.J. Kox, M. Janssen & J. Illy. Princeton: Princeton University Press, 1998.

Einstein, Albert. "On the Present State of the Problem Of Gravitation." *The Genesis of General Relativity: Volume 3*. Ed. J. Renn. Springer: Dordrecht, 2007: 543–566.

Freundlich, Erwin. *The Foundations of Einstein's Theory of Gravitation*. Cambridge: Cambridge University Press, 1920 – translated by Henry Brose.

Haldane, Richard. *The Reign of Relativity*. New Haven: Yale University Press, 1921.

Janssen, Michel. "The Einstein-De Sitter-Weyl-Klein Debate." Cf. Einstein 1998: 351–357.

Lindemann, Frederick. "Einstein's Theory: A Revolution in Thought." *The Times, Educational Supplement* of January 29, 1920: 59.

Lowe, Victor. *Understanding Whitehead*. Baltimore: The John Hopkins Press, 1966.

Lowe, Victor. *Alfred North Whitehead: The Man and His Work Volume I: 1861–1910*. Baltimore: The John Hopkins University Press, 1985.

Lowe, Victor. *Alfred North Whitehead: The Man and His Work; Volume II: 1910–1947*. Ed. J.B. Schneewind. Baltimore: The John Hopkins University Press, 1990.

McCausland, Ian. "Anomalies in the History of Relativity." *Journal of Scientific Explorations* Vol. 13(2), 1999: 271–290.

Minkowski, Hermann. "Die Grundgleichungen für die elektromagnetischen Vorgänge in bewegten Körpern." *Mathematische Annalen* 68, 1910: 472–525.

Minkowski, Hermann. "Space and Time." *The Principle of Relativity*. Ed. A. Sommerfeld. New York: Dover, 1952: 73–96.

Minkowski, Hermann. "Mechanics and the Relativity Postulate." *The Genesis of General Relativity: Volume 3*. Ed. J. Renn. Springer: Dordrecht, 2007: 273–285.

Nunn, T. Percy. *Relativity and Gravitation*. London: University of London Press, 1923.

Pais, Abraham. *'Subtle is the Lord...' The Science and Life of Albert Einstein*. Oxford: Oxford University Press, 1983.

Realdi, Matteo. "The Universe of Willem de Sitter." 2007. http://www.phil-inst.hu/~szekely/PIRT_BUDAPEST.

Russell, Bertrand. *My Philosophical Development*. 1959. London: George Allen and Unwin, 1975.

Sanchez-Ron, José M. "The Reception of General Relativity Among British Physicists and Mathematicians (1915–1930)." *Studies in the History of General Relativity*. Eds. J. Eisenstaedt & A.J. Kox. Boston: Birkhäuser, 1992: 57–88.

Silberstein, Ludwik. *The Theory of Relativity*. London: Macmillan and co., 1914.

Silberstein, Ludwik. "General Relativity Without the Equivalence Hypothesis." *Philosophical Magazine*. 36, 1918: 94–128.

Stachel, John. *Einstein from 'B' to 'Z'*. Boston: Birkhäuser, 2002.

Temple, George. "A Generalisation of Professor Whitehead's Theory of Relativity." *Proceedings of the Physical Society of London*. 36, 1923: 176–193.

Walter, Scott. "The Non-Euclidean style of Minkowskian Relativity." 1999a. *The Symbolic Universe: Geometry and Physics 1890–1930*. Ed. J. Gray. Oxford: Oxford University Press, 1999: 91–127.

Walter, Scott. "Minkowski, Mathematicians, and the Mathematical Theory of Relativity." 1999b. *The Expanding Worlds of General Relativity*. Eds. H. Groenner, J. Renn, J. Ritter & T. Sauer. Boston: Birkhäuser, 1999: 45–86.

Walter, Scott. "Breaking in the 4-Vectors: The Four-Dimensional Movement In Gravitation, 1905–1910." *The Genesis of General Relativity: Volume 3*. Ed. J. Renn. Dordrecht: Springer, 2007: 193–252.

Warwick, Andrew. *Masters of Theory: Cambridge and the Rise of Mathematical Physics*. Chicago: The University of Chicago Press, 2003.

Whitehead, Alfred North. In chronological order:

UA. *A Treatise on Universal Algebra, with Applications*. 1898. New York: Hafner Publishing, 1960.

MCMC. "On Mathematical Concepts of the Material World." 1905/1906. *Alfred North Whitehead: An Anthology*. Eds. F.S.C. Northrop & M.W. Gross. New York: MacMillan, 1961: 7–82.

OT. *The Organisation of Thought*. 1917. Westport Connecticut: Greenwood Press, 1974.
PNK. *An Enquiry Concerning the Principles of Natural Knowledge*. 1919. New York: Dover, 1982.
RS. "A Revolution of Science." *The Nation* of November 15, 1919: 232–233.
CN. *Concept of Nature*. 1920. Cambridge: Cambridge University Press, 1986.
R. *The Principle of Relativity with Applications to Physical Science*. Cambridge: Cambridge University Press, 1922.
SMW. *Science and the Modern World*. 1925. New York: The Free Press, 1967.
AI. *Adventures of Ideas*. Cambridge: Cambridge University Press, 1933.
ESP. *Essays in Science and Philosophy*. 1947. Westport Connecticut: Greenwood Press, 1968.
IS. *Interpretation of Science: Selected Essays*. Ed. A.H. Johnson. New York: Bobbs-Merril Company, 1961.
Wilson, Edwin Bidwell & Lewis, Gilbert Newton. "The Space-Time Manifold of Relativity." *Proceedings of the American Academy of Arts and Sciences*. 48(11), 1912: 389–507.

The Experimental Verdict on Spacetime from Gravity Probe B

James Overduin

Abstract Concepts of space and time have been closely connected with matter since the time of the ancient Greeks. The history of these ideas is briefly reviewed, focusing on the debate between "absolute" and "relational" views of space and time and their influence on Einstein's theory of general relativity, as formulated in the language of four-dimensional spacetime by Minkowski in 1908. After a brief detour through Minkowski's modern-day legacy in higher dimensions, an overview is given of the current experimental status of general relativity. Gravity Probe B is the first test of this theory to focus on spin, and the first to produce direct and unambiguous detections of the geodetic effect (warped spacetime tugs on a spinning gyroscope) and the frame-dragging effect (the spinning earth pulls spacetime around with it). These effects have important implications for astrophysics, cosmology and the origin of inertia. Philosophically, they might also be viewed as tests of the propositions that spacetime acts on matter (geodetic effect) and that matter acts back on spacetime (frame-dragging effect).

1 Space and Time Before Minkowski

The Stoic philosopher Zeno of Elea, author of Zeno's paradoxes (c. 490-430 BCE), is said to have held that space and time were unreal since they could neither act nor be acted upon by matter [1]. This is perhaps the earliest version of the *relational view* of space and time, a view whose philosophical fortunes have waxed and waned with the centuries, but which has exercised enormous influence on physics. The opposing *absolutist view*, that space and time do possess independent existence apart from matter, has an equally distinguished history that might be traced back to the Stoics' philosophical rivals, the Epicureans, whose founder Leucippus of Abdera (active c. 450 BCE) introduced the concept of a pre-existing void as the "emptiness between atoms" [2]. The earliest explicit statement of the absolutist view has been attributed by Max Jammer to the Pythagorean philosopher Archytas (428-347 BCE): "Since everything which is moved into a certain place, it is plain that the place where the thing moving or being moved shall be, must exist first" [3].

V. Petkov (ed.), *Space, Time, and Spacetime*, Fundamental Theories of Physics 167, DOI 10.1007/978-3-642-13538-5_2, © Springer-Verlag Berlin Heidelberg 2010

Aristotle (384-322 BCE) constructed a hybrid of the absolute and relational views. He accepted arguments similar to that of Archytas, but was deeply unhappy with the atomistic idea of void, "since no preference can be given to one line of motion more than to another, inasmuch as the void, as such, is incapable of differentiation ... how [then] can there be any natural movement in the undifferentiated limitless void?" To get around this difficulty Aristotle developed the arguably relational idea that space is defined by that which contains it. He was led in this way (in the *Physics*) to his influential picture of a cosmos pinned simultaneously to the center of the earth and the firmanent of fixed stars: "The center of the universe and the inner surface of the revolving heavens constitute the supreme 'below' and the supreme 'above'; the former being absolutely stable, and the latter constant in its position as a whole." Such was Aristotle's authority that few questioned it for two millenia. An exception was John Philoponus (c. 490-570), who argued for a more purely absolute picture and reacted in particular against the idea that space is somehow defined by that which contains it: "Place is *not* the adjacent part of the surrounding body ... It is a given interval, measurable in three dimensions; it is distinct from the bodies in it, and is, by its very nature, incorporeal. In other words, it is the dimensions alone, devoid of any body."

Claudius Ptolemy (c. 85-165) elaborated on Aristotle's system, using only circular motions and uniform speeds so as to "save the phenomena" in the face of increasingly accurate observations. However, the way in which he did so points up the limited extent to which Aristotle's thinking can truly be considered relational. The fact that the "firmanent of fixed stars" and "center of the earth" defined the rest frame of Aristotle's cosmos did not mean that space was *physically* anchored to the matter making up the earth or stars. Rather it so happened that these referents stood still in a background space that was more properly conceived as existing absolutely. Thus, adopting an earlier idea of Hipparchus, Ptolemy first detached the sun's "orbit" from the center of the earth (giving it an "eccentricity"). Later he added planetary "deferents," "epicycles" and finally "equants"–all reference points or paths in *empty space* (some of them even with inherent motions of their own). These so-called "void points" make sense only with respect to absolute space–or perhaps to "matter" of a divine kind, as hinted at in the *Almagest*: "The first cause of the first motion of the universe, if one considers it simply, can be thought of as an invisible and motionless deity." Here Ptolemy anticipated Newton, who would later refer to absolute space (in the *Opticks*) as the "sensorium" of God.

The nature of time as well as space was eagerly debated in this way by the ancients. The Epicurean philosopher Lucretius (c. 99-55 BCE) may have been the first to argue explicitly for a relational view of time, writing in *The Nature of the Universe* that: "Time by itself does not exist ... It must not be claimed that anyone can sense time by itself apart from the movement of things." Saint Augustine (354–430) put a theological twist on this argument in his *Confessions*, emphasizing that "God created the world *with* time, not in time."

Nicolaus Copernicus (1473–1543) relocated the center of Aristotle's universe from the earth to the sun. This step was not quite so daring as often thought, for Hipparchus and Ptolemy had already nudged the sun's "orbit" away from the center of the earth by introducing "eccentricity." As Copernicus himself noted near the beginning of *De Revolutionibus*: "Nothing prevents the earth from moving ... For, it is not the center of all the revolutions." Furthermore, although he re-centered the cosmos kinematically on the sun, Copernicus did not attach space dynamically to the rest frame of the sun or any other physical body, but followed Aristotle in associating it with the metaphysical "sphere of the fixed stars," which (he wrote): "contains itself and everything, and is therefore immovable. It is unquestionably the place of the universe, to which the motion and position of all the other heavenly bodies are compared."

Fifty years later, the notion of rigid planetary spheres could no longer be reconciled with astronomical observations, leading Johannes Kepler (1571–1630) to declare: "From henceforth the planets follow their paths through the ether like the birds in the air. We must therefore philosophize about these things differently." Thoughts such as these led him to the radical idea of attaching the rest frame of space to *physical bodies* rather than a metaphysical construct such as absolute space (he conceived of forces extending outward from the sun and sweeping the planets along in their orbits). The laws of planetary motion that he subsequently derived have been wonderfully characterized by Julian Barbour as a "pre-Machian triumph of Mach's Principle" [2].

A similar shift in thinking is apparent in Galilei Galileo (1564–1642). Rather than identifying the fixed stars with the rest frame of space in an abstract sense, he asserted (in the *Dialogo*) that they are *physically* at rest in space: "The fixed stars (which are so many suns) agree with our sun in enjoying perpetual rest." However, Galileo did not further define this state of "rest," and appears to have implicitly adopted the absolutist view. In fact he was the first to use the actual term "absolute motion," in his theory of the tides. Réné Descartes (1596–1650) also relied on the concept of absolute space (which he referred to as a "plenum") in arriving at something similar to Newton's eventual first law of motion. After learning of Galileo's trial by the Inquisition, however, he put off publishing his results by more than a decade and eventually prefaced them (in the *Principia Philosophiae*) by a disclaimer stating that all motion was, after all, relative. He may have been the first to hold both absolutist and relational views at the same time.

This inconsistency irritated Isaac Newton (Fig. 1), who complained in *De Gravitatione* that if all motion was really relative as Descartes said, then "it follows that a moving body has no determinate velocity and no definite line in which it moves." It was partly to do away with any such confusion that he expressed himself so

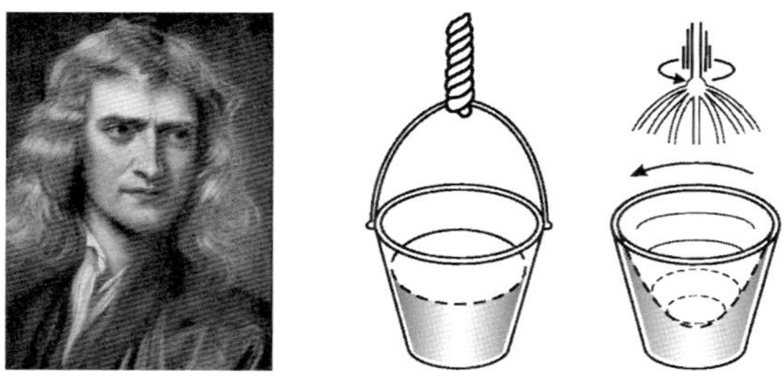

Fig. 1 Isaac Newton (1643–1727) and his bucket experiment: the concavity of the water's surface indicates that the water is rotating with respect to "absolute space"

categorically in the famous opening of his *Principia*: "Absolute, true and mathematical time, of itself and from its own nature, flows equably without relation to anything external ... absolute space, in its own nature, without relation to anything external, remains always similar and immoveable." He added that the existence of absolute space could be demonstrated by watching the water in a spinning bucket. The fact that the water's surface gradually assumed a concave shape showed that it was spinning with respect to *something*; how else would it know what to do? Proof of the reality of space, in other words, could be found in the inertia of matter.

Newton's most formidable relational critic was the mathematician and philosopher Gottfried Wilhelm Leibniz (1646–1716), who retorted (in a letter to Christiaan Huygens): "If there are 1,000 bodies, I still hold that ... each separately could be considered as being at rest ... Mr. Newton recognizes the equivalence of hypotheses in the case of rectilinear motion, but with regard to circular motion he believes that the effort which revolving bodies make to recede from the axis of rotation enables one to know their absolute motion. But I have reasons for believing that *nothing* breaks this general law of equivalence" [4]. The philosopher Bishop George Berkeley (1685–1753) went even farther, writing in *De Motu* that the very concept of two bodies "moving" around a common center is meaningless in empty space, since a co-rotating observer will not see anything change. "Suppose," however, "that the sky of fixed stars is created; suddenly from the conception of the approach of the globes to different parts of the sky the motion will be conceived."

If Newton's was the definitive statement of the absolutist view of space, then his most notorious relational counterpart was Ernst Mach (Fig. 2), who addressed himself directly to Newton's bucket argument, writing in *The Science of Mechanics*: "No one is competent to say how the experiment would turn out if the sides of the vessel increased in thickness and mass until they were ultimately several leagues thick." A sufficiently large or massive bucket, in other words, might carry the local inertial frame of the water around *with it* and leave the water's surface flat. This was perhaps the first explicit, though physically incomplete suggestion of a phenomenon now generally referred to as *frame-dragging*.

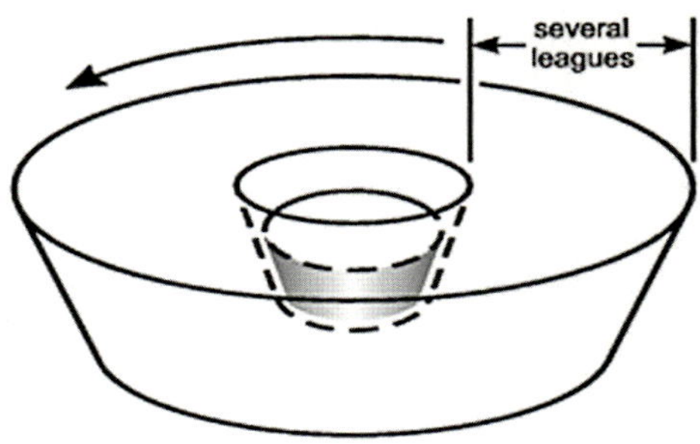

Fig. 2 Ernst Mach (1838–1916) and his revision of Newton's bucket experiment: would the water still climb up the walls if the bucket were *arbitrarily large and massive?*

Mach's principle, as his rather vague suggestion has come to be known, has proved stubbornly difficult to formulate in a precise physical way, and even more difficult to test experimentally. At a conference on this subject in Tübingen in 1993, leading experts discussed at least 21 different versions of "Mach's principle" in the scientific literature, some of them mutually contradictory [5]. It is probably for this reason that Mach's relational ideas have proved to be more inspirational than fruitful in physics. Nevertheless they led to some fascinating experimental investigations, even before Einstein's time. In 1894 the German vulcanologist Immanuel Friedländer (1871–1948) and his brother Benedict (1866–1908) looked for evidence that heavy rotating millstones could exert a Mach-type force on a sensitive torsion balance, and confessed (in *Absolute or Relative Motion?*) that they could find no definite results either way. In 1904, fellow German physicist August Föppl (1854–1924) published the results of an experiment designed to detect the influence of the rotating Earth on the angular momentum of a pair of heavy flywheels whose spin axis could be aligned along either lines of latitude or longitude (Fig. 3). He too found nothing, but noted that his accuracy was limited to about 2%.

Experiments like Gravity Probe B should not be seen as tests of Mach's principle (which is ill-defined as it stands), but rather as tests of specific theories of gravity (which may or may not incorporate well-defined "Machian" features such as frame-dragging). Nevertheless, it is possible to think of Gravity Probe B as a realization of the experiment suggested by Mach (and actually attempted by Föppl) in which the role of the "bucket" is played by the earth and the dragging of local inertial

Fig. 3 Early experimenters in frame-dragging: Benedict Friedländer (*top left*), August Föppl (*bottom left*), Föppl's experimental apparatus (*right*)

frames is measured not by water but by orbiting gyroscopes over a million times more sensitive than the best navigational gyros on earth.

2 Spacetime After Minkowski

Albert Einstein (1879–1955) radically re-ordered the traditional priorities of metaphysics when he showed in 1905 that there is a quantity more fundamental than either space or time, namely the speed of light c. Space and time are interconvertible, and must be so in order to preserve the constancy of c for all observers. The geometrical inference that space and time could be seen as components of a single four-dimensional spacetime fabric came from Hermann Minkowski (Fig. 4), who announced it in Cologne 100 years ago with the words:

> "Henceforth space by itself, and time by itself, are doomed to fade away into mere shadows, and only a kind of union of the two will preserve an independent reality."

Einstein initially dismissed Minkowski's four-dimensional interpretation of his theory as "superfluous learnedness" [6]. To his credit, he quickly changed his mind. The language of spacetime (tensor calculus) proved to be essential in making the transition from special to general relativity.

This transition required two main steps, a physical one and a mathematical one, and both relied crucially on Minkowski's spacetime picture. The physical step occurred in 1907 when, in the patent office in Bern, Einstein was struck by what he later called his "happiest thought": a man falling off the side of a building feels no gravity. The significance of this observation lies in the fact that the *same* choice of accelerated coordinates suffices to transform away the earth's gravitational field, *regardless of who or what is dropped*. If gravity were like any other force–electromagnetism, say–differently charged objects would "fall" quite differently, some of them even accelerating upward. By contrast, gravity appears matter-blind. From this observational fact (now known as the *equivalence principle*) Einstein leapt to the spectacular inference that gravitation must originate, not in any property of matter, but in spacetime itself. He eventually identified the relevant property of

Fig. 4 Hermann Minkowski (1864–1909)

spacetime as its curvature. This idea is the physical foundation of general relativity, succinctly summarized by John Wheeler as: "Spacetime tells matter how to move; matter tells spacetime how to curve." But while the resulting theory has been very successful, Einstein himself saw it as incomplete. In particular, he was unhappy with its dualistic division of physical reality into "spacetime" and "matter," describing these in 1936 as being like two wings of the same building, one made of "fine marble ... the other of low-grade wood." In the 1956 edition of *The Meaning of Relativity*, published in the year after his death, he still expressed the belief that this distinction would prove to be a temporary one: "In reality space will probably be of a uniform character and the present theory be valid only as a limiting case." If indeed matter and spacetime could be described as aspects of a single, unified field (as many physicists still hope), the very philosophical distinction between "relational" and "absolute" points of view might lose its meaning.

The second, more mathematical step toward general relativity was the search for a way to describe the dynamics of curved spacetime in a way that would hold for all observers–even accelerating ones–regardless of their choice of coordinates. By contrast with the equivalence principle, this principle (known as *general covariance*) did not arrive in a flash but required years of difficult slogging through the forest of tensor analysis (Fig. 5). Einstein memorably described the goal of expressing physical laws without coordinates as "equivalent to describing thoughts without words."

Today it is commonplace to speak of equivalence and general covariance as the two foundations of general relativity. In 1918, however, Einstein himself identified a third, philosophical pillar of his theory: *Mach's principle*. This characterization is now widely regarded as wishful thinking. Einstein was undoubtedly inspired by Mach's relational views, and initially hoped that his new theory of gravitation would "secure the relativization of inertia" by binding spacetime so tightly to matter that one could not exist without the other. In fact, however, the equations of general relativity are perfectly consistent with spacetimes that contain no matter at all. Flat (Minkowski) spacetime is a trivial example, but empty spacetime can also be curved, as demonstrated by Willem de Sitter in 1916. There are even spacetimes whose distant reaches rotate endlessly around the sky relative to an observer's local inertial frame, as demonstrated by Kurt Gödel in 1949. The bare existence of such solutions

Fig. 5 Albert Einstein in 1916

in Einstein's theory shows that it cannot be Machian in any strong sense; matter and spacetime remain logically independent. The term "general relativity" is thus something of a misnomer, as emphasized by Minkowski and others since. The theory does *not* make spacetime more relational than it was in special relativity. Just the opposite is true: the absolute space and time of Newton are retained. They are merely amalgamated and endowed with a more flexible mathematical skeleton (the metric tensor). When this became clear, Einstein's interest in Mach faded, and he wrote to a colleague in 1954: "As a matter of fact, one should no longer speak of Mach's principle at all."

Nevertheless, Einstein's theory of gravity represents a major swing back toward the relational view of space and time, in that it answers the objection of the ancient Stoics. Space and time *do* act on matter, by guiding the way it moves. And matter *does* act back on spacetime, by warping and twisting it. Perhaps nowhere is this more strikingly illustrated than in the two effects Gravity Probe B is designed to detect directly for the first time: the *geodetic effect*, in which curved spacetime around the massive earth causes an orbiting gyroscope to precess about an axis perpendicular to the plane of its motion; and the *frame-dragging effect*, in which the rotating earth pulls spacetime around with it, twisting the gyroscope's spin axis along the equatorial plane (Fig. 6). In that sense, general relativity is indeed nearly as relational as Mach might have wished. Some physicists, most notably Julian Barbour, have asserted that general relativity is in fact perfectly Machian, at least for closed (i.e. finite) spacetimes [7]. Key to this claim is the argument that allegedly "un-Machian" empty spacetimes (like those of Minkowski and de Sitter) are idealizations that do not take gravitational degrees of freedom into account. (The idea that gravitational radiation is responsible for transmitting inertia between mutually accelerated masses has been explored by Dennis Sciama [8], John Wheeler [9] and

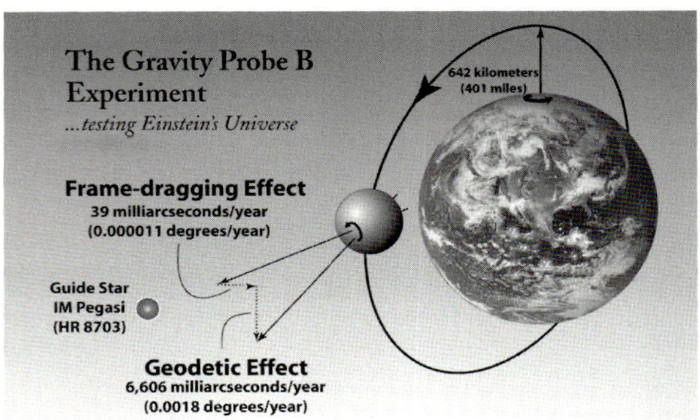

Fig. 6 In Einstein's theory of general relativity, spacetime acts on matter through its curvature, causing the spin axis of a gyroscope in orbit around a large mass like the earth to "fall into" the direction of travel (geodetic effect). Matter acts back on spacetime, not only by curving it, but also by *pulling spacetime with it*, causing the spin axis of a gyroscope to precess in the direction of the earth's rotation (frame-dragging effect). Gravity Probe B is designed to detect both of these effects directly and unambiguously for the first time

others [10].) In the context of modern quantum field theory, the distinction between absolute and relational views of spacetime breaks down as "empty space" becomes populated not only by gravitational waves but also by matter in the form of virtual particles, zero-point fields, etc. [11]. Within the classical world of Minkowski and Einstein, however, the majority view might best be summed up as follows: *spacetime behaves relationally but exists absolutely.*

3 Minkowski's Legacy in Higher Dimensions

By uniting space and time in a common metrical framework Minkowski shattered the prejudice, going back to the ancient Pythagoreans, that geometry applies only to *lengthlike quantities*. He was the first to make such a proposal in the context of a fully realized physical theory (special relativity) and it is entirely appropriate to consider him the father of spacetime. Nevertheless there were intriguing precursors for such a union before 1908, and these may have helped to prepare the conceptual ground for the eventual acceptance of relativity theory.

Possibly the first to refer to time as a fourth dimension was the French mathematician Jean d'Alembert, in an article in the *Encyclopédie* that he co-edited with Denis Diderot in 1754. Mysteriously, d'Alembert attributed the idea to "an enlightened man of my acquaintance" [12]. This unnamed source is thought to be the French-Italian mathematician Joseph-Louis Lagrange, who though only 18 years old at the time of the publication of the *Encyclopédie*, later observed in *Theory of Analytical Functions* (1797) that with time as a fourth coordinate "one can regard mechanics as four-dimensional geometry."

The German philosopher Arthur Schopenhauer referred repeatedly to matter, motion and causation as equivalent to the "union of space and time" in *The World as Will and Representation* (1818). He was, however, not concerned with physics, but rather with staking out a philosophical position relative to his predecessor Immanuel Kant. By equating these concepts Schopenhauer aimed to reduce the number of mental categories that Kant had argued were necessary for the mind to make sense of experience. For both thinkers space and time were "united" mainly in the sense that they existed more as forms of perception than as features of any external reality. Schopenhauer in turn exerted tremendous influence on the composer Richard Wagner, whose opera *Parsifal* (1877) contains this fascinating exchange between two knights on their way to the temple of the holy grail: "I barely tread, yet seem already to have come so far ... You see, my son, time here becomes space."[1]

---

[1] Much scholarly ink has been spilt on this passage by Wagner; see for instance Hans Melderis' *Space-Time-Myth: Richard Wagner and modern science* [14]. The composer's debt to Kant and Schopenhauer is suggested by a letter he wrote while working on *Parsifal* in 1860: "Since time and space are merely our way of perceiving things, but otherwise have no reality, even the greatest tragic pain must be explicable to those who are truly clear-sighted as no more than the error of the individual" [13].

Ideas of unifying space and time were not restricted to Europe, as evidenced by this line from the prose poem *Eureka* (1848) by the American author Edgar Allan Poe: "... the considerations through which, in this Essay, we have proceeded step by step, enable us clearly and immediately to perceive that Space and Duration are one." This has sometimes been interpreted as a prophetic anticipation of relativity theory,[2] but it is likely that Poe was merely stressing in a literary way that the size and age of the visible universe are correlated via the speed of light (a fact that he used elsewhere in *Eureka* to present the germ of the first scientifically correct solution to Olbers' paradox in astronomy [16]).

A more mathematical precursor to the spacetime concept is found in the generalizations of complex numbers known as quaternions, invented by the Irish physicist and mathematician William Rowan Hamilton in 1843. The fact that these objects consist of one real (scalar) component plus an imaginary (three-vector) component led Hamilton to argue as follows: "Time is said to have only one dimension, and space to have three dimensions ... The mathematical quaternion partakes of both these elements; in technical language it may be said to be 'time plus space,' or 'space plus time' " [17]. But probably the most explicit anticipation of Minkowski came from "S.," an anonymous contributor to the British journal *Nature* in 1885, who wrote: "... there is a new three-dimensional space for each successive instant of time; and, by picturing to ourselves the aggregate formed by the successive positions in time-space of a given solid during a given time, we shall get the idea of a four-dimensional solid ..." [18]. "S." was likely the English mathematician James Joseph Sylvester [19]. In the wake of articles such as this, the idea of time as a fourth dimension seeped into public awareness, culminating in novels like H.G. Wells' *The Time Machine* (1895), whose hero opens the book by telling his listeners that "there is no difference between Time and any of the three dimensions of Space except that our consciousness moves along it."[3] One final illustration of the extent to which spacetime was in the air prior to Minkowski's pronouncement is the "New theory of space and time" (1901) of Hungarian philosopher Menyhért Palágyi, in which space and time were combined in a four-dimensional "flowing space" by means of mixed coordinates $x + it, y + it, z + it$ [20].[4]

Minkowski's 1908 geometrization of time via the relation $x^0 = ct$ was, of course, physically motivated by Einstein's successful union of Newtonian mechanics and Maxwellian electromagnetism in the form of special relativity. Given our present mania for further kinds of unification in higher dimensions, it is surprising that more physicists have not taken Minkowski's example to heart and attempted to expand the domain of geometry *beyond space and time*.

---

[2] Einstein was apparently familiar with Poe's *Eureka*, referring to it in 1934 as "a beautiful achievement of an unusually independent mind" [15].

[3] Hubert Goenner [20] makes the interesting observation that Minkowski could have read Wells' *Time Machine*, as it appeared in German translation in 1904.

[4] After learning of Minkowski's speech in 1908, Palágyi attempted unsuccessfully to claim priority for the discovery of spacetime.

Historically, the reluctance to consider new kinds of coordinates is a practical one: we see no evidence for extra dimensions at experimentally accessible scales of length, time and energy. The same objection applied in Minkowski's day. The reason why time and space appeared independent until 1908 is that the size of the dimension-transposing constant "c" that converts one into the other is many orders of magnitude larger than the characteristic speeds of everyday life. The main effect of the new coordinate in four-dimensional (4D) special relativity is to multiply familiar (non-relativistic) quantities by the factor

$$\gamma_{4D} = \frac{1}{\sqrt{1-(dx/c\,dt)^2}}. \tag{1}$$

When $dx/dt \ll c$, as is true nearly everywhere on earth outside modern particle accelerators, then $\gamma_{4D} \approx 1$ and spacetime looks like space.

Inspired by the unification of mechanics and electromagnetism in four dimensions, the Finnish physicist Gunnar Nordström (1914) and the German mathematician Theodor Kaluza (1921) hit upon the idea of further unifying electromagnetism and *gravity* by means of a fifth lengthlike coordinate $x^5 = \ell$ (Fig. 7).[5] Nordström's was a scalar theory of gravity that was soon proven incompatible with observation. Kaluza's, however, was a five-dimensional (5D) extension of Einstein's tensor theory. The resulting theory turned out to contain both standard general relativity *and* Maxwell's electromagnetism in four dimensions, a miracle that is nowadays understood as arising from the fact that U(1) gauge invariance is "added onto" Einstein's theory in the guise of invariance with respect to coordinate transformations along the extra dimension. To explain why this new coordinate is not seen in nature, Kaluza imposed a "cylinder condition" whereby 4D physics is essentially independent of ℓ by fiat. The Swedish physicist Oskar Klein (1926) showed that this independence could arise in a more natural way if the new coordinate had a circular topology

Fig. 7 First to consider extending Minkowski's spacetime with a fifth dimension: Gunnar Nordström (*left*), Theodor Kaluza (*center*) and Oskar Klein (*right*)

[5] The superscript "4" is generally reserved for an imaginary version of Minkowski's fourth coordinate x^0, written as $x^4 = ict$, which allows the metric of flat Minkowski space to be written in Euclidean form.

and a compact scale (below $\sim 10^{-18}$ cm). Compactified Kaluza-Klein theory was picked up by Einstein and Bergmann (1938), Jordan (1947) and others, and eventually re-emerged as the basis for nearly all higher-dimensional unified theories today, including string and M-theory [21].

However, while it has been immensely influential, Nordström, Kaluza and Klein's idea is less radical than Minkowski's in that the proposed new coordinate shares the lengthlike character of ordinary three-space. Philosophically, this represents a return to the Pythagorean prejudice that geometry should deal only in quantities that can be measured with a meter-stick. Others have been bolder. The remainder of this section is intended as a brief and undoubtedly incomplete introduction to some of the *non*-lengthlike coordinates that have been considered in the literature. Among the earliest such proposals were those of W. Band (1939) and O. Hara (1959) relating x^5 to particle *spin* [22, 23], and that of Y.B. Rumer, who proposed in 1949 a fifth coordinate based on *action* via $x^5 = S/mc$ [24]. Rumer applied this idea to what he termed "5-optics" and imposed a restriction (called the "requirement of physical admissibility") similar to Kaluza's cylinder condition [25]. Related work has been done more recently by Yu and Andreev [26].

At its most fundamental, physics deals with dimensions of length [L], time [T] and mass [M], so the most natural choice for a new "post-Minkowskian" coordinate is arguably one related to *mass* via either $x^5 = Gm/c^2$ or $x^5 = h/mc$. Newton's gravitational constant G (or alternatively Planck's constant h) is thereby promoted to the same dimension-transposing role as "c" in 4D special relativity. This proposal is most closely associated with P.S. Wesson and his collaborators beginning in 1983 [27, 28], though related ideas were discussed as early as 1967 by de Vos and Hilgevoord [29] and 1974 by Edmonds [30, 31].[6] In the context of *non*-compactified or Space-Time-Matter (STM) theory, where Kaluza's cylinder condition is relaxed in principle, the identification of x^5 with rest mass is suggested by several lines of argument including the fact that the 4D relativistic energy-momentum relation $p^\alpha p_\alpha = E^2 - c^2 p^2 = m^2 c^4$ reduces simply to $p^A p_A = 0$ in 5D (where $\alpha = 0, 1, 2, 3$ and $A = 0, 1, 2, 3, 5$); and that the 4D free-particle action principle $\delta(\int m ds) = 0$ is contained in the simpler 5D one $\delta(\int dS) = 0$ (where $ds^2 = g_{\alpha\beta} dx^\alpha dx^\beta$ and $dS^2 = g_{AB} dx^A dx^B$) [21, 33, 34]. The size of the dimension-transposing constant G/c^2 provides a natural explanation for the fact that no mass-like fifth dimension has yet been detected. "Velocity" along such a direction means *varying rest mass*, and the 5D generalization of (1) reads:

$$\gamma_{5D} = \frac{1}{\sqrt{1 - (1/c^2)(dx/dt)^2 \pm (G^2/c^6)(dm/dt)^2}}. \tag{2}$$

The factor $1/c^2$ is already small enough that time was not recognized as part of space until 1908. The factor G^2/c^6 is so much smaller yet (by 54 orders of magni-

[6] The idea that x^5 might be related to mass has also been independently attributed to M.A. Neacsu in 1981 [32].

tude in SI units) that it is no surprise that possible variation in, say, the rest masses of elementary particles has yet to be observed. Such variation, if it exists, likely takes place on cosmological scales. STM theory is consistent with the classical tests of general relativity in the solar system as well as cosmological and other experimental data [35–37]. The status of mass as a fifth coordinate has been most recently reviewed by Wesson in 2003 [38].

Additional *time*like dimensions in the context of Kaluza-Klein theory lead to the wrong sign for the Maxwell action, and to the appearance of tachyons (negative-mass eigenstates) in the theory. More generally, extra temporal dimensions raise the specter of causality violation via closed timelike curves (CTCs). For these reasons few physicists have taken Minkowski's example literally enough to posit additional coordinates with the dimensions of time (i.e., $x^5 = c\tau$). The earliest such proposal may be that of A.D. Sakharov in 1984 [39]. Sakharov considered even numbers of additional compact time dimensions and argued that causality could be preserved for macroscopic processes if the radius of compactification were suitably small. This work was further developed by Aref'eva and Volovich [40]. Other "two-time" theories have been propounded by Burakovsky and Horwitz [41], Bars and Kounnas [42], Wesson [43], Kociński and Wierzbicki [44], Erdem and Ün [45] and Quiros [46].

Other ways of extending Minkowski's four-dimensional spacetime have been considered as well. Fukui [47] studied the possibility of extra coordinates proportional to both mass and *charge* via $x^5 = \sqrt{G/c^4}\,q$. Such an identification goes somewhat against the spirit of Kaluza's original theory, in which electric charge arises in the form of momentum along $x^5 = \ell$. M. Carmeli, in his theory of cosmological special relativity [48], proposed a fifth coordinate proportional to *cosmological recession velocity* via $x^5 = v/H$, where H =Hubble's constant (the expansion rate of the universe). The resulting theory is intended to supplant general relativity on cosmological scales where, for example, it has been claimed to predict cosmic acceleration [49] and obviate the need for dark matter [50]. Additional kinds of "post-Minkowskian" coordinates have been explored by Redington [32], Matute [51] and Delbourgo [52] and others.

It is too early to say whether any of these candidate coordinates will eventually prove to be as useful as Minkowski's in unifying the laws of physics. If and when that happens, we may find ourselves amending his speech of 100 years ago only slightly so that, for instance: "Space and time by themselves, *and mass by itself*, are doomed to fade away into mere shadows, and ony a kind of union of the *three* will preserve an independent reality."

4 Experimental Tests: An Unfinished Job

General relativity, based on a flexible and animated version of Minkowski's four-dimensional spacetime, has survived over 90 years of experimental test. Nevertheless there are at least four good reasons to think that the theory is incomplete and must be overthrown just as Newton's was. First, general relativity predicts its own demise; it breaks down in *singularities*, regions where the curvature of spacetime becomes infinite and the field equations can no longer be applied. These cannot be dismissed as mere academic curiosities, because they do apparently occur in the real universe if general relativity holds. Theoretical work by S. Hawking, R. Penrose and others has proven that singularities must form within a finite time (the universe is necessarily "geodesically incomplete"), given only very generic assumptions such as the positivity of energy. Two places where we expect to find them are at the big bang, and inside black holes like the one at the center of the Milky Way. If we are to fully understand these phenomena, then general relativity must be modified or extended in some way.

Second, there is the question of cosmology. Under the reasonable assumptions that the universe on large scales is homogeneous and isotropic (the same in all places and in all directions), as implied by observation in combination with the Copernican principle, general relativity has led to a cosmological theory known as the big bang theory. This theory has had some spectacular successes; for instance, the prediction of the cosmic microwave background radiation, the calculation of the abundances of light elements, and a basis for understanding the origin of structure in the universe. It also has some weaknesses, notably involving finely tuned initial conditions (the "flatness" and "horizon problems"). More troublingly, in recent decades it has become impossible to match the predictions of big-bang cosmology with observation unless the thin density of matter observed in the universe (i.e. that which can be seen by emission or absorption of light, or inferred from consistency with light-element synthesis) is supplemented by much larger amounts of unseen *dark matter* and *dark energy* that cannot consist of anything in the standard model of particle physics. The observations are quite clear: the required exotic dark matter has a density some five times that of standard-model matter, and the required dark energy has an energy density some three times greater still. To date, there is no direct experimental evidence for the existence of either component, and there are strong theoretical reasons (the "cosmological constant problem") to be suspicious of dark energy in particular. There is also no convincing explanation of why two new and as-yet unobserved forms of matter-energy should be so closely matched in energy density (the "coincidence problem"). While the majority of cosmologists seem prepared to accept both dark matter and dark energy as necessary, if inelegant facts of life, others are beginning to interpret them as possible evidence of a breakdown of general relativity at large distances and/or small accelerations.

Third, existing tests of general relativity have been restricted to *weak gravitational fields* (or arguably moderate ones in the case of the binary pulsar). Major surprises in this regime would have been surprising, since Einstein's theory goes over to Newton's in the weak-field limit, and we know that Newtonian gravity works reasonably well. But surprises are quite possible, and even likely, in the strong-field regime, where we hope to see hints about the ways in which general relativity must be modified in order to unify it with the other forces of nature.

Fourth, Einstein's theory as it stands is *incompatible with the rest of physics* (i.e. the "standard model" based on quantum field theory). The problem stems from the fact that the gravitational field carries energy and thus "attracts itself" (by contrast the electromagnetic field, for example, carries no charge). In field-theory language, the quantization of gravity requires an infinite number of renormalization parameters. It is widely believed that our present theories of gravity and/or the other interactions are only approximate "effective field theories" that will eventually be seen as limiting cases of a unified theory in which all four forces become comparable in strength at very high energies. But there is no consensus as to whether it is general relativity or particle physics–or both–that must be modified, let alone how. Experimental input may be our only guide to unification.

Gravitational experiments can be divided into two kinds: those that test fundamental principles, and those that test individual theories (including general relativity). The fundamental principles include such basic axioms as local position invariance (or LPI; the outcome of any experiment should be independent of where or when it is performed) and local Lorentz invariance (or LLI; the outcome of any experiment should be independent of the velocity of the freely-falling reference frame in which it is performed). The fundamental principle of most direct physical relevance to general relativity is the *equivalence principle*, which predicts that different test bodies should accelerate the same way in the same gravitational field, independent of their mass or internal structure, provided they are small enough not to disturb the environment or to be affected by tidal forces. The *approximate* validity of this statement has been known to some since at least the sixth century, when John Philoponous noted in a critique of Aristotle that "the ratio of the times of fall of the bodies does not depend on the ratio of their weights." It is most famously associated with Galileo at the leaning tower of Pisa. Historians of science are divided on whether that particular event actually took place, and similar ones were reported decades earlier by other people such as the Flemish engineer Simon Stevin in 1586. However, Galileo was the first to understand the significance of the measurement, and pushed it further by using a variety of different materials including gold, lead, copper and stone. He also refined the experiment by rolling his test masses down inclined tables and eventually by using pendulums.

Many people have improved on these tests since, most notably Newton and Loránd Eötvös. Newton refined Galileo's pendulum experiments, and brilliantly perceived that celestial bodies could also serve as test masses (he checked that the earth and moon, as well as Jupiter and its satellites, fall at the same rate toward the sun). This idea was reintroduced as a test of the equivalence principle by K. Nordtvedt in the 1970s, and is now applied together with laser ranging to the moon to set an upper limit on any difference in lunar and terrestrial accelerations toward the sun at less than three parts in 10^{13} (this is particularly significant because the earth has a nickel-iron core while the moon is largely composed of silicates). Eötvös pioneered the use of the torsion balance, enabling a six-order-of-magnitude advance in sensitivity over pendulum tests. Torsion balances are still the basis for the best equivalence-principle tests today; these limit any difference between the accelerations of different kinds of test masses in the gravitational field of the sun (or in a component of the field of the earth) to less than two parts in 10^{13}.

Fig. 8 The proposed Satellite Test of the Equivalence Principle (STEP)

It may be possible to reach even higher accuracy in the future through the use of laser atom interferometry to measure the rates of fall of isotopes of the same element with slightly different atomic weight. In general, however, gravitational experiments on earth are subject to inherent limitations due to factors such as seismic noise, and it is likely that further significant increases in precision will require going into space. One such proposal, the Satellite Test of the Equivalence Principle (STEP), is currently under development at Stanford University (Fig. 8). STEP is conceptually a return to Galileo's free-fall method, but one in which pairs of test masses are continuously "dropped" inside an orbiting spacecraft, allowing for a longer integration time and a periodic rather than quadratic signal. It inherits key technologies from Gravity Probe B, including drag-free control and a cryogenic readout system. STEP's design sensitivity of one part in 10^{18} would make it a true test, not only of the foundation of general relativity, but also of theories that attempt to unify gravity with the standard model of particle physics [53].

The "three classical tests" of general relativity were historically inaugurated by Einstein's derivation of *gravitational redshift*. In fact, this effect follows from the equivalence principle alone, so it is not a test of general relativity per se and is more properly grouped with the fundamental tests. (Some have called it the "half" in Einstein's "two and a half classical tests.") A clock in a gravitational field is, by the equivalence principle, indistinguishable from an identical one in an accelerated frame of reference. The gravitational redshift is thus equivalent to a Doppler shift between two accelerating frames. The most precise measurement of this shift to date was carried out by R. Vessot and M. Levine in 1976 (Fig. 9). Known as Gravity Probe A, their experiment compared a hydrogen maser clock on earth to an identical one in orbit at about 10,000 km and confirmed expectations based on the equivalence principle to an accuracy of 0.02%. The modern-day Global Positioning System (GPS) also functions as a de facto confirmation of gravitational redshift.

Fig. 9 Robert Vessot and Martin Levine with the Gravity Probe A payload (1976)

GPS satellites must coordinate their time signals to about 30 ns in order to reach their specified civilian accuracy of about 10 m. This required precision in time is more than a thousand times smaller than the discrepancy between clocks on the surface and those aboard GPS satellites due to gravitational redshift, which must consequently be correct to at least 0.1% for GPS trackers to work.

The first true "classical test" of general relativity came with its successful explanation of the anomalous *perihelion shift* of the planet Mercury (the rate at which its orbit slews around the sun, as measured by its point of closest approach). This effect (along with most of the other gravitational tests) is now described in terms of a formalism invented by A. Eddington and later developed by K. Nordtvedt and C. Will into what is known as the Parametrized Post-Newtonian (PPN) framework. Here, weak, spherically-symmetric gravitational fields like that around the sun are modeled with two parameters γ (describing the warping of space) and β (describing the warping of time, or the nonlinearity of the theory). General relativity predicts that β and γ are both equal to one, and most of the experimental tests effectively place upper limits on $|\beta - 1|$ and/or $|\gamma - 1|$. Mercury's anomalous perihelion shift is proportional to $(2 + 2\gamma - \beta)/3$, which is equal to one in general relativity. Initial measurements relied on optical telescopes; modern ones are based on radar data and constrain any departure from general relativity to less than 0.3%. An important early source of systematic error came from uncertainty in solar oblateness (quadrupole moment), but this has now been well constrained from helioseismology. Perihelion shift has also been observed using radio telescopes in distant binary pulsar systems, where it is known as periastron shift.

Perihelion shift led to the rapid acceptance of general relativity among Einstein's peers but *light deflection*, the last of the three classical tests, brought him public fame. He had already found in 1911 that the equivalence principle implies some light deflection, since a beam of light sent horizontally across a room will appear to bend toward the floor if the room is accelerating upwards. In 1915, however, Einstein realized that space curvature doubles the size of the effect, and that it might be possible to detect it by observing the bending of light from background stars around the sun during a solar eclipse. Teams led by Eddington and A. Crommelin were able to confirm this prediction to an accuracy of about 30% during the eclipse of May 1919. The light deflection angle is proportional to $(1+\gamma)/2$, which is equal to one in general relativity. Constraints on γ from optical telescopes were superseded in the 1960s by the use of linked arrays of radio telescopes (Very Long-Baseline Interferometry or VLBI) to measure the deflection around the sun of radio waves from distant quasars. By 1995 these observations had confirmed general relativity to an accuracy of 0.04%. In cosmology, light deflection (better known as gravitational lensing) is used to weigh dark matter, measure the expansion rate of the universe and even function as a cosmic "magnifying glass" to bring the faintest and most distant objects into closer view.

The space age made possible what is sometimes known as a "fourth classical test" based on the *time delay* of light signals in a gravitational field. I.I. Shapiro realized in 1964 that if general relativity was correct, then a light signal sent past the sun to a planet or spacecraft would be slowed in the sun's gravitational field by an amount proportional to the light-bending factor, $(1+\gamma)/2$, and that it would be possible to measure this effect if the signal were reflected back to earth. Typical time delays are on the order of several hundred microseconds. Passive radar reflections from Mercury and Mars were consistent with general relativity to an accuracy of about 5%. Use of the Viking Mars lander as an active radar retransmitter in 1976 confirmed Einstein's theory at the 0.1% level. The most precise of all time delay experiments to date has involved Doppler tracking of the Cassini spacecraft on its way to Saturn in 2003; this limits any deviations from general relativity to less than 0.002%–the most stringent test of Einstein's theory so far.

Radio astronomy provided a fifth test in the form of the *binary pulsar*. General relativity predicts that a non-spherically-symmetric system (such as a pair of masses in orbit around each other) will lose energy through the emission of gravitational waves. While these waves themselves have not yet been detected directly, the loss of energy has. The evidence comes from binary systems containing at least one pulsar. Pulsars are rapidly rotating neutron stars that emit regular radio pulses from their magnetic poles. These pulses can be used to reconstruct the pulsar's orbital motions. The fact that these objects are neutron stars makes them particularly valuable as experimental probes because their gravitational fields are much stronger than those of the sun (thus providing arguably "moderate-field" tests of general relativity, if not strong-field in the sense that $Gm/c^2r \sim 1$). The first binary pulsar was discovered by R.A. Hulse and J.H. Taylor in 1974. Timing measurements produce three constraints on the two unknown masses plus one more quantity; when applied to the general-relativistic energy loss formula, the results are consistent at the 0.2% level. Several other relativistic binary systems have since been discovered, including one whose orbital plane is seen almost edge-on and another in which the companion is probably a white dwarf rather than a neutron star. Most compelling is a *double pulsar* system, in which radio pulses are detected from both stars. This imposes six constraints on the two unknown masses and allows for four independent tests of general relativity. The fact that all four are mutually consistent is itself impressive confirmation of the theory. After two and a half years of observation, the most precise of these tests (time delay) verifies Einstein's theory to within 0.05%.

The perihelion-shift, light-deflection and time-delay tests firmly establish the validity of general relativity in the slow-velocity, weak-field limit within the solar system. The binary pulsar provides extra-solar confirmation of these tests and also goes some way toward extending them into the "moderate-field" regime. But there is, as yet, little accurate confirmation of Einstein's theory for strong fields such as those found near neutron-star surfaces or black-hole horizons, or over distances on the scale of the galaxy or larger. Both these difficulties may be addressed by experimental efforts aimed at the *direct* detection of gravitational waves. Most of these efforts employ interferometers to measure the difference in displacement between the lengths of two perpendicular "arms" as they are alternately stretched and compressed by the waves' passage. No gravitational waves have been detected to date. This null result does not yet impose a meaningful constraint on general relativity because of the astrophysical uncertainties inherent in predicting the strength and number of gravitational wave sources in the universe, as well as the computational challenges in modeling the characteristics of the expected signals. The strongest limits so far, from the Laser Interferometry Gravity-wave Observatory (LIGO), imply that the most frequent source events (binary neutron-star mergers) occur no more than approximately once per year per galaxy. The best theoretical estimates imply that they would not be expected more than once per $10^4 - 10^5$ years per galaxy. An upgraded version of LIGO (Advanced LIGO) is currently under construction with at least ten times the initial sensitivity.

Ground-based detectors are sensitive primarily to the high-frequency gravitational waves produced by transient phenomena (explosions, collisions, inspiraling binaries). A complementary Laser Interferometer Space Antenna (LISA) is being planned jointly by NASA and ESA; this will use a trio of spacecraft arranged in an equilateral triangle with 5 million km-long arms to look for lower-frequency waves from quasi-periodic sources, like compact objects well before coalescence and mergers between the supermassive black holes thought to lie at the centers of galaxies. LISA will rely crucially on the drag-free technology proven by Gravity Probe B. If successful, it will go a long way toward confirming the validity of general relativity, not only for strong fields but also throughout the universe.

5 The Geodetic and Frame-Dragging Effects

There is one other regime in which general relativity has been poorly tested to date: *spin*. Einstein's theory predicts that the spin axis of a rotating test body will precess in a gravitational field (*geodetic effect*), and that it will undergo an additional precession if the source of the gravitational field is itself rotating (*frame-dragging effect*). These phenomena might be termed the sixth and seventh tests of general relativity. Gravity Probe B is designed to confirm or disprove them directly and unambiguously for the first time. However, it is important to note that the significance of these effects goes beyond testing Einstein's theory. The question of spin is particularly important from a fundamental point of view, because it is the intrinsic spin of elementary particles that poses one of the greatest obstacles to the geometrization of standard-model fields via higher spacetime dimensions, the most promising route to unification of these fields with gravity. Thus Nobel prizewinner C.N. Yang commented in 1983 that, general relativity, "though profoundly beautiful, is likely to be amended ... whatever [the] new geometrical symmetry will be, it is likely to entangle with spin and rotation, which are related to a deep geometrical concept called torsion ... The proposed Stanford experiment [Gravity Probe B] is especially interesting since it focuses on the spin. I would not be surprised at all if it gives a result in disagreement with Einstein's theory."

The physical content of both the geodetic and frame-dragging effects can be understood in terms of analogies with electromagnetism. (Such analogies go back to Michael Faraday's experiments with "gravitational induction" beginning in 1849.) When gravitational fields are weak and velocities are low compared to c, then it becomes feasible to perform a "3+1 split" and decompose 4D spacetime into a scalar or 0-dimensional "time–time" component, a vector or 1-dimensional "time-space" component and a tensor or 2-dimensional "space–space" component. If one calls the scalar component a "gravito-electric potential" and the vector one a "gravito-magnetic potential," then the "gravito-electric field" $\boldsymbol{g}$ and "gravito-magnetic field" $\boldsymbol{H}$ constructed in the usual way from the divergence and curl of these potentials turn out to obey equations that are nearly identical to Maxwell's equations and the Lorentz force law of ordinary electrodynamics. Based on this analogy, the geodetic and frame-dragging effects are sometimes referred to as "gravito-electricity" and "gravito-magnetism" respectively. However, such an identification must be used with care because the distinction between gravito-electricity and gravito-magnetism depends on the frame in which it is observed, just like its counterpart in Maxwell's theory. This means that observers using different coordinate systems as, for example, one centered on the earth and another on the barycenter of the solar system, may disagree on the relative size of the effects they are discussing.

It is possible to argue that these effects have already been observed *indirectly* in the solar system, since gravito-electromagnetic fields are a necessary manifestation of Einstein's gravitational field in the low-velocity, weak-field limit, and the validity of general relativity is now routinely assumed in, for instance, updating the ephemeris of planetary positions. In this sense, it would be surprising if an experiment like Gravity Probe B, which is designed to observe gravito-electromagnetic effects *directly*, did not see them. Such a result would suggest that general relativity needs to be extended or modified in some way such that terms involved in the geodetic and/or frame-dragging effects are strongly affected while leaving predictions for other post-Newtonian effects unchanged. Nevertheless, surprises do occur in science, and a surprise here would have major implications for unification of gravity with the rest of physics. On such fundamental questions, history has shown that there is no substitute for the direct test.

Symmetry considerations dictate that the earth's gravito-electric field must be radial and its gravito-magnetic one dipolar (Fig. 10). From these facts one can immediately write down the precessions due to the geodetic effect ($\boldsymbol{\Omega}_{\mathrm{g}}$) and

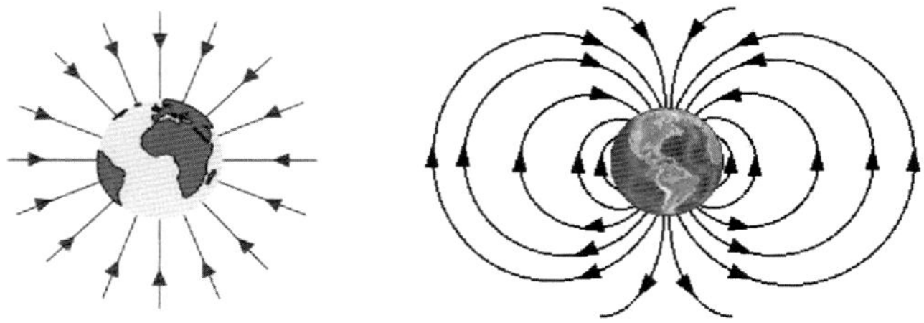

Fig. 10 The earth's radial gravito-electric field (*left*) and dipolar gravito-magnetic field (*right*)

frame-dragging effect ($\boldsymbol{\Omega}_{\mathrm{fd}}$) by referring to standard formulae governing the motion of a test charge in an external electromagnetic field, and replacing the electric and magnetic fields by $\boldsymbol{g}$ and $\boldsymbol{H}$ respectively. The result is:

$$\boldsymbol{\Omega}_{\mathrm{GR}} = \boldsymbol{\Omega}_{\mathrm{g}} + \boldsymbol{\Omega}_{\mathrm{fd}} = \frac{3GM}{2c^2r^3}(\boldsymbol{r} \times \boldsymbol{v}) + \frac{GI}{c^2r^3}\left[\frac{3\boldsymbol{r}}{r^2}(\boldsymbol{S} \cdot \boldsymbol{r}) - \boldsymbol{S}\right], \tag{3}$$

where M, I and $\boldsymbol{S}$ refer to the mass, moment of inertia and angular momentum of the central body and $\boldsymbol{r}$ and $\boldsymbol{v}$ are the radial position and instantaneous velocity of the test body. Equation (3) is sometimes referred to as the *Schiff formula* after Leonard I. Schiff, who derived it in 1959.

The geodetic or $\boldsymbol{r} \times \boldsymbol{v}$ term in (3) arises from the way that angular momentum is transported through a gravitational field. Einstein's Dutch friend and colleague Willem de Sitter (1872–1934; Fig. 11) began to study this problem in 1916 when general relativity was less than a year old. He found that the orbital angular momentum of the earth-moon system precesses in the field of the sun, a special case now referred to as the de Sitter or "solar geodetic" effect (although "heliodetic" might be more descriptive). De Sitter's calculation was extended to the spin angular momentum of rotating test bodies by two of his countrymen: in 1918 by the mathematician Jan Schouten (1883–1971) and in 1920 by the physicist and musician Adriaan Fokker (1887–1972). Eddington brought these results to the attention of the wider community in *The Mathematical Theory of Relativity* (1923), writing that "If the earth's rotation could be accurately measured ... by gyrostatic experiments, the result would differ from the rotation relative to the fixed stars." This was the germ of the idea that would eventually grow into Gravity Probe B.

In the framework of the gravito-electromagnetic analogy, the geodetic effect can be seen partly as a spin-orbit interaction between the spin of the gyroscope and the "mass current" of the rotating earth. This is the analog of Thomas precession in electromagnetism, where the electron experiences an induced magnetic field due to the apparent motion of the nucleus around it (in its rest frame). In the gravitomagnetic case, the gyroscope "feels" the massive earth orbiting around it (in its rest frame) and experiences an induced *gravito*-magnetic torque, causing its spin vector to precess. This spin-orbit interaction accounts for one third of the total geodetic precession; the other two thirds arise due to space curvature alone and cannot be easily interpreted

Fig. 11 Discoverers of the geodetic effect in general relativity: Willem de Sitter (*left*), Jan Schouten (*center*) and Adriaan Fokker (*right*)

gravito-electromagnetically. They can however be easily understood geometrically (Fig. 12). The gyroscope's spin vector remains always perpendicular to its plane of motion (arrows), and in flat space its direction remains constant as the gyroscope completes an orbit (left). If, however, space is folded into a cone to simulate the effect of curvature, then part of the area of the circle (shaded) must be removed and the gyroscope's spin vector no longer lines up with itself after one making one complete circuit (right). The angle between the spin vectors "before" and "after" produces the other two thirds of the geodetic effect. In the case of Gravity Probe B this is sometimes referred to as the phenomenon of the "missing inch" because space curvature shortens the circumference of the spacecraft's orbital path around the earth by 1.1 in. In polar orbit at an altitude of 642 km the total geodetic effect (comprising both the spin-orbit and space curvature effects) causes a gyroscope's spin axis to precess in the north-south direction by 6,606 milliarcseconds over the course of a year–an angle so small that it is comparable to the average angular size of the planet Mercury as seen from earth.

Experimental limits on geodetic precession place new constraints on a broad class of alternatives to Einstein's theory of gravity known as "metric theories" (loosely speaking, theories that differ from Einstein's but still respect the equivalence principle). These are characterized by the PPN parameters β and γ, both equal to one in general relativity [55]. The geodetic effect is proportional to $(1+2\gamma)/3$, so an experimental detection translates directly into a constraint on γ. It also probes other kinds of "generalizations of general relativity" such as those involving extra spacetime dimensions [35], scalar fields [56], torsion [57, 58] and violations of Lorentz invariance, the conceptual foundation of special relativity [59, 60].

The frame-dragging or $\boldsymbol{S}$-dependent term in (3) is smaller in magnitude than the geodetic one, but reveals more clearly the Machian aspect of Einstein's theory. In fact, it is curious that Einstein did not discover this effect himself, given that he had explicitly looked for dragging phenomena in his earlier attempts at gravitational field theories, and that he still attached enough importance to Mach's principle to refer to it as a pillar of general relativity in 1918. For whatever reason, frame-dragging within general relativity was first discussed that same year by Austrian physicists Hans Thirring (Fig. 13; 1888–1976) and Josef Lense (1890–1985); it is

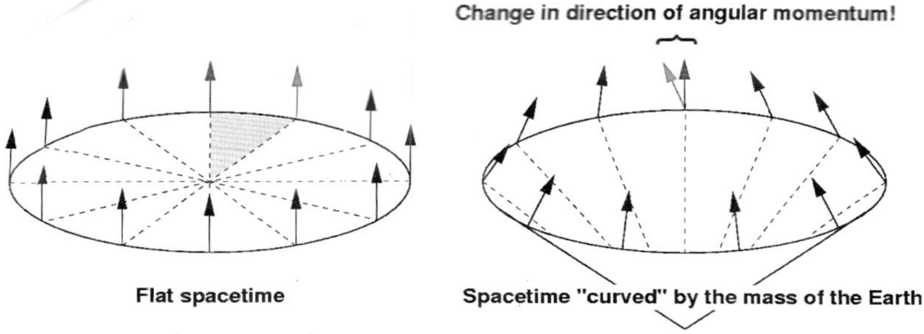

Fig. 12 Geodetic precession and the "missing inch"

often referred to as the Lense-Thirring effect. Thirring originally approached this problem as an experimentalist; he hoped to look for Mach-type dragging effects inside a massive rotating cylinder. Unable to raise the necessary financing, he reluctantly settled down to solve the problem theoretically instead [61]. It is his second calculation (with Lense) involving the field outside a slowly rotating solid sphere that forms the basis for modern gyroscopic tests. But both his results are "Machian" in the sense that the inertial reference frame of the test particle is influenced by the motion of the larger mass (the cylinder or sphere). This is completely unlike Newtonian dynamics, where local inertia arises entirely due to motion with respect to "absolute space" and is unaffected by the distribution of matter.

In terms of the gravito-electromagnetic analogy, frame-dragging is a manifestation of the *spin–spin* interaction between the test body and central mass. It is analogous to the interaction of a magnetic dipole $\boldsymbol{\mu}$ with a magnetic field $\boldsymbol{B}$ (the basis of nuclear Magnetic Resonance Imaging or MRI). Just as a torque $\boldsymbol{\mu} \times \boldsymbol{B}$ acts in the magnetic case, so a test body with spin $\boldsymbol{s}$ experiences a torque proportional to $\boldsymbol{s} \times \boldsymbol{H}$ in the gravitational case. In the case of Gravity Probe B, this torque causes the gyroscope spin axes to precess in the east-west direction by 39 milliarcseconds per year–an angle so tiny that it is equivalent to the average angular width of the dwarf planet *Pluto* as seen from earth.

The orbital plane of an artificial satellite is also a kind of "gyroscope" whose nodes (the points where it intersects a reference plane) will exhibit a similar frame-dragging precession (the de Sitter effect). Such an effect has been reported in the case of the earth-orbiting Laser Geodynamic Satellites (LAGEOS and LAGEOS II) by Ignazio Ciufolini and colleagues using laser ranging [62, 63]. This method of looking for frame-dragging is elegant and complementary to the more direct gyroscopic test. It is not definitive on its own because the general-relativistic effect (31 milliarcseconds per year at the LAGEOS altitude of 59,000 km) is swamped by Newtonian contributions that are as much as a billion times larger. To model or otherwise remove these terms necessarily involves systematic uncertainties whose magnitude is still a subject of debate [64–66].

In principle, frame-dragging imposes another new constraint on alternative metric theories of gravity. Lense-Thirring precession is proportional to the combination

Fig. 13 Discoverers of the frame-dragging effect in general relativity: Josef Lense (*left*) and Hans Thirring (*right*)

of PPN parameters $(\gamma+1+\alpha_1/4)/2$ where γ describes the warping of space and α_1 is a "preferred-frame" parameter that allows for a possible dependence on motion relative to the rest frame of the universe, taking the value zero in general relativity [55]. In practice, frame-dragging in the solar system is so weak that the experimental bounds it places on these parameters are not likely to be competitive with those from other tests. Seen purely as a testbed for distinguishing between alternative theories of gravity, therefore, frame-dragging has sometimes been dismissed as being of little practical interest.

That way of thinking has largely disappeared with the realization that frame-dragging takes on its true importance in the strong-field and cosmological regimes. Astrophysicists now invoke gravitomagnetism as the engine and alignment mechanism for the vast jets of gas and magnetic field ejected from quasars and galactic nuclei like the radio source NGC 6251 (Fig. 14, left). These jets are generated by compact objects at the centers of galactic nuclei that are almost certainly supermassive black holes (right). The megaparsec length scale of the jets implies that their direction is held constant over time scales as long as tens of millions of years. This can only be accomplished by the gyroscopic spin of the black hole, and the only way the direction of that spin can be communicated to the jet is via the black hole's gravitomagnetic field $\boldsymbol{H}$ [54]. The field causes the accretion disk to precess around the black hole, and that precession combines with the disk's viscosity to drive the inner region of the disk into the hole's equatorial plane, gradually forcing the jets to align with the north and south poles of the black hole. This phenomenon, known as the Bardeen-Petterson effect, is widely believed to be the physical mechanism responsible for jet alignment.

Gravitomagnetism is also thought to lie behind the generation of the astounding energy contained in these jets in the first place. The event horizon of the black hole can act like a gigantic "battery" where the gravitomagnetic potential of the black hole interacts with the tangential component of the ordinary magnetic field $\boldsymbol{B}$ to produce a drop in electric potential [54]. This phenomenon, known as the Blandford-Znajek mechanism, effectively converts the immense gravitomagnetic, rotational energy of the supermassive black hole into an outgoing stream of ultra-relativistic charged particles. Gravity Probe B has thus become a test of the mechanism that powers the most violent explosions in the universe.

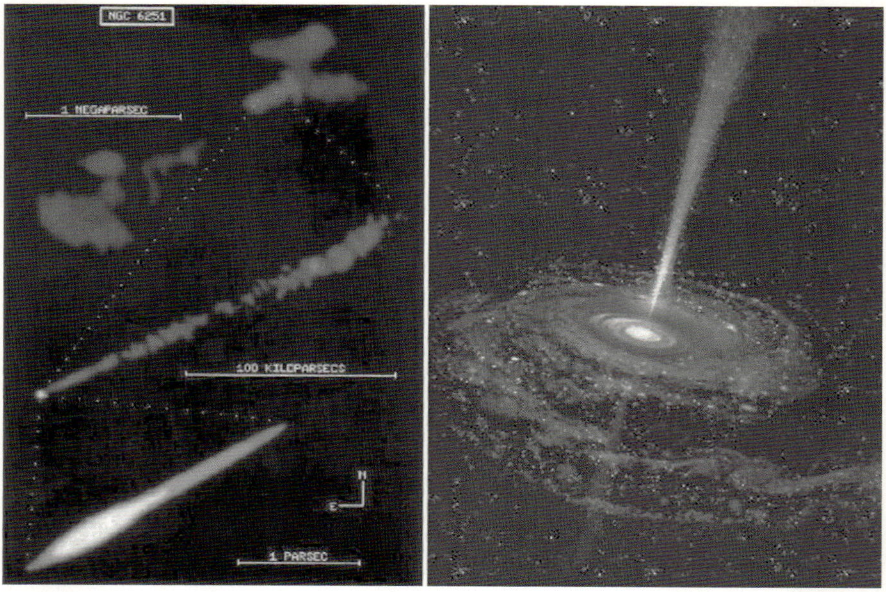

Fig. 14 Megaparsec-scale jet associated with the strong radio source NGC 6251 (*left*); such jets are now thought to be aligned and powered by the gravitomagnetic fields of rotating supermassive black holes (*right*)

It is on cosmological scales, however, that frame-dragging may take on its deepest significance, as part of the explanation for the origin of inertia. Consider the earth's equatorial bulge. Textbooks teach us that this phenomenon is due to rotation with respect to the "inertial frame" in which the universe as a whole happens to be at rest (Fig. 15, left). But what if it were the *earth* that stood still, and the rest of the universe that rotated (right)? Would the equator still bulge? Newton would have said "No." For him, inertial frames were tied ineluctably to absolute space which "in its own nature, without relation to anything external, remains always similar and immoveable." We know that the concept of absolute space (time) is retained in general relativity, so we might have expected that the same answer would carry over to Einstein's theory as well. However, it does not. As demonstrated by Thirring in his original calculation of 1918, and amplified by many others since, general-relativistic frame-dragging goes over to "perfect dragging" *when the dimensions of the large mass become cosmological*. That is, if the entire universe were to rotate, it would drag the inertial frame of the earth around with it. On this basis, Einstein would have had to answer "Yes" to the question posed above. In this respect general relativity is indeed more relativistic than its predecessors, as Mach would have wished. Early calculations were flawed in many ways, but the phenomenon of perfect dragging has persisted through a series of increasingly sophisticated treatments, notably those of Sciama [8], Brill and Cohen [67,68], Lindblom and Brill [69], Pfister and Braun [70] and Klein [71,72]. Pfister sums up the situation as follows [61]: "Although Einstein's

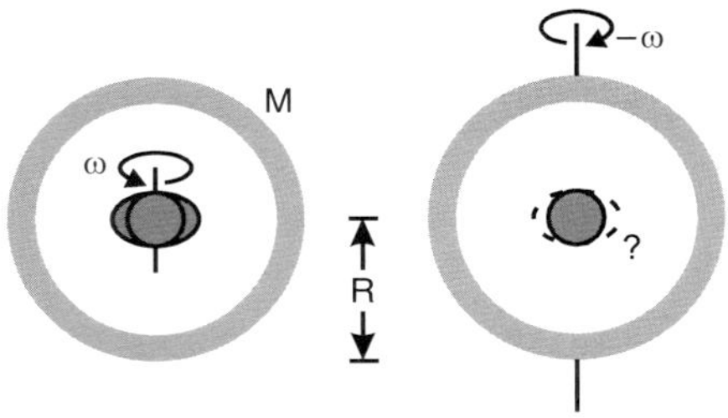

Fig. 15 Would the earth still bulge, if it were standing still and the rest of the universe were rotating around *it?*

theory of gravity does not, despite its name 'general relativity,' yet fulfil Mach's postulate of a description of nature with only relative concepts, it is quite successful in providing an intimate connection between inertial properties and matter, at least in a class of not too unrealistic models for our universe." The apparently *instantaneous* nature of the connection is particularly mysterious. Much remains to be learned. What is clear, however, is that direct detection of frame-dragging by Gravity Probe B will do much to give us confidence in what has been a largely theoretical enterprise to date; namely, to understand how a relational explanation for inertia may be possible within a theory of absolute spacetime.

6 Gravity Probe B

The English physicist P.M.S. Blackett reportedly considered the idea of looking for the de Sitter effect with a laboratory gyroscope as early as the 1930s [73]. The smallness of the signal, however, put such an experiment far out of reach until after post-World War II improvements in gyroscope technology and the dawning of the space age. To measure a yearly precession of order 10 milliarcseconds to 1% accuracy requires a gyroscope with drift rate less than 10^{-18} rad/s. On earth, where (for instance) density inhomogeneities contribute to this drift rate with the full force of the earth's gravitational acceleration $a \sim g$, the gyro rotor would have to be homogeneous to a part in $\sim 10^{17}$–a hopelessly unattainable number. A similar argument holds for rotor asphericity. The only way around such fundamental limitations is to go into space, where unwanted accelerations can be suppressed–with a great deal of work–so that $a \sim 10^{-11} g$. The rotor need then be homogeneous, for example, to "only" one part in $\sim 10^6$, a level that can be achieved, with great effort, using the best materials on earth [73].

Considerations of this kind led two American physicists to take a new look at gyroscopic tests of general relativity independently within months of each other in late 1959 and early 1960. George E. Pugh (b. 1928; Fig. 16) was spurred by a talk given by the Turkish-American theoretical physicist Huseyin Yilmaz on the possible use of an artificial satellite to distinguish his theory of gravity from Einstein's. He noted that such an experiment "would be, in the most literal sense a direct measurement of space itself" [74]. Leonard. I. Schiff (1915–1971) was inspired at least in

Fig. 16 Leonard Schiff c. 1970 (*top left*), George Pugh in 2007 (*bottom left*) and Dan Debra, Bill Fairbank, Francis Everitt and Bob Cannon with a model of Gravity Probe B in 1980 (*right*)

part by a magazine advertisement for a new "Cryogenic Gyro … with the possibility of exceptionally low drift rates" [73]. Schiff had a longstanding interest in both general relativity and Mach's principle, and went so far as to refer to his proposal as "an experimental test of Mach's principle" [75]. He was joined by low-temperature experimentalist Bill Fairbank and guidance and control specialist Bob Cannon, and together the three men set Gravity Probe B on the path to reality. Under its original name (the Stanford Relativity Gyroscope Experiment) the project received its first NASA funding in 1964.

Pugh's paper attracted less notice at the time but is now recognized as the birth of the concept of *drag-free motion*. This is a critical element of the Gravity Probe B mission, whereby any one of the gyroscopes can be isolated from the rest of the experiment and protected from non-gravitational forces (such as those caused by solar radiation pressure and atmospheric drag); the rest of the spacecraft is then made to "chase after" the reference gyro by means of helium boiloff vented through a revolutionary porous plug and specially designed thrusters. Gravity Probe B's demonstration that cross-track accelerations can be suppressed in this way to less than $10^{-11}g$ paves the way for the development of future gravitational experiments such as the Satellite Test of the Equivalence Principle (STEP) and the Laser Interferometer Space Antenna (LISA). The porous plug has already proved vital to other cryogenic NASA missions including COBE, IRAS, WMAP and Spitzer.

The experimental concept is illustrated in Fig. 17. In principle it is simplicity itself: a gyroscope, a readout mechanism to monitor the spin axes, and a telescope to compare these axes with the line of sight to a distant guide star. In practice, Gravity Probe B evolved into one of the most complex experiments ever flown, requiring at least a dozen new technologies that did not exist when it was conceived. How, for instance, is one to locate the spin axis of a perfectly spherical, perfectly homogeneous gyroscope, suspended in vacuum (Fig. 18)? This is the readout problem; another, closely related challenge is how to spin up such a gyroscope in the first place. Various possibilities were considered in the early days, until 1962 when

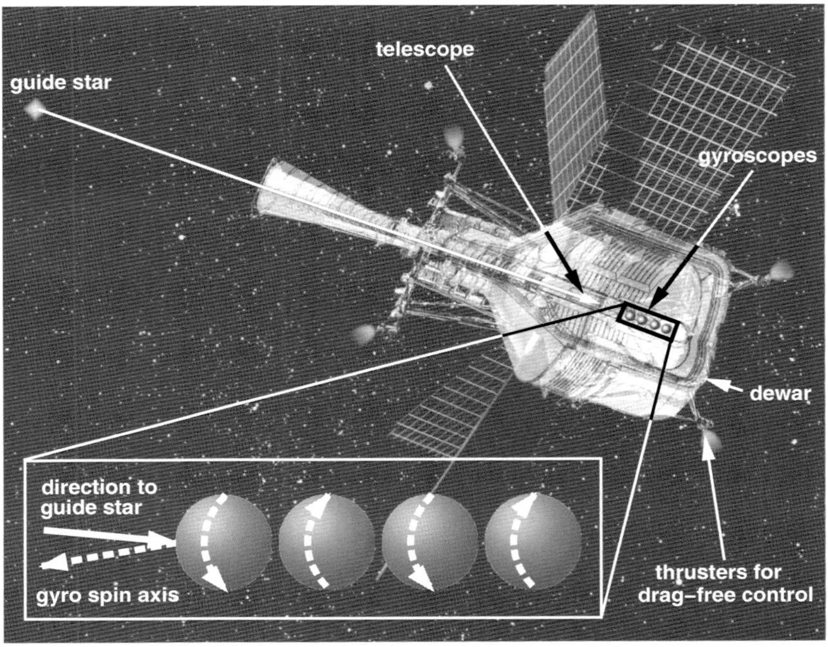

Fig. 17 The Gravity Probe B concept, described by William Fairbank, one of the project founders, as "just a star, a telescope and a spinning sphere"

C.W. Francis Everitt (now in charge of the experiment) hit on the idea of exploiting what had until then been a small but annoying source of unwanted torque in magnetically levitated gyroscopes. Spinning superconductors develop a magnetic moment, known as the *London moment*, which is aligned with the spin axis and proportional to the spin rate. If the rotors were levitated electrically instead of magnetically, this tiny effect could be used to monitor their spin axes. (Measuring it would require magnetic shielding orders of magnitude beyond anything available in 1962, another story in itself.) Thus was born the London moment readout, which in its modern incarnation uses niobium-coated quartz spheres as rotors and SQUIDs (Superconducting QUantum Interference Devices) as magnetometers. So sensitive are these devices that they register a change in spin-axis direction of 1 milliarcsecond in just five hours of integration time.

How can one meaningfully compare the spin-axis direction (from the SQUID, in volts) with the direction to the guide star (from an onboard telescope, in radians)? The answer is to exploit nature's own calibration in the form of *stellar aberration*. This phenomenon, an apparent back-and-forth motion of the guide star position due to the orbit of the earth around the sun, is entirely Newtonian and inserts "wiggles" into the data whose period and amplitude are exquisitely well known. (Such is the precision of the experiment that this calibration requires terms of second, as well as

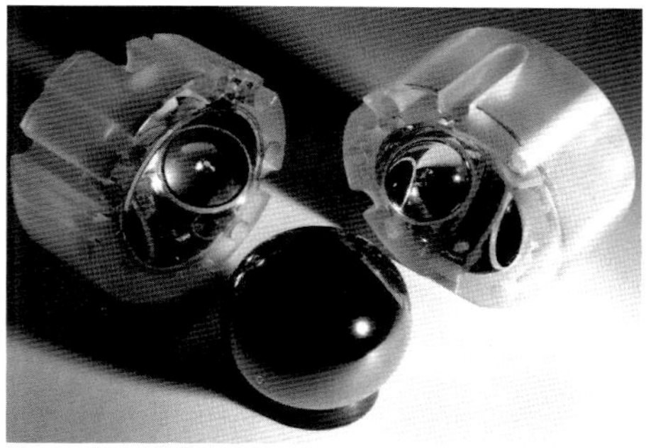

Fig. 18 Gravity Probe B gyroscope rotor and housing. Note suspension electrodes (circular patterns) and gas spin-up channel (groove)

first order in the earth's speed v/c.) What about the fact that the guide star *itself* has an unknown proper motion large enough to obscure the predicted relativity signal? This apparent liability is turned into an advantage by designing the experiment in classic "double-blind" fashion: a separate team of radio astronomers uses VLBI to monitor the movements of the guide star relative to even more distant quasars. Only at the conclusion of the experiment are the two sets of data to be compared; this helps to prevent the physicists from finding what they expect to see.

Necessity in the form of a 10^{-18} rad/s precession rate was the mother of many more marvels. Among these are the roundest man-made objects in the world and a suspension system capable of keeping them within microns of their housings, at spin rates averaging 4,000 rpm, over a dynamic range of eight or more orders of magnitude in force. A beam splitter and image divider assembly was created to increase the resolution of the onboard telescope (inherently limited by the size of the spacecraft) by three orders of magnitude over existing star trackers. A novel optical bonding technique had to be devised to fasten the telescope (sculpted out of a single lump of quartz) to the quartz block containing the science instrument. Expandable nested lead shields were employed to reduce the strength of the magnetic field inside the dewar to less than one-millionth that of the Earth, the lowest level ever achieved in space. New techniques were invented to spin up the gyros, reduce vacuum pressure and remove charge buildup on the rotors. Many of these innovations have led to engineering and commercial spinoffs.[7]

Gravity Probe B was launched from Vandenberg Air Force Base in California on 20 April 2004 (Fig. 19). Once in orbit, it underwent an initial orbit checkout phase, during which the attitude and control system was tuned and the gyroscopes suspended, spun up, calibrated and aligned with the guide star. These tasks required 129 days. The science or data-collecting phase of the mission lasted from 27 August 2004 until 14 August 2005, or 353 days, just under the original goal of one full year. The mission concluded with a final post-flight calibration phase, which continued until 29 September 2005, when there was no longer enough liquid helium in the dewar to maintain the experiment at cryogenic temperatures.

---

[7] For more details, see the Gravity Probe B website at http://einstein.stanford.edu.

Fig. 19 Launch of Gravity Probe B at 09:57:24 PDT, April 20, 2004

Figure 20 shows approximately 140 days of science data from one of the gyroscopes (points) superimposed on the predictions of general relativity (lines). North-south (geodetic) precession is plotted in the upper panel, while east-west (frame-dragging) precession is plotted in the lower panel. These plots give us *our first direct look at the warping and twisting of spacetime around the earth.* If Newton were correct, the data would fall on horizontal lines.

As might be expected in an experiment that pushes gyroscope performance six orders of magnitude beyond existing limits, unexpected complications have cropped up in the data analysis. First, it became apparent during the science phase of the mission that there were variations in the polhode rate of the gyros. (Polhode motion had been expected, but its period had not been expected to change appreciably over the mission lifetime, given characteristic rotor spin-down periods on the order of 10,000 years). It is critical to understand and model these polhode variations in order to match the data from successive orbits and thereby attain integration times long enough to realize the full precision of the SQUID readout system. Second, two larger-than-expected forms of Newtonian torque, known as the "misalignment" and "roll-polhode resonance" torques, were discovered during post-flight calibration. Misalignment torques were proportional to the angle between the gyroscope spin axis and the spacecraft roll axis, while resonance torques acted on individual gyroscopes during times when there was a high-order resonance between the slowly changing polhode period and the satellite roll period. All three phenomena have been traced to larger-than-anticipated electrostatic patch effects. In essence, while both the gyro rotors and housings achieved almost perfect *mechanical* sphericity, they were not quite spherical *electrically*. The anomalous torques are due to

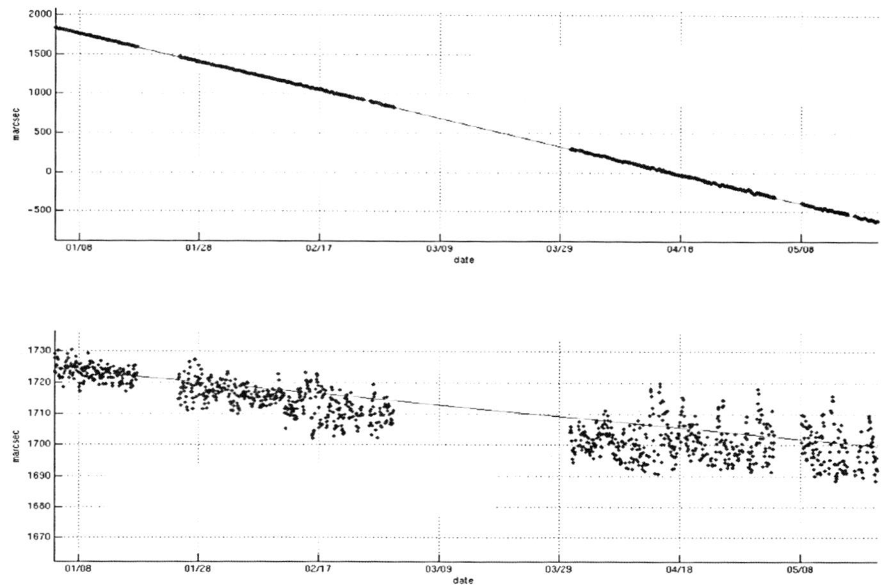

Fig. 20 Preliminary results for the precession of one of the Gravity Probe B gyroscopes in the north-south or geodetic direction (*top*) and the east–west or frame-dragging direction (*bottom*)

interactions between patches on the gyro rotors and housings, and the time-varying polhode periods are caused by the fact that these interactions extract energy from the spinning rotors.

Fortunately, Gravity Probe B was designed to take various kinds of "superfluous" data, and these are now proving their worth. In particular, real-time snapshots of trapped flux on the rotors have enabled the data analysis team to reconstruct the polhode phase of each of the gyros to within $\sim 1°$ over the entire mission. Spin speeds are known to $\sim$1 nHz, and spin-down rates to $\sim$1 pHz/s. With this data, in combination with a complete physical understanding of all three (fully Newtonian) effects, it has been possible to develop a more comprehensive method of data analysis that is expected to lead to final accuracies close to those originally envisioned for the experiment. Table 1 summarizes interim results from all four gyroscopes as of December 2008 [76]. These numbers are preliminary and do not include all sources of systematic error or model sensitivity analysis. Nevertheless it is possible at this stage to state that geodetic precession has been directly observed at better than 1%, and that frame-dragging has been directly observed with an accuracy of about 15%.

Final results from Gravity Probe B are to be announced in 2010.

Table 1 Preliminary Gravity Probe B results (milliarcsec/yr) [76]

		Solar geodetic	Guide star proper motion	Net predicted (Ω_{GR})	Observed (Ω_{obsd})
North–south:	−6606	+7	$+28 \pm 1$	-6571 ± 1	-6550 ± 14
East–west:	−39	−16	-20 ± 1	-75 ± 1	-69 ± 6

7 Summary

The detection of geodetic precession and frame-dragging by Gravity Probe B can be seen as the culmination of a debate that stretches back to Greek antiquity. That debate was originally philosophical: do space and time exist absolutely, or only in relation to matter? As natural philosophy evolved into natural science, it began to take on a physical character. The absolute picture, advocated most forcefully by Newton, was physically simpler but carried with it uncomfortable metaphysical baggage (inertia as resistance to motion with respect to "absolute space," which itself could neither be observed nor acted upon in any way). The relational view, most strongly associated with Mach, was philosophically more elegant but troublingly vague in the physical sense (what kind of relation, exactly, gives rise to inertia?) Attempts were already made to distinguish between the two points of view by experimentalists such as Föppl before the time of Einstein and Minkowski.

With the advent of general relativity, it became possible to frame the debate in precise physical terms. It turned out that Minkowski's spacetime, as shaped and animated in the presence of matter according to Einstein's gravitational field equations, took neither side in the debate–or rather, took them both. The spacetime of general relativity exists absolutely and behaves relationally, as exemplified by the geodetic and (especially) frame-dragging effects.

Ninety years of experiment have solidified the case for Einstein's theory. However, most of the evidence so far is limited to the solar system where fields are weak and velocities low. Gravitational-wave astronomy has the potential to improve the situation, as do experiments that challenge the foundations (as opposed to predictions) of general relativity, like tests of the equivalence principle.

The geodetic and frame-dragging effects test Einstein's theory in another direction by focusing on the *spin* of the central mass and test body, with important implications for astrophysics, cosmology and the origin of inertia. Such is their subtlety, however, that detecting them with confidence has required 40 years of scientific and engineering ingenuity and perseverance. That story is not quite finished, and may yet fulfil the original aim of the Gravity Probe B mission: to provide the "most rigorously validated of all tests of Einstein's theory" [76]. Preliminary data are consistent with general relativity. These results tighten constraints on alternative theories of gravity, improve confidence in astrophysical models of the jets and accretion disks associated with supermassive black holes, and suggest that we may be close to understanding why our local compass of inertia is aligned with the rest frame of the distant galaxies.

They also settle an old debate in metaphysics. It would be hard to imagine a more direct demonstration that spacetime acts on matter than the geodetic effect (warped spacetime twists a spinning gyroscope), or a more convincing proof that matter acts back on spacetime than the frame-dragging effect (the spinning earth pulls spacetime around with it). In that sense Gravity Probe B shows how a physics experiment–when pushed to the furthest possible extremes of near-zero temperature, pressure, electric charge, magnetic field and acceleration–can also become a test of philosophy.

Acknowledgements The author thanks Ron Adler, Francis Everitt, Hans-Jörg Fahr, Bob Kahn, Holger Mueller, Alex Silbergleit, Martin Tajmar and Paul Wesson for helpful discussions.

References

1. Hahm, D.E., *The origins of stoic cosmology* (Ohio State University Press, Ohio, 1977)
2. Barbour, J.B., *The discovery of dynamics* (Oxford University Press, Oxford, 2001)
3. Jammer, M., *Concepts of space* (Harvard University Press, Cambridge, 1954)
4. Alexander, H.G., *The Leibniz-Clarke correspondence* (Manchester University Press, Manchester, 1956)
5. Barbour, J.B. and Pfister, H. (eds.), *From Newton's bucket to quantum gravity* (Birkhäuser, Boston, 1995), p. 530
6. Pais, A., *'Subtle is the Lord... '* (Oxford University Press, Oxford, 1982)
7. Barbour, J.B., in Barbour, J.B. and Pfister, H. (eds.), *From Newton's bucket to quantum gravity* (Birkhäuser, Boston, 1995), p. 214
8. Sciama, D., *The unity of the universe* (Faber and Faber, London, 1959)
9. Wheeler, J.A., in Infeld, L. (ed.), *Relativistic theories of gravitation* (Pergamon, Oxford, 1964), p. 223
10. Berry, M., *Principles of cosmology and gravitation* (Cambridge University Press, Cambridge, 1976)
11. Overduin, J.M. and Fahr, H.-J., *Naturwissenschaften* **88**, 491 (2001)
12. Archibald, R.C., "Time as a fourth dimension," *Bull. Am. Math. Soc.* **20**, 409 (1914)
13. May, A., "Parsifal as proto-SF," http://www.andrew-may.com/parsifal.htm, accessed April 1, 2009
14. Melderis, H., *Raum-Zeit-Mythos. Richard Wagner und die modernen Naturwissenschaften* (Hamburg, Europische Verlagsanstalt, 2001)
15. Cappi, A., "Edgar Allan Poe and his cosmology," http://www.bo.astro.it/~cappi/poe.html, accessed April 1, 2009
16. Overduin, J.M. and Wesson, P.S., *The light/dark universe* (World Scientific, Singapore, 2008)
17. Graves, R.P., *Life of Sir William Rowan Hamilton*, Vol. 3 (1889), p. 635
18. S., "Four-dimensional space," *Nature* **31**, 481 (1885)
19. Halpern, P., *The great beyond: higher dimensions, parallel universes and the extraordinary search for a theory of everything* (Wiley, Hoboken, 2004), p. 57
20. Goenner, H., "On the history of the geometrization of space-time" (2008); arXiv:0811.4529 [gr-qc]
21. Overduin, J.M. and Wesson, P.S., "Kaluza-Klein gravity," *Phys. Rep.* **283**, 303 (1997)
22. Band, W., "Klein's fifth dimension as spin angle," *Phys. Rev.* **56**, 204 (1939)
23. Hara, O., "A study of charge independence in terms of Kaluza's five dimensional theory," *Prog. Theor. Phys.* **21**, 919 (1959)
24. Rumer, Y., *Zh. Eksp. Teor. Fiz.* **19**, 86 (1949); Rumer, Y., *Zh. Eksp. Teor. Fiz.* **19**, 207 (1949)
25. Rumer, Y.B., "Action as a space coordinate," *Sov. Phys. JETP* **36**, 1348 (1959)

26. Tsipenyuk, D.Y. and Andreev, V.A., "Structure of extended space," *Kratk. Soobshch. Fiz.* **6**, 23 (2000); arXiv:gr-qc/0106093
27. Wesson, P.S., "A new approach to scale-invariant gravity," *Astron. Astrophys.* **119**, 145 (1983)
28. Wesson, P.S. et al., *Int. J. Mod. Phys.* **A11**, 3247 (1996)
29. de Vos, J.A. and Hilgevoord, J., "Five-dimensional aspect of free particle motion," *Nucl. Phys.* **B1**, 494 (1967)
30. Edmonds, J.D., "Five-dimensional space-time: mass and the fundamental length," *Int. J. Theor. Phys.* **11**, 309 (1974)
31. Edmonds, J.D., "Extended relativity: mass and the fifth dimension," *Found. Phys.* **5**, 239 (1975)
32. Redington, N., "On the significance of the fifth coordinate in Wesson's version of Kaluza-Klein theory," unpublished preprint (1997); arXiv:gr-qc/9701062
33. Wesson, P.S., *Space-time-matter* (World Scientific, Singapore, 1999)
34. Wesson, P.S., *Five-dimensional physics: classical and quantum consequences of Kaluza-Klein cosmology* (World Scientific, Singapore, 2006)
35. Liu, H. and Overduin, J.M., "Solar system tests of higher-dimensional gravity," *Astrophys. J.* **548**, 386 (2000); arXiv:qr-qc/0003034
36. Overduin, J.M., "Solar system tests of the equivalence principle and constraints on higher-dimensional gravity," *Phys. Rev.* **D62**, 102001 (2000); arXiv:gr-qc/0007047
37. Overduin, J.M., Wesson, P.S. and Mashhoon, B., "Decaying dark energy in higher-dimensional cosmology," *Astr. Astrophys.* **473**, 727 (2007); arXiv:0707.3148 [astro-ph]
38. Wesson, P.S., "The equivalence principle as a symmetry," *Gen. Rel. Grav.* **35**, 307 (2003); arXiv:gr-qc/0302092
39. Sakharov, A.D., "Cosmological transitions with changes in the signature of the metric," *Sov. Phys. JETP* **60**, 214 (1984)
40. I.Y. Aref'eva and I.V. Volovich, "Kaluza-Klein theories and the signature of space-time," *Phys. Lett.* **164B**, 287 (1985)
41. Burakovsky, L. and Horwitz, L.P., "5D generalized inflationary cosmology," *Gen. Rel. Grav.* **27**, 1043 (1995)
42. Bars, I. and Kounnas, C., "Theories with two times," *Phys. Lett.* **B402**, 25 (1997); arXiv:hep-th/9703060
43. Wesson, P.S., "Five-dimensional relativity and two times," *Phys. Lett.* **B538**, 159 (2002); arXiv:gr-qc/0205117
44. Kociński, J. and Wierzbicki, M., "The Schwarzschild solution in a Kaluza-Klein theory with two times," *Rel. Grav. Cosmol.* **1**, 19 (2004); arXiv:gr-qc/0110075
45. R. Erdem and C.S. Ün, "Reconsidering extra time-like dimensions," *Europhys. J.* **C47**, 845 (2006); arXiv:hep-ph/0510207
46. I. Quiros, "Causality and unitarity might be preserved in higher-dimensional space-times with compact extra dimensions, arXiv:0706.2400 [hep-ph]
47. Fukui, T., "Vacuum cosmological solution in a 6D universe," *Gen. Rel. Grav.* **24**, 389 (1992)
48. Carmeli, M., *Cosmological special relativity* (Singapore: World Scientific, 1997)
49. Carmeli, M., "Fundamental approach to the cosmological constant issue," *Int. J. Mod. Phys.* **A17**, 4219 (2002); arXiv:astro-ph/0205395
50. Hartnett, J.G., "Spiral galaxy rotation curves determined from Carmelian general relativity," *Int. J. Theor. Phys.* **45**, 2118 (2006); arXiv:astro-ph/0511756
51. Matute, E.A., "Geometry-matter duality and electromagnetism in higher dimensions," *Class. Quant. Grav.* **14**, 2771 (1997)
52. Delbourgo, R., "The flavour of gravity," *J. Phys.* **A39**, 5175 (2006); arXiv:hep-th/0512173
53. Overduin, J.M. et al., "Advances in Space Research", 43, 1532 (2009); arXiv:0902.2247
54. Thorne, K., "Gravitomagnetism, jets in quasars and the Stanford gyroscope experiment," in Fairbank, J.D., Deaver, B.S., Everitt, C.W.F. and Michelson, P.F. (eds.), *Near zero: new frontiers of physics* (W.H. Freeman, New York, 1988), p. 572
55. Will, C.M., *Theory and experiment in gravitational physics* (Cambridge University Press, Cambridge, 1993)
56. Nandi, K. et al., "Brans-Dicke corrections to the gravitational Sagnac effect," *Phys. Rev.* **D63**, 084027 (2001); arXiv:gr-qc/0006090

57. Halpern, L., "A geometrical theory of spin motion," *Int. J. Theor. Phys.* **23**, 843 (1984)
58. Mao, Y. et al., "Constraining torsion with Gravity Probe B," *Phys. Rev.* **D76**, 104029 (2007); arXiv:gr-qc/0608121
59. Bailey, Q. and Kostelecky, A., "Signals for Lorentz violation in post-Newtonian gravity," *Phys. Rev.* **D74**, 045001 (2006); arXiv:gr-qc/0603030
60. Overduin, J.M., "Constraints on Lorentz violation from Gravity Probe B," in Kostelecky, A. (ed.), *Fourth meeting on CPT and Lorentz symmetry* (World Scientific, Singapore, 2008), p. 199
61. Pfister, H., "Dragging effects near rotating bodies and in cosmological models," in Barbour, J.B. and Pfister, H. (eds.), *From Newton's bucket to quantum gravity* (Birkhäuser, Boston, 1995), p. 315
62. Ciufolini, I. et al., "Test of general relativity and measurement of the Lense-Thirring effect with two earth satellites," *Science* **279**, 2100 (1998)
63. Ciufolini, I. and Pavlis, E.C., "A confirmation of the general relativistic prediction of the Lense-Thirring effect," *Nature* **431**, 958 (2004)
64. Will, C.W., "The search for frame-dragging," *Matters of gravity* **10** (1997)
65. Will, C.W., "Frame-dragging in the news in 2004," *Matters of gravity* **25** (2005); arXiv:gr-qc/0503086
66. Iorio, L., "Conservative evaluation of the uncertainty in the LAGEOS-LAGEOS II Lense-Thirring test," *Cent. Eur. J. Phys.* 1 (2009); arXiv:0710.1022 [gr-qc]
67. Brill, D.R. and Cohen, J.M., "Rotating masses and their effect on inertial frames," *Phys. Rev.* **143**, 1011 (1966)
68. Cohen, J.M. and Brill, D.R., "Further examples of 'Machian' effects of rotating bodies in general relativity," *Nuovo Cim.* **56B**, 209 (1968)
69. Lindblom, L. and Brill, D.R., "Inertial effects in the gravitational collapse of a rotating shell," *Phys. Rev.* **D10**, 3151 (1974)
70. Pfister, H. and Braun, K., "Induction of correct centrifugal force in a rotating mass shell," *Class. Quant. Grav.* **2**, 909 (1985)
71. Klein, C., "Rotational perturbations and frame dragging in a Friedmann universe," *Class. Quant. Grav.* **10**, 1619 (1993)
72. Klein, C., "Second-order effects of rotational perturbations of a Friedmann universe," *Class. Quant. Grav.* **11**, 1539 (1994)
73. Everitt, C.W.F., "The Stanford relativity gyroscope experiment (A): history and overview," in Fairbank, J.D., Deaver, B.S., Everitt, C.W.F. and Michelson, P.F. (eds.), *Near zero: new frontiers of physics* (New York: W.H. Freeman, 1988), p. 587
74. Pugh, G.E., "Proposal for a satellite test of the coriolis prediction of general relativity," U.S. Dept. of Defense Weapons Systems Evaluation Group Research Memorandum No. 11 (1959); reprinted in Ruffini, R.J. and Sigismondi, C., *Nonlinear gravitodynamics: the Lense-Thirring effect* (Singapore: World Scientific, 2003), p. 414
75. Schiff, L.I., "Possible new experimental test of general relativity theory," *Phys. Rev.* **4**, 215 (1960)
76. *Gravity Probe B final NASA report* (Dec. 2008); http://einstein.stanford.edu/content/final_report/GPB_Final_NASA_Report-020509-web.pdf

Rigidity and the Ruler Hypothesis

Stephen N. Lyle

Abstract In special relativity, one often speaks of rigid rods, looking at them in one inertial frame or another and observing that they do not always have the same length, despite their rigidity. This paper is about what happens to the rod as it gets from one inertial frame to another, i.e., as it accelerates. The problem is not entirely academic. For those who would like to model extended charge distributions and their fields, and in particular the forces they exert upon themselves via these electromagnetic effects, when they are accelerating, some hypothesis must be made about the way the charge distribution shifts around in the relevant spatial hypersurfaces of Minkowski's spacetime. A notion of rigidity is indeed usually applied and that is discussed here (Sect. 1), in connection with frames of reference adapted to accelerating observers in the spacetime of special relativity. The physical legitimacy of adapted frames of reference is discussed in some detail throughout the paper, but particularly in the context of the Pound–Rebka experiment in Sect. 1.10. The aim is to elucidate the roles of what are usually referred to as the clock and ruler hypotheses. One would also like to consider rigid motions of any material medium in a more general framework, even in the context of general relativity. The notion of rigidity can be extended (Sect. 2) in a simple but perhaps questionable way. The aim here will indeed be to cast a critical glance.

1 Rigid Rods and Rigid Spheres

1.1 A Toy Electron

If one wanted to make a model of the electron in which the electric charge were no longer all concentrated at a mathematical point in any spacelike hypersurface of Minkowski spacetime, one might get the idea of dividing the charge into two equal amounts labelled A and B, each one occupying such a point, and spaced apart by some distance D when the system is moving inertially and observed from an inertial frame moving with it. This would certainly be a toy electron (see Fig. 1). It would

V. Petkov (ed.), *Space, Time, and Spacetime*, Fundamental Theories of Physics 167,
DOI 10.1007/978-3-642-13538-5_3,

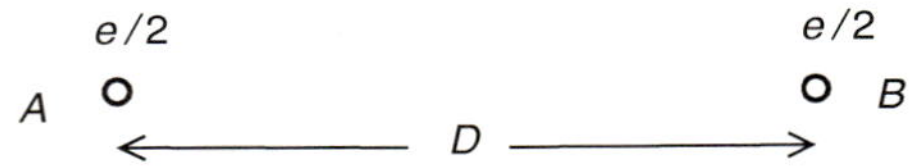

Fig. 1 A toy electron at rest in an inertial frame. The system is in equilibrium under the forces between A and B

Fig. 2 System in motion in an inertial frame $\mathcal{I}$. If we impose some motion on A, what is the motion of B ?

appear to a large extent to defeat the object of giving the electron a spatial extent since there are now two mathematical points of charge instead of just one. But it does nevertheless bring out some of the advantages and some of the difficulties. We shall only be concerned with one of the difficulties here.

The idea of such a model is to allow it to move, then work out the electromagnetic fields due to each point of charge using the Lienard–Wiechert potential, and calculate the force that each charge can thereby exert on the other. One is of course interested in the net force that such a system might exert upon itself when accelerated. In a first approach, one does not worry about the force required to hold the system together. For it should not be forgotten that the two charges are alike and will repel one another. And yet this very question raises another, more urgent one.

For suppose point A, on the left, has a one-dimensional motion given by $x_A(t)$ along the axis from A to B, as observed relative to some inertial frame $\mathcal{I}$ (see Fig. 2). What will be the motion $x_B(t)$ of the right-hand end of the system? Let the coordinate speed of A in $\mathcal{I}$ be

$$v_A(t) = \dot{x}_A(t) := \frac{\mathrm{d}x_A}{\mathrm{d}t}.$$

If the system were rigid in the pre-relativistic sense, the speed of B would be

$$v_B(t) = v_A(t).$$

There is an obvious problem with this: if A and B always have the same coordinate speed, the separation of A and B will always be the same in $\mathcal{I}$, viz., D, whereas we expect lengths of objects to contract in special relativity when viewed from frames in which those objects are moving.

If the system had been set in uniform motion, with A moving at constant coordinate speed v_A in frame $\mathcal{I}$, and if it were rigid in the sense of special relativity, we would expect the separation of A and B to be the contracted length $D\sqrt{1 - v_A^2/c^2}$

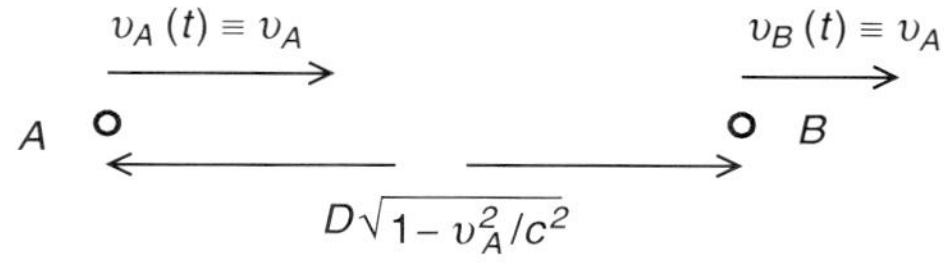

Fig. 3 System in uniform motion $v_A(t) \equiv v_A$ in an inertial frame $\mathcal{I}$. We expect $v_B(t) \equiv v_A$ and the length to be contracted as shown

as observed from $\mathcal{I}$ (see Fig. 3). There would then be no problem for point B to have the same coordinate speed as point A.

The difficulty occurs, of course, when v_A is changing with time, i.e., when there is acceleration, and the system has to adjust all the time. This means that v_B looks as though it might be a very complicated function of time indeed. Perhaps we require some physical assumption about the relaxation time of the system. One immediately wonders how B is supposed to adjust. Indeed, what are the forces on the system? What is accelerating it? Since we started by attributing a motion $x_A(t)$ to A, we might imagine some external force applied to A and ask how the effects of this force might be transmitted to B to make it too accelerate. On the other hand, an external force might equally be applied to both points, as would happen for example if it were due to a force field. In addition, we know that there are mutually repulsive electromagnetic forces between A and B due to their own fields, and we said there had to be binding forces to oppose them, to hold the thing together. How will these react to the changes?

In the pre-relativistic context, there was no problem because the appropriate notion of rigidity automatically delivered the motion of point B given the motion of point A. But this hypothesis was unequivocally pre-relativistic: if the system was accelerated by applying a force to point A, the effects had to propagate instantaneously to point B, and in such a way that the net force on B was always identical. Of course, if the system was accelerated by a uniform field acting simultaneously on both A and B, then one had the advantage of not having to consider the binding forces at all.

So what could be the separation of A and B as observed from $\mathcal{I}$ when $v_A(t)$ changes with time? One idea is that, since A has speed $v_A(t)$, the separation of A and B at time t could just be $D\sqrt{1 - v_A(t)^2/c^2}$, the contracted length for that speed. This might be a good approximation in some cases, but there is an obvious problem with it: because there is some contraction going on, this too will induce a motion of B that ought to be included. And if B does not have the same speed as A, why not use the speed of B to work out the contraction? Or some average of the speeds?

Although the situation looks rather hopeless, we do appear to have an approximation. This looks especially promising if D is small. Is there some way of making D infinitesimal and taking a limit? Let us switch to a material rod under acceleration.

1.2 A Rigid Rod

Of course, we know what happens to a rigid rod when it has uniform motion relative to an inertial frame $\mathcal{I}$. In other words, we know what we want rigidity to mean in that context. But can we say how a rigid rod should behave when it accelerates? Can we still have some kind of rigidity?

Let A and B be the left- and right-hand ends of the rod and consider motion $x_A(t)$ and $x_B(t)$ along the axis from A to B (see Fig. 4). Let us first label the particles in the rod by their distance s to the right of A when the system is stationary in some inertial frame (see Fig. 5). This idea of labelling particles will prove extremely useful when considering continuous media later on. In the present case, we imagine the rod as a strictly one-dimensional, continuous row of particles.

Now let A have motion $x_A(t)$ relative to an inertial frame $\mathcal{I}$ (see Fig. 6) and let $X(s,t)$ be a function giving the position of particle s at time t as

$$x_s(t) = x_A(t) + X(s,t),$$

where naturally we require

$$X(0,t) = 0, \qquad X(D,t) = x_B(t) - x_A(t).$$

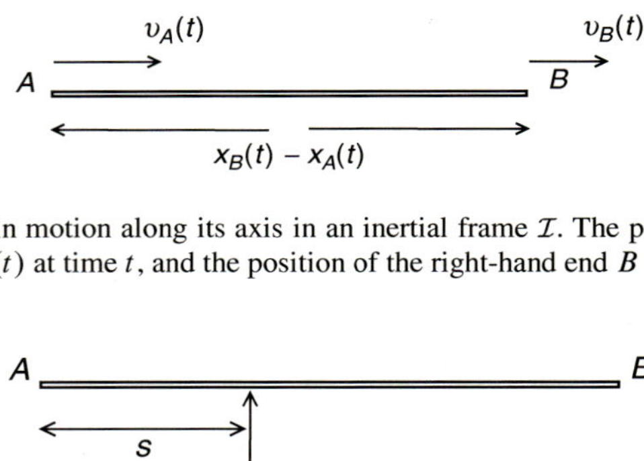

Fig. 4 Material rod in motion along its axis in an inertial frame $\mathcal{I}$. The position of the left-hand end A is given by $x_A(t)$ at time t, and the position of the right-hand end B is given by $x_B(t)$

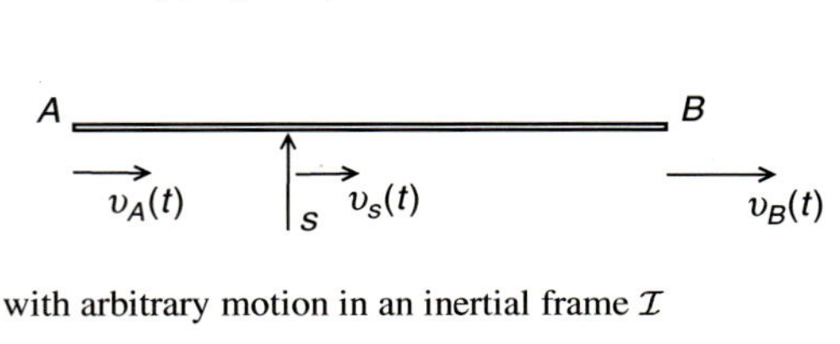

Fig. 5 Stationary material rod in an inertial frame $\mathcal{I}$. Labelling the particles in the rod by their distance s from A, so that $s \in [0, D]$

Fig. 6 Material rod with arbitrary motion in an inertial frame $\mathcal{I}$

Let us require the element between s and $s + \delta s$ to have coordinate length

$$\left[1 - \frac{v(s,t)^2}{c^2}\right]^{1/2} \delta s, \tag{1}$$

where $v(s,t)$ is its instantaneous coordinate velocity, with $v(0,t) = v_A(t)$. This is precisely the criterion suggested by Rindler [1, pp. 39–40]. We can integrate to find

$$X(s,t) = \int_0^s \left[1 - \frac{v(s',t)^2}{c^2}\right]^{1/2} \mathrm{d}s'. \tag{2}$$

This implies that

$$X_B = \int_0^D \left[1 - \frac{v(s',t)^2}{c^2}\right]^{1/2} \mathrm{d}s'. \tag{3}$$

Note the highly complex equation this gives for the speed function $v(s,t)$, viz.,

$$v(s,t) = v_A(t) + \frac{\partial X(s,t)}{\partial t}. \tag{4}$$

Let us observe carefully that we are not assuming any simple Galilean addition law for velocities here. This is a straightforward differentiation with respect to t of the formula for the coordinate position of atom s at time t, viz., $x_A(t) + X(s,t)$. The partial time derivative of X is not the velocity of s relative to A, that is, it is not the velocity of s measured in a frame moving with A.

Now (2) seems to embody the idea of the rod being rigid. For surely this rod could no longer be elastic, in the sense that (1) only allows the element δs to relativistically contract for the value of its instantaneous speed, forbidding any other contortions. One could well imagine the rod undergoing a very complex deformation along its length, in which relativistic contraction effects were quite negligible compared with a certain looseness in the molecular bonding, but we are not talking about this. In fact we are seeking a definition of rigidity that does not refer to microscopic structure.

Let us just note what assumption is expressed by the earlier idea that

$$x_B(t) = x_A(t) + D\sqrt{1 - v_A(t)^2/c^2}, \tag{5}$$

which would lead to

$$v_B(t) = v_A(t) - \gamma(v_A) v_A(t) \ddot{x}_A(t) D/c^2, \tag{6}$$

where $\gamma(v_A)$ is the usual function of the speed. Thinking of A as a kind of base point, to which a force is perhaps applied, it says that the relativistic contraction is instantaneous: when A moves at speed v_A, the rod immediately has coordinate length $D/\gamma(v_A)$ for that value of v_A. We are now improving on this, accounting for the fact that, if the adjustment takes time, that time will depend how long the rod is, and that in turn depends on what we are trying to establish, namely the instantaneous length of the rod. We might imagine that a signal leaves A to tell B where it should be, and as it moves across, the speed of B is changing in response to past messages of the same kind. But using (2) and (4), can we tell when the new signal will get there?

1.3 Equation of Motion for Points on the Rod

So far the main equations for the atom labelled s on the rod are (2) and (4), viz.,

$$X(s,t) = \int_0^s \left[1 - \frac{v(s',t)^2}{c^2}\right]^{1/2} \mathrm{d}s' \tag{7}$$

and

$$v(s,t) = v_A(t) + \frac{\partial X(s,t)}{\partial t}. \tag{8}$$

The first implies that

$$\frac{\partial X(s,t)}{\partial s} = \left[1 - \frac{v(s,t)^2}{c^2}\right]^{1/2}. \tag{9}$$

We can write one nonlinear partial differential equation for $X(s,t)$ by eliminating $v(s,t)$ to give

$$c^2\left(\frac{\partial X}{\partial s}\right)^2 + \left[\frac{\partial X}{\partial t} + v_A(t)\right]^2 = c^2. \tag{10}$$

This is effectively the equation that we have to solve to find the length of our rod. It is important to see that there is a boundary condition too, viz.,

$$0 = \left.\frac{\partial X(s,t)}{\partial t}\right|_{s=0}, \tag{11}$$

because we do require $v(0,t) = v_A(t)$ in conjunction with (8).

We shall find a solution to this problem, although not by solving (10) directly. Instead we shall follow a circuitous but instructive route and end up guessing the relevant solution.

1.4 A Frame for an Accelerating Observer

Let AO be the name for an observer moving with the left-hand end A of the proposed rod. AO is an accelerating observer and it is well known [3] that such a person can find well-adapted coordinates y^μ with the following properties (where the Latin index runs over $\{1,2,3\}$):

- First of all, any curve with all three y^i constant is timelike and any curve with y^0 constant is spacelike.
- At any point along the worldline of AO, the zero coordinate y^0 equals the proper time along that worldline.
- At each point of the worldline of AO, curves with constant y^0 which intersect it are orthogonal to it where they intersect it.
- The metric has the Minkowski form along the worldline of AO.
- The coordinates y^i are Cartesian on every hypersurface of constant y^0.
- The equation for the worldline of AO has the form $y^i = 0$ for $i = 1, 2, 3$.

Such coordinates could be called semi-Euclidean.

Let us consider a 1D acceleration and temporarily drop the subscript A on the functions $x_A(t)$ and $v_A(t)$ describing the motion of AO in the inertial frame $\mathcal{I}$. The worldline of the accelerating observer is given in inertial coordinates by

$$t = \sigma, \qquad x = x(\sigma), \qquad \frac{\mathrm{d}x}{\mathrm{d}\sigma} = v(\sigma), \tag{12}$$

$$\frac{\mathrm{d}^2x}{\mathrm{d}\sigma^2} = a(\sigma), \qquad y(\sigma) = 0 = z(\sigma), \tag{13}$$

using the time t in $\mathcal{I}$ to parametrise. The proper time $\tau(\sigma)$ of AO is given by

$$\frac{\mathrm{d}\tau}{\mathrm{d}\sigma} = (1 - v^2/c^2)^{1/2}. \tag{14}$$

The coordinates y^μ are constructed on an open neighbourhood of the AO worldline as follows (see Fig. 7). For an event (t, x, y, z) not too far from the worldline, there is a unique value of τ and hence also the parameter σ such that the point lies in the hyperplane of simultaneity (HOS) of AO when its proper time is τ. This hyperplane of simultaneity is given by

$$t - \sigma(\tau) = \frac{v\big(\sigma(\tau)\big)}{c^2}\big[x - x\big(\sigma(\tau)\big)\big], \tag{15}$$

which solves, for any x and t, to give $\sigma(\tau)(x, t)$.

The semi-Euclidean coordinates attributed to the event (t, x, y, z) are, for the time coordinate y^0, (c times) the proper time τ found from (15) and, for the spatial coordinates, the spatial coordinates of this event in an instantaneously comoving inertial frame at proper time τ of AO. In fact, every other event in this instantaneously comoving inertial frame is attributed the same time coordinate $y^0 = c\tau$ and the appropriate spatial coordinates borrowed from this frame. Of course, the HOS of AO at time τ is also the one borrowed from the instantaneously comoving inertial frame.

There is just one detail to get out of the way: there are many different instantaneously comoving inertial frames for a given τ, and there are even many different ways to choose these frames as a smooth function of τ as one moves along the AO worldline, rotating back and forth around various axes in the original inertial frame

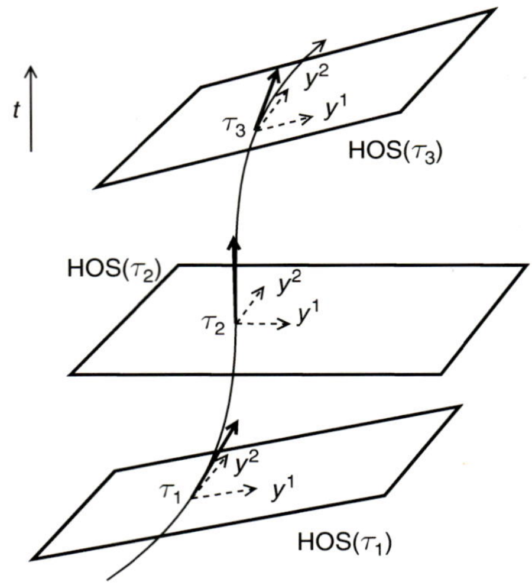

Fig. 7 Constructing a semi-Euclidean (SE) frame for an accelerating observer. View from an inertial frame with time coordinate t. The curve is the observer worldline given by (12). Three hyperplanes of simultaneity (HOS) are shown at three successive proper times τ_1, τ_2, and τ_3 of the observer. These hyperplanes of simultaneity are borrowed from the instantaneously comoving inertial observer, as are the coordinates y^1, y^2, and y^3 used to coordinatise them. Only two of the latter coordinates can be shown in the spacetime diagram

$\mathcal{I}$ as τ progresses. We choose a sequence with no rotation about any space axis in the instantaneous local rest frame. It can always be done by solving the Fermi–Walker transport equations (see Sect. 2.5). The semi-Euclidean coordinates are then given by

$$\begin{cases} y^0 = c\tau, \\ y^1 = \dfrac{[x - x(\sigma)] - v(\sigma)(t - \sigma)}{\sqrt{1 - v^2/c^2}}, \\ y^2 = y, \\ y^3 = z, \end{cases} \tag{16}$$

where $\sigma = \sigma(t, x)$ as found from (15). The inverse transformation, from semi-Euclidean coordinates to inertial coordinates, is given by

$$\begin{cases} t = \sigma(y^0) + \dfrac{v(y^0)}{c^2} y^1 \left[1 - \dfrac{v(y^0)^2}{c^2}\right]^{-1/2}, \\ x = x(y^0) + y^1 \left[1 - \dfrac{v(y^0)^2}{c^2}\right]^{-1/2}, \\ y = y^2, \\ z = y^3, \end{cases} \tag{17}$$

where the function $\sigma(y^0)$ is just the expression relating inertial time to proper time for the accelerating observer, and the functions $x(y^0)$ and $v(y^0)$ should really be written $x\big(\sigma(y^0)\big)$ and $v\big(\sigma(y^0)\big)$, respectively.

The above relations are not very enlightening. They are only displayed to show that the idea of such coordinates can be made perfectly concrete. One calculates the metric components in this frame, viz.,

$$g_{00} = 1/g^{00} = \left[1 + \frac{a(\sigma)[x - x(\sigma)]/c^2}{1 - v(\sigma)^2/c^2}\right]^2, \tag{18}$$

where $\sigma = \sigma(t, x)$ as found from (15), and

$$g_{i0} = 0 = g_{0i}, \qquad g_{ij} = -\delta_{ij}, \qquad i, j \in \{1, 2, 3\}, \tag{19}$$

and checks the list of requirements for the coordinates to be suitably adapted to the accelerating observer.

Although perfectly concrete, the coordinates are not perfectly explicit: the component g_{00} of the semi-Euclidean metric has been expressed in terms of the original inertial coordinates! This can be remedied as follows. One observes that, with the help of (15),

$$y^1 = [x - x(\sigma)]\sqrt{1 - v^2/c^2}. \tag{20}$$

One calculates the four-acceleration in the inertial frame to be

$$a^\mu := \frac{\mathrm{d}^2 x^\mu}{\mathrm{d}\tau^2} = a(1 - v^2/c^2)^{-2}\left(\frac{v}{c}, 1, 0, 0\right), \tag{21}$$

and transforms this by Lorentz transformation to the inertial frame instantaneously comoving with the observer to find only one nonzero four-acceleration component in that frame, which is called the absolute acceleration of the observer:

$$a_{01} := \text{absolute acceleration} = a(1 - v^2/c^2)^{-3/2}. \tag{22}$$

The notation a_{01} for the 1-component of the absolute acceleration will appear again on p. 94. One now has the more comforting formula

$$g_{00} = 1/g^{00} = \left[1 + \frac{a_{01}(\sigma)y^1}{c^2}\right]^2. \tag{23}$$

To obtain still more explicit formulas, one needs to consider a specific motion $x(\sigma)$ of AO, the classic example being uniform acceleration:

$$x(\sigma) = \frac{c^2}{g}\left[\left(1 + \frac{g^2\sigma^2}{c^2}\right)^{1/2} - 1\right], \qquad t = \sigma, \tag{24}$$

where g is some constant with units of acceleration. This does not look like a constant acceleration in the inertial frame:

$$\frac{\mathrm{d}x}{\mathrm{d}\sigma} = \frac{g\sigma}{(1+g^2\sigma^2/c^2)^{1/2}}, \qquad \frac{\mathrm{d}^2x}{\mathrm{d}\sigma^2} = \frac{g}{(1+g^2\sigma^2/c^2)^{3/2}}. \tag{25}$$

However, the 4-acceleration defined in the inertial frame $\mathcal{I}$ by

$$a^\mu = \frac{\mathrm{d}^2x^\mu}{\mathrm{d}\tau^2}, \tag{26}$$

where τ is the proper time, has constant magnitude. It turns out that

$$a^2 := a_\mu a^\mu = -g^2,$$

with a suitable convention for the signature of the metric.

In this case, the transformation from inertial to semi-Euclidean coordinates is

$$y^0 = \frac{c^2}{g}\tanh^{-1}\frac{ct}{x+c^2/g}, \tag{27}$$

$$y^1 = \left[\left(x+\frac{c^2}{g}\right)^2 - c^2t^2\right]^{1/2} - \frac{c^2}{g}, \qquad y^2 = y, \qquad y^3 = z, \tag{28}$$

and the inverse transformation is

$$t = \frac{c}{g}\sinh\frac{gy^0}{c^2} + \frac{y^1}{c}\sinh\frac{gy^0}{c^2}, \tag{29}$$

$$x = \frac{c^2}{g}\left(\cosh\frac{gy^0}{c^2} - 1\right) + y^1\cosh\frac{gy^0}{c^2}, \qquad y = y^2, \qquad z = y^3. \tag{30}$$

One finds the metric components to be

$$g_{00} = \left(1+\frac{gy^1}{c^2}\right)^2, \qquad g_{0i} = 0 = g_{i0}, \qquad g_{ij} = -\delta_{ij}, \tag{31}$$

for $i, j \in \{1,2,3\}$, in the semi-Euclidean frame. Interestingly, this metric is static, i.e., g_{00} is independent of y^0. It is the only semi-Euclidean metric that is [7].

It is worth pausing to wonder why AO should adopt such coordinates. It must be comforting to attribute one's own proper time to events that appear simultaneous. But what events are simultaneous with AO? In the above construction, AO borrows the hyperplane of simultaneity of an inertially moving observer, who does not have the same motion at all. AO also borrows the lengths of this inertially moving observer. But if AO were carrying a rigid measuring rod, what lengths would be measured with it?

1.5 Lengths Measured by the Rigid Rod

In fact the rigid rod of Sect. 1.2 measures the spatial coordinates of AO when this observer uses semi-Euclidean coordinates. Let us prove this for the case of a uniform acceleration g, where formulas are explicit.

We write down the path of a point with some fixed spatial coordinate s along the axis of acceleration (putting the other spatial coordinates equal to zero). The formula we have for the path of the origin of the SE frame as expressed in Minkowski coordinates is

$$x_A(t) = \frac{c^2}{g}\left(\sqrt{1+\frac{g^2t^2}{c^2}}-1\right), \tag{32}$$

giving a coordinate velocity

$$v_A(t) = \frac{gt}{\sqrt{1+\dfrac{g^2t^2}{c^2}}}. \tag{33}$$

The formula for the path of the point at fixed SE spatial coordinate s from the origin as expressed in Minkowski coordinates is

$$x_s(t) = X(s,t) + x_A(t) = \frac{c^2}{g}\left[\sqrt{\left(1+\frac{gs}{c^2}\right)^2+\frac{g^2t^2}{c^2}}-1\right]. \tag{34}$$

We are going to show that the function $X(s,t)$ defined by the last relation actually satisfies our equation of motion (10) in the case where the function $x_A(t)$ gives the path of the left-hand end A of the rod, i.e., when the point A is uniformly accelerated by g.

Proof That (34) Is a Solution for (10)

We begin with the partial derivatives:

$$\frac{\partial X}{\partial t} = \frac{gt}{\sqrt{\left(1+\dfrac{gs}{c^2}\right)^2+\dfrac{g^2t^2}{c^2}}} - v_A(t), \tag{35}$$

$$\frac{\partial X}{\partial s} = \frac{1+gs/c^2}{\sqrt{\left(1+\dfrac{gs}{c^2}\right)^2+\dfrac{g^2t^2}{c^2}}}. \tag{36}$$

Hence,

$$\left[\frac{\partial X}{\partial t} + v_A(t)\right]^2 = \frac{g^2t^2}{\left(1+\frac{gs}{c^2}\right)^2 + \frac{g^2t^2}{c^2}} \tag{37}$$

and

$$c^2\left(\frac{\partial X}{\partial s}\right)^2 = \frac{c^2(1+gs/c^2)^2}{\left(1+\frac{gs}{c^2}\right)^2 + \frac{g^2t^2}{c^2}}. \tag{38}$$

Adding the last two equations together, it is clear that we just get c^2, as required by (10). The boundary condition (11) on p. 66 is obviously satisfied too. ■

For a rod with arbitrary 1D acceleration, the formulas are much more involved, due to the lack of explicitness, but the proof is nevertheless straightforward [2]. So not only have we found the length of our rigid rod when it is accelerating along its own axis, but we discover that any AO with 1D motion could use it to measure semi-Euclidean coordinates along the direction of acceleration. This means that the rigid rod automatically satisfies what is sometimes called the ruler hypothesis, namely, it is at any instant of time ready to measure lengths in an instantaneously comoving inertial frame, since this is precisely the length system used by the semi-Euclidean coordinates.

The accelerating observer would not necessarily have to be holding one end of the rod. It could be lying with one end held fixed at some semi-Euclidean coordinate value $y^1 = s_1$ and the other end would then remain at a constant coordinate value $y^1 = s_2 > s_1$. This is shown by exactly the same kind of analysis as above. In other words, if the rod always manages to occupy precisely this interval on the axis of the SE coordinate system, its length as viewed in the original inertial frame $\mathcal{I}$ will satisfy the rigidity (10) on p. 66. Hence, a rigid rod whose left-hand end is compelled to follow the worldline $y^1 = s_1$ will always appear to have the same length $s_2 - s_1$ to the SE observer.

Before taking a look at some of the remarkable features of the semi-Euclidean coordinate frame, let us just note in passing that, to first order in the rest length D of the rod, one finds

$$x_B(t) = x_A(t) + \left[1 - \frac{v_A(t)^2}{c^2}\right]^{1/2} D + O(D^2), \tag{39}$$

which is precisely our original approximation (5) back on p. 65.

1.6 Properties of a Semi-Euclidean Frame

Since the notion of rigidity expressed by (10) fits in so nicely with the semi-Euclidean frame of the associated observer, it is worth summarising some of the features of these frames. We consider any point B with fixed spatial semi-Euclidean coordinates $(y^1, y^2, y^3) = (s, 0, 0)$ and varying y^0. Then we have:

- Viewed from the inertial frame $\mathcal{I}$, B follows an accelerating worldline, but generally with a different acceleration to the observer at the origin of the semi-Euclidean frame. In the case of a uniformly accelerating observer, it turns out that such a point also has uniform acceleration, but a smaller one than the observer at the origin, and ever smaller as s increases [see (79) on p. 97].
- Viewed from the inertial frame $\mathcal{I}$, if B is simultaneous with the accelerating observer A at the origin of the semi-Euclidean frame as judged by that observer, i.e., A and B have semi-Euclidean coordinates $(c\tau, 0, 0, 0)$ and $(c\tau, s, 0, 0)$, respectively, for some τ, then they have the same 4-velocity at those two events. Of course, they are not then simultaneous for the original inertial observer with frame $\mathcal{I}$ (except at the coincident origins of the two frames). But since the accelerating observer borrows the hyperplane of simultaneity of an instantaneously comoving inertial observer, the two events in question will be simultaneous for the latter. In other words, when an inertial observer instantaneously comoving with A looks at B, that observer will have the same 4-velocity as B.

The first of these is not difficult to show from the definitions of the semi-Euclidean coordinates. The second follows immediately from the expression

$$\frac{\partial X}{\partial t} = v_A(y^0) - v_A(t), \tag{40}$$

which can be proven for general 1D accelerations of the observer. For then, by (8) on p. 66,

$$\begin{aligned} v_B(t) &= v(s, t) \\ &= v_A(t) + \frac{\partial X}{\partial t} \\ &= v_A(t) + v_A(y^0) - v_A(t) \\ &= v_A(y^0). \end{aligned} \tag{41}$$

So to find the speed of B at some event on its worldline, we must draw the HOS of the accelerating observer A which contains that event and find the proper time y^0 at which the HOS intersects the worldline of A. The point B has the speed which A had at that proper time. If we draw the worldlines of A and B on the Minkowski diagram, this result about their speeds tells us that any HOS through the A worldline intersects the two worldlines at points where they have the same gradient in the Minkowski (t, x) plane.

1.7 Behaviour of a Rigid Rod

From the properties in the last section, we may deduce something about the behaviour of a rigid rod under acceleration, i.e., we may deduce something about what the material points of the rod must do if the rod is to satisfy our rigidity criterion (1) on p. 65.

According to the first observation, all points of the rod have different coordinate accelerations and indeed different 4-accelerations relative to the inertial frame $\mathcal{I}$ when the rod is viewed in any hyperplane of simultaneity of AO. It turns out that all points of the rod have different 4-accelerations relative to the inertial frame $\mathcal{I}$ when the rod is viewed in any hyperplane of simultaneity of $\mathcal{I}$. We may deduce that the 4-forces on different material points of the rod must always be different at any instant of time for any inertial observer, and in a specific way that depends on the 4-force at A.

According to the second observation, all points of the rod have the same coordinate velocity and indeed the same 4-velocity relative to the inertial frame $\mathcal{I}$ when the rod is viewed in any hyperplane of simultaneity of AO.

What we have then here is a rather complex system of 4-forces within the rod. We might say that they conspire in such a way that, if AO carries one end of it (labelled A) and uses the semi-Euclidean frame to judge simultaneity, the points of the rod will always have the same speed relative to $\mathcal{I}$. They also conspire in such a way that the rod will always instantaneously have the right length to measure semi-Euclidean coordinate lengths for A, which are also proper lengths for AO in the semi-Euclidean system.

Note, however, that, apart from the first instant when the rod is at rest in $\mathcal{I}$, the rod will never have the relativistically contracted length

$$\left[1-\frac{v_A(\sigma)^2}{c^2}\right]^{1/2} D \tag{42}$$

when observed from $\mathcal{I}$. Its length according to $\mathcal{I}$ will be the quantity $X(D,t)$ defined by inserting $s=D$ in [2]

$$X(s,t) = x_A(y^0) - x_A(t) + s\left[1-\frac{v_A(y^0)^2}{c^2}\right]^{-1/2}, \tag{43}$$

where $y^0 = y^0(s,t)$ is given by

$$t = \sigma(y^0) + \frac{v_A(y^0)}{c^2}s\left[1-\frac{v_A(y^0)^2}{c^2}\right]^{-1/2}. \tag{44}$$

These things are illustrated in Fig. 8 for the case of a uniform acceleration of magnitude g, where explicit formulas are possible. Axes t and x are those of the inertial frame $\mathcal{I}$. Of course we have dropped two space dimensions. The rigid rod is the four-dimensional region between the two worldlines [see (32) and (34) on p. 71]

$$x_A(t) = \frac{c^2}{g}\left(\sqrt{1+\frac{g^2t^2}{c^2}}-1\right)$$

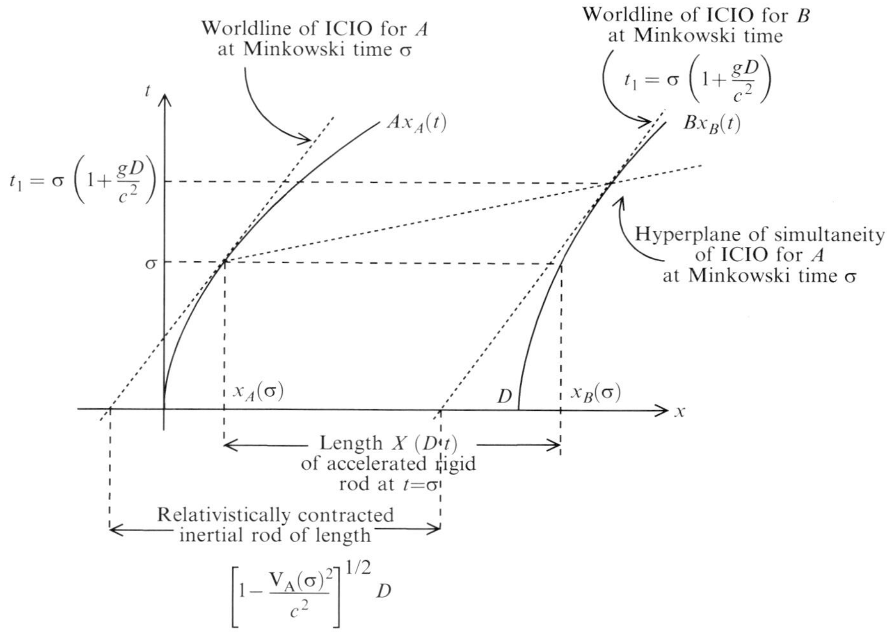

Fig. 8 Uniformly accelerating rigid rod. *Slanting dotted axes* are those of the instantaneously comoving inertial observer (ICIO) for A at Minkowski time σ or for B at Minkowski time $t_1 = \sigma(1 + gD/c^2)$

and

$$x_B(t) = \frac{c^2}{g}\left[\sqrt{\left(1 + \frac{gD}{c^2}\right)^2 + \frac{g^2t^2}{c^2}} - 1\right].$$

Sloping dotted axes are those of the instantaneously comoving inertial observer (ICIO) for A at Minkowski time σ. The hyperplane of simultaneity for this ICIO intersects the worldline of B at an event $\big(t_1, x_B(t_1)\big)$ where that worldline has the same gradient as the worldline of A at the event $\big(\sigma, x_A(\sigma)\big)$, i.e., the same speed relative to $\mathcal{I}$ as A at the event $\big(\sigma, x_A(\sigma)\big)$. The Minkowski time t_1 of this event on the worldline of B is found to be

$$t_1 = \sigma\left(1 + \frac{gD}{c^2}\right). \tag{45}$$

Note that, because A has the same speed relative to $\mathcal{I}$ at the event $\big(\sigma, x_A(\sigma)\big)$ as B at the event $\big(t_1, x_B(t_1)\big)$, they have the same 4-velocity components relative to $\mathcal{I}$, and hence also relative to the ICIO for A at $\big(\sigma, x_A(\sigma)\big)$. The two 4-velocities (of A and B) are located at different events in spacetime, and the last conclusion follows because the Lorentz transformation from $\mathcal{I}$ to the frame of the ICIO is constant in spacetime. But if we transform the two vectors at different events by a spacetime-dependent transformation, such as the transformation to SE coordinates, we would not expect to end up with the same sets of components, and indeed we do not.

Concerning the length of the rod:

- AO always considers the rod to have length D when using the semi-Euclidean system.
- The instantaneously comoving inertial observers with A, or indeed with B, always consider the rod to have length D, but only at the event where they are instantaneously comoving with A or B. As mentioned above, this is precisely what is meant by saying that the rod satisfies the ruler hypothesis.
- The inertial observer with frame $\mathcal{I}$ considers the rod to have length

$$X(D,t) = x_B(t) - x_A(t) = \frac{c^2}{g}\left[\sqrt{\left(1+\frac{gD}{c^2}\right)^2 + \frac{g^2t^2}{c^2}} - \sqrt{1+\frac{g^2t^2}{c^2}}\right]. \tag{46}$$

- A rod represented by the 4D region between the two dotted time axes tangent to the worldline of A at Minkowski time σ and the worldline of B at Minkowski time t_1 would have the relativistically contracted length

$$\left[1-\frac{v_A(\sigma)^2}{c^2}\right]^{1/2} D$$

for $\mathcal{I}$, but there is no such rod here.

1.8 Rigid Spheres and Instantaneous Transmission of Motion

We can see what our rod is doing in a spacetime picture, drawn in the inertial frame in which it is originally at rest. It sweeps out a region of spacetime and we are saying that the SE observer is using it to measure length. It would be easy to become euphoric about such calculations, particularly the fact that the rigidity criterion (10) is satisfied by an expression like (46) set up for rather different reasons. But perhaps we should be asking what we would have to do to get the rod to move like that. Could the SE observer just accelerate the left-hand end and let the rest of the rod adapt somehow to what is happening via its rigidity?

After all we paid no attention to microscopic structure. If we think about the toy electron with its two point charge components, we avoided making any detailed model of the binding forces. In fact, we appear to have gone a long way without doing any real physics.

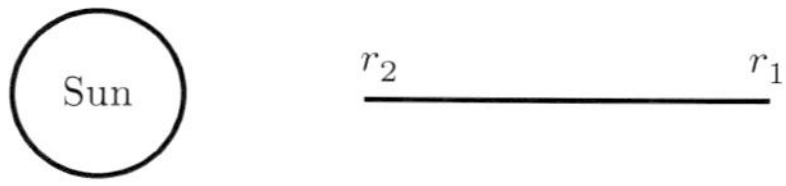

Fig. 9 Measuring stick in Schwarzschild spacetime, lying along a fixed radial coordinate interval. The length is given by the usual formula $L = \int_{r_2}^{r_1} \mathrm{d}r\, (1-2m/r)^{-1/2}$, where r is the Schwarzschild radial coordinate. But is this really the length of a measuring stick? Could a measuring stick really have this motion? Or put another way, what would one have to do to get it to behave in this way?

To show the generality of the problem, we find this in an elementary course on general relativity (see Fig. 9) [4]:

> To get a more quantitative feel for the distortion of the geometry produced by the gravitational field of a star, consider a long stick lying radially in the gravitational field, with its endpoints at the [Schwarzschild] coordinate values $r_1 > r_2$. To compute its length L, we have to evaluate
>
> $$L = \int_{r_2}^{r_1} \mathrm{d}r\, (1 - 2m/r)^{-1/2}.$$

Since this set of points lies in a hyperplane of simultaneity for the Schwarzschild coordinates, a Schwarzschild observer would call this the proper distance between the two endpoints. But is it really the length of a stick? What would we have to do to get a stick to do this? For example, none of the points of it are in free fall, so they all have some kind of 4-acceleration, and in fact, they all have different 4-accelerations, exactly as we have found for the accelerating rigid rod in a flat spacetime.

Of course, we cannot say whether the measuring stick in the above quote is rigid until we label the material particles in it and extend our definition of rigidity to the curved spacetimes of general relativity. A step is taken in this direction in Sect. 2.7. However, it is clear that if real rods do behave like this, there must be some physical reason for it. On the other hand, in pre-relativistic mechanics, rigidity was always an ideal concept, at best a convenient approximation that no one would really have expected to be possible.

One finds the same attitude in calculations of self-force on small charge distributions. This is discussed in the recent book by Yaghjian [5]. Calculations are made for a relativistically rigid spherical shell of charge of radius a, whose center has an arbitrary motion:

> 'Relativistically rigid' refers to the particular model of the electron, proposed originally by Lorentz, that remains spherical in its proper (instantaneous rest) frame, and in an arbitrary inertial frame is contracted in the direction of velocity to an oblate spheroid with minor axis equal to $2a/\gamma$.

This is exactly the kind of rigidity we have been talking about. Like our rod, the sphere always has the same dimensions to the instantaneously comoving inertial observer. Above all, this makes it possible to carry out the self-force calculation. Like so many approximations in physics, it is largely motivated by mathematical convenience. Note, however, that the value of $2a/\gamma$ for the minor axis of the spheroid is only an approximation, as we have been at pains to show [see (42) on p. 74].

Yaghjian goes on to say [5]:

> Even a relativistically rigid finite body cannot strictly exist because it would transmit motion instantaneously throughout its finite volume. Nonetheless, one makes the assumption of relativistically 'rigid motion' to avoid the possibility of exciting vibrational modes within the extended model of the electron.

He imputes the last remark to Pauli. But is there really any sense in which motion is transmitted instantaneously? That did seem to be the assumption with pre-relativistic rigidity: if a force was applied to one end of a rod, the same force had to be transmitted instantaneously to all the particles in the rod, so that all particles would always have the same acceleration and the same speed.

Are we assuming something like this in the present case? Viewing from the inertial frame $\mathcal{I}$, imagine the left-hand end A of the rod as being accelerated in some active way, whilst the rest of the rod follows suit in some sense. As the left-hand end moves faster, the other points on the rod pick up speed too. In the case of a uniform acceleration of A, we know that each point of the rigid rod has uniform acceleration, but always lesser, until we come to the right-hand end B, which has the smallest value. However, the end B eventually reaches the speed that A had some time previously. In this view of things, we may be thinking that speed somehow propagates along the rod, with a delay that we ought to be able to calculate.

On p. 73, we showed the following result. If we consider a point $x_A(\sigma)$ on the worldline of A, when it has speed $v_A(\sigma)$, and draw the HOS of the ICIO, this HOS will intersect the worldline $x_B(t)$ of B at a Minkowski time t_1 where $v_B(t_1) = v_A(\sigma)$. In the case of a uniform acceleration g, where we have explicit formulas, we know from (45) that

$$t_1 = \sigma\left(1 + \frac{gD}{c^2}\right).$$

We therefore know how much Minkowski time is required for the speed of A to propagate through to the other end of the rod, if indeed there is propagation, viz., $\sigma g D/c^2$. The Minkowski observer will consider that the signal, if indeed it is one, has propagated from $x_A(\sigma)$ to $x_B(t_1)$, so that it has travelled a distance $x_B(t_1) - x_A(s)$. We can calculate this from the formulas in the last section, and the result is

$$x_B(t_1) - x_A(\sigma) = D\left(1 + \frac{g^2\sigma^2}{c^2}\right)^{1/2}.$$

We know this anyway, because it is the projection onto the x axis of the imaginary inertial rod mentioned earlier.

If we now divide the distance travelled by the putative signal by the time it has taken, as reckoned in the Minkowski frame, we find the value

$$\frac{D(1 + g^2\sigma^2/c^2)^{1/2}}{g\sigma D/c^2} = c\left(1 + \frac{c^2}{g^2\sigma^2}\right)^{1/2} > c \tag{47}$$

for the speed of propagation. If speed propagates, it does so faster than light.

Is this why our rigid rod is instantaneously ready to measure lengths when accelerated to a new speed? When we look back at the formulation of our equation of motion (10) for the material particles making up the rod, it is clear that we never explicitly introduced any delays. We merely hoped that the more sophisticated model would cater for this.

Alternatively, it may be that we should not consider speed as propagating in the rod. After all, the result (47) is most catastrophic when $\sigma = 0$, simply because the two ends of the rod happen to have the same speed at the same Minkowski time. The explanation of whatever paradox there seems to be here is more likely to be this. When the motion begins from rest (all points of the rod being at rest when $\sigma = 0$), each one has to instantaneously have the appropriate four-acceleration. This may be a problem in itself, but once that is accomplished, there is no need for the speed to propagate. Each point of the rod is subject to its appropriate four-acceleration and so acquires the required speed locally as it were. The real problem is: how can each point be subject to the appropriate four-acceleration. If a force is applied at one end, it is indeed four-acceleration (or four-force, or just force) that has to take a little time to transmit to the various points of the rod. The rod has to adjust in some way.

So could rigidity be equivalent to instantaneous transmission of four-acceleration? This too looks unnecessary since the rod may have been forever undergoing this motion, at least theoretically. More realistically, one could always wait until the required distribution of four-accelerations has set itself up within the rod and thereafter describe its motion as rigid. Yaghjian says that the assumption of relativistically rigid motion is made to avoid the possibility of exciting vibrational modes within the extended electron. This corresponds to the idea that, in reality, there must always be some time of adjustment after the motion is initiated.

Let us return to the question posed at the beginning of this section, viz., could one just accelerate the left-hand end of a rigid rod and let the rest of the rod adapt somehow to what is happening via its rigidity? We can investigate this idea by applying the result (41) on p. 73. We imagine a rod that is stationary in an inertial frame $\mathcal{I}$ and to which an external acceleration is applied to the left-hand end A at some time t_{acc} (reckoned in that frame).

On the Minkowski diagram, the worldline of A is represented by a vertical line which we may take to be the time axis, up to $t = t_{\mathrm{acc}}$, where it begins to curve over to the right. The hyperplanes of simultaneity of an observer moving with A are at first horizontal, but begin to slant upwards after $t = t_{\mathrm{acc}}$, slanting up more and more as A moves faster. This effectively determines what the right-hand end B of the rod will do, if we recall the simple result (41) on p. 73. Because the rod is rigid, B always moves with speed equal to the speed of A at the event on the worldline of A that the instantaneously comoving inertial observer moving with A considers to be simultaneous. So the worldline of B is clearly vertical up to the time $t = t_{\mathrm{acc}}$. But what happens next?

As soon as $t > t_{\mathrm{acc}}$, the relevant HOS of A through the worldline of B must be one of those that are beginning to slant upwards, no matter how soon after t_{acc} we look at the worldline of B. So B will have the speed of A at a slightly earlier time (as reckoned in the frame $\mathcal{I}$), but nevertheless at some time after t_{acc}, when A

was already accelerating. This means that B will have started moving. Its worldline curves over to the right after t_{acc}. It curves over more slowly, but the important thing is that it does so immediately (in the $\mathcal{I}$ reckoning) after the time t_{acc}.

We conclude that the acceleration of B is indeed instantaneous, i.e., simultaneous with the acceleration of A in this frame. And, of course, this means that in some other inertial frame, B will begin to accelerate before A, throwing out the idea that the acceleration of A could be the cause of the acceleration of B. Put another way, if one really were applying an external force only at A, one would not expect B to be able to react for at least the time it takes light to propagate along the length of the rod. This suggests another notion of rigidity, wherein a rigid rod is one in which the speed of sound in the rod is equal to the speed of light [6].

Presumably this shows that one cannot expect any rod to have our kind of rigid motion when tampered with in this way. So rigid motion is not an easy behaviour to achieve, whatever one's medium is made of. However, one could imagine that some other means of accelerating the medium could result in its having rigid motion, e.g., applying different external forces to all particles making up the rod, perhaps by means of a force field like gravity or electromagnetism. Indeed, the accelerations of all particles in the rod are completely determined when it has the rigid motion specified by the criterion (1) on p. 65.

The above considerations suggest that one should investigate rigid motion, rather than rigid spheres or rods. One might then have to conclude that rigid motion cannot strictly occur when caused by an external force applied at just one point of the object. This would still leave open the question of motions due to fields of force. We consider the idea of rigid motions of a general medium in Sect. 2.

1.9 Rigid Electrons and Rigid Atoms

We asked above how each point could be subject to the appropriate four-acceleration when the rigid rod or sphere is made to move. One could envisage an interplay of repulsive and binding forces within these objects conspiring to move each material point in the appropriate way. It does look possible a priori. Indeed, this idea is effectively applied to the spherical shell of charge in the self-force calculations discussed by Yaghjian in his book [5]. The binding forces are precisely those required to keep the shell of charge spherical in its proper frame.

Of course, it could be that small particles like electrons are rigid in this sense. In any case, one has to assume something and this seems to be as convenient an assumption as one could hope for within the framework of relativity theory. Another small particle is the atom. In a pre-quantum model, an electron orbits a nucleus under an electromagnetic attraction so one only has the binding force to worry about. One might, like Bell in his paper *How to Teach Special Relativity* [10], treat the nucleus as an accelerating point charge for which the exact electromagnetic potential (the Lienard–Wiechert potential) is known from Maxwell's theory, then calculate the exact orbit of the electron in this field as the nucleus accelerates. In principle, this would give a perfect description of the way the length of the atom would change in the direction of acceleration.

One might then be forgiven for forgetting the complexity of the rigid rod with its long string of atoms and the unfortunate way they might interfere with one another by being bound together in some manner. One could just suppose that the way measuring sticks contract is just, or should be just, if they are any good for measuring, the way their constituent atoms contract, according to the simple idea of the last paragraph. Would this then give us a different notion of rigidity? It certainly looks likely on the face of it, if one could carry out precise calculations.

However, one might give the following argument for supposing the Bell atom to be at least approximately rigid. Figure 10 shows our solution for accelerating a rigid rod from one state of uniform velocity to another. During the acceleration, the length of the rod is always the same for an instantaneously comoving inertial observer. This is precisely what one expects for the radius of the Bell atom if the acceleration is slow enough, i.e., if the electron gets in plenty of revolutions around the nucleus before the acceleration ever has time to change very much. In an accurate calculation, one would expect to require some adjustment time, but to a certain level

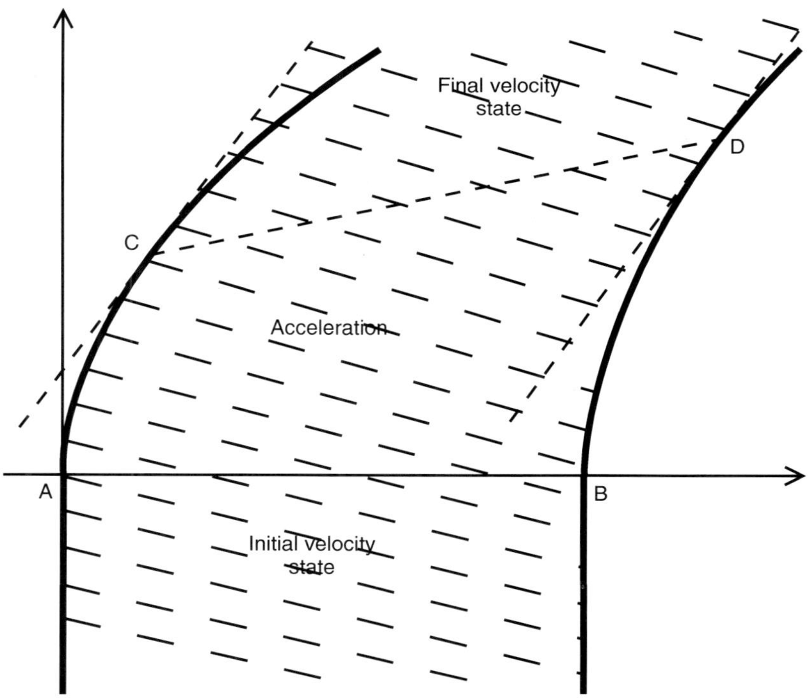

Fig. 10 Rigid motion of a rod from one uniform velocity state to another. The rod is represented by the *shaded region* of the spacetime diagram. The initial velocity state continues up to the horizontal line AB, while the acceleration occupies the region up to the slanting line CD

of approximation, the atom will behave rigidly. This is discussed further at the end of Sect. 2.7.

The aim of the second part of this discussion (Sect. 2) will be to consider rigid motion in a more general context, for any material medium. We are trying here to avoid awkward questions about how the motion is brought about, or how the medium manages to get into or remain in a state of rigid motion.

1.10 A Note on the Pound–Rebka Experiment

This is just a simplified overview to identify some hidden assumptions. We imagine an emitter E at the bottom of a tower and a receiver R at the top (in fact, iron nuclei emitting and absorbing gamma rays). The latter is going to detect the gravitational redshift predicted by general relativity. We may suppose that the gravitational field is perfectly uniform here, by which we mean that there are coordinates relative to which the metric takes the form (31) on p. 70. (This assumption in itself is worth examining much more closely [7].) Of course, this spacetime is flat, i.e., there is a global inertial frame, often called the freely falling frame. By the strong equivalence principle, electromagnetic effects relating to E and R can be examined using the ordinary Maxwell equations, or ordinary quantum electrodynamics, in the freely falling frame.

In his book [9, Sect. 5.7], Brown mentions the need to assume the clock hypothesis, which declares that a clock worthy of the name will measure the proper time along whatever worldline it happens to be following. This could be taken as the definition of an ideal clock, or a hypothesis, to be tested, that some particular putative clock approximates to ideality. Let us understand this in the context of the Pound–Rebka experiment. E emits waves that leave at precise intervals, but relative to what time scale? The proper time associated with its worldline? If the iron nuclei used as emitter and receiver satisfy the clock hypothesis, we would answer affirmatively there.

As an aside, which is nevertheless quite relevant to the general ethos of this book, we may well ask why this should be. What is the physical explanation? It is suggested here that the Bell approach may show that this is just a good approximation [10], but that detailed calculations with the relevant theories in special relativity (this spacetime is flat) or minimally extended from flat spacetime in general relativity where necessary [8], would give a better answer. So we are suggesting that the clock 'hypothesis' is necessary insofar as one needs to know when the waves or photons are emitted, but that one could also prove that this is a good approximation, so that the only assumption needed is the assumption that one has good theories for the emission process.

But there is already an interesting problem with identifying the emitter and receiver worldlines in a global inertial frame when the tower is uniformly accelerating. In fact, there are at least two obvious possibilities:

1. The emitter E follows the usual hyperbola in spacetime of a uniformly accelerating point and the receiver R likewise, with the same uniform acceleration, i.e.,

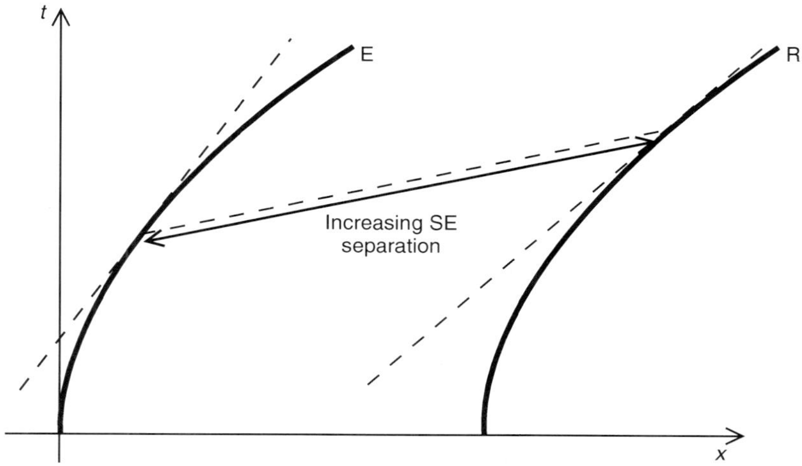

Fig. 11 Case 1. The emitter E follows the usual hyperbola in spacetime of a uniformly accelerating point and the receiver R likewise, with the same uniform acceleration. With a ruler satisfying the ruler hypothesis, E considers R to recede

with the same hyperbola but shifted along the space axis (see Fig. 11). If E uses semi-Euclidean (Rindler) coordinates, i.e., rigid rulers as described in this paper, then R recedes from it.

2. The emitter follows the usual hyperbola of a uniformly accelerating point and the receiver likewise, but a different one, viz., the hyperbola of a point at fixed semi-Euclidean distance from E (see Fig. 12). If that distance is fixed, it must have a lower uniform acceleration.

In general relativity, which is what we are doing here (even though Brown is considering a case where one is still trying to do special relativity but finding that the results of the Pound–Rebka experiment create a problem with the notion of inertial frame), the emitter and receiver are accelerating because they are *not* being allowed to fall freely. But should they have the same acceleration for some reason related to the fact that the gravitional field is uniform, as in (1) above (it has zero curvature, but then that does not tell us how strong the field is, only that there are no tidal effects, hence no variation in it); or should they have constant spatial separation as in 2, if indeed that is what we should mean by separation (ruler hypothesis)?

If the receiver is supported by the roof of the tower, then it is indeed the structure of the tower that determines the motion of the receiver. In fact it is usually assumed that the tower is rigid, i.e., case 2 above. If that is so, it is important to see that one is assuming that the emitter and receiver have different accelerations, i.e., they are being supported differently against the uniform gravitational field. This might look surprising when one considers that the gravitational field is supposed to be uniform. But it just illustrates the fact that supporting something in a gravitational field in general relativity introduces effects that are quite different from the gravitational field,

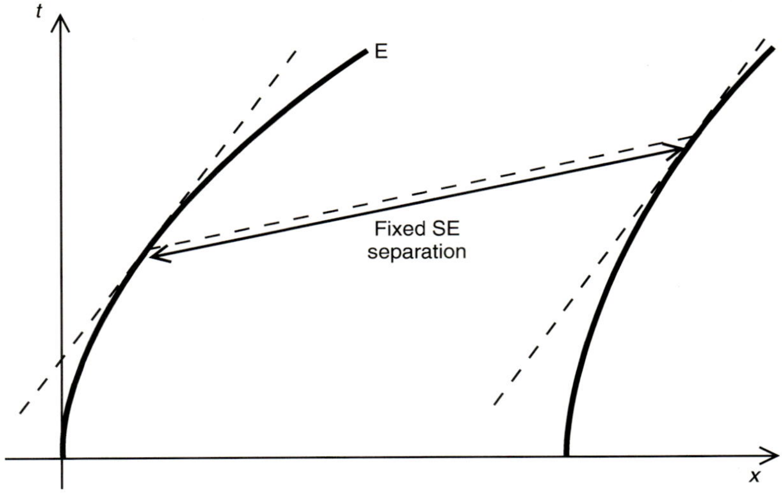

Fig. 12 Case 2. The emitter follows the usual hyperbola of a uniformly accelerating point and the receiver likewise, but a different one, viz., the hyperbola of a point at fixed semi-Euclidean distance from E

viz., a supporting force which causes an acceleration (while freely falling objects have no acceleration). In fact, supporting something is a rather arbitrary thing to do in a certain sense. The notion of supporting is just specified by saying that a thing is not allowed to move relative to some coordinates one happens to be using, and coordinates are not fundamental in general relativity. A classic case would be an object held at some fixed values of the usual coordinates for a Schwarzschild spacetime.

So what coordinates are being used in the Pound–Rebka experiment and how would they be set up? Presumably one wants to say that the distance between the emitter and receiver is constant. But what distance is this? Suppose one measures distances up the tower using a ruler held by an observer sitting with the emitter. If it satisfies the ruler hypothesis, then it measures semi-Euclidean (Rindler) spatial coordinates. In case 1 above, where the receiver is supported in such a way that it has the same uniform acceleration as the emitter, the receiver would be measured to recede according to such measurements. But if the tower is rigid (in the usually accepted sense, discussed in this paper) and the receiver is fixed relative to a point of the tower, then the receiver would be considered to remain at a fixed distance from the emitter.

The point about mentioning this is just to say that, just as one discusses the clock hypothesis in this context, there is a similar consideration of the ruler hypothesis. If one did measure the emitter–receiver separation with a ruler and wanted to say that this gave the semi-Euclidean spatial coordinate (because the constancy of the separation would allow us to do the redshift calculation in the usual way), one would effectively be assuming that the ruler satisfied the ruler hypothesis, i.e., that despite the acceleration of the observer holding one end of it, it is always precisely ready to give the proper distance of an instantaneously comoving inertial (freely falling) observer. This is also the rigidity assumption, viz., the ruler is rigid, or at least undergoing what will be called rigid motion in Sect. 2 of this paper. In short, the rigidity assumption is actually precisely the ruler hypothesis.

As an aside, it was mentioned above that there can be no such thing as a rigid object because, if the external force is applied at one point of the object, it cannot remain rigid. Hence the discussion of rigid motion in Sect. 2, without consideration of how one might achieve the rigid motion of an object. But here one has a case where one might actually achieve rigid motion, i.e., the tower might actually be undergoing rigid motion, because the gravitational effect on it is *not* applied at just one point.

In an analogous way, one assumes that the emitter and receiver satisfy the clock hypothesis, i.e., that despite their accelerations, they emit and receive exactly as instantaneously comoving inertial emitters and receivers would. In other words, used as clocks, they would deliver proper time as it is usually defined. Of course, proper time is perfectly well defined in a mathematical sense along arbitrary worldlines in special relativity, without the need to mention any clocks. The only hypothesis one needs there in a context like this is the hypothesis that what one is actually hoping to use as a clock does read proper time. Of course, if it did not, it would not be regarded as a clock. What we would like to add here is Bell's idea that one should be able to show theoretically that any particular device is or is not a clock, or is a good or bad approximation to a clock, and this entirely within special relativity if one is using special relativity. The only extra assumption in the latter case is that one's theories about how the clock is working (e.g., electromagnetism for an electron going round an atom, or quantum electrodynamics for a better model) are actually valid theories in that context.

In this context nothing is really different in general relativity, except that one has to add the strong equivalence principle in order to be able to apply non-gravitational bits of physics when the spacetime is curved. In a certain sense one can consider special relativity as a special case of general relativity, viewing special relativity as general relativity with no gravitational effects (and saying, of course, that special relativity treats gravity very differently when there is any gravity). Moreover, general relativity adds nothing as far as acceleration is concerned. One can perfectly well consider accelerating test particles in special relativity, as in general relativity. But if some process is occurring in the particle, e.g., an electron orbiting a central nucleus, we do not know a priori whether that process is going just as it would for an instantaneously comoving inertial particle of the same kind, insofar as the two processes could be compared. It seems unlikely, but presumably a detailed calculation with the relevant theories would allow one to estimate the discrepancy. Presumably it would also show that the discrepancy is very small for most things we use as clocks, and the scale of accuracy on which no physical process fits with the theoretical proper time would be the one where we would have to admit that general relativity was beginning to fail.

Why do some people claim that special relativity cannot treat accelerated motions? Perhaps they are thinking, not of accelerated test particles, but accelerated observers. The view here is that one has exactly the same problem with accelerating observers in general relativity as in special relativity. If an observer is uniformly accelerating in special relativity, what coordinates would this observer set up? Everyone seems to use the semi-Euclidean (Rindler) coordinates as though there were something special about them. Of course, the observer remains at the spatial origin of those coordinates, the time coordinate is the proper time of the observer, and other obvious things like that. But are those the coordinates the observer would set up? If we are thinking about using clocks and rulers to set them up in a real world, it would seem that we do not actually know. The clock and ruler hypotheses merely assert that they would be in that context. Whether our actual physical clocks and rulers would fit the bill is another matter.

But would general relativity help here? Of course there are nice coordinates for any timelike worldline, in which the worldline remains at the spatial origin and the time coordinate is the proper time, etc. But are those the coordinates that an observer following that worldline would set up using clocks and rulers? It would seem that we are in exactly the same situation as in the last paragraph.

Both of the above cases 1 and 2 lead to redshift. The point here is just to see that the receiver with the same uniform acceleration as the observer will still detect a redshift (case 1), since the other case is the standard one. Consider the situation in the local inertial frame, which happens to be global for a uniform gravitational field. In the spacetime diagram, we have two identically shaped curves, curving over to the right, translates of one another along the space axis. The one on the left is the emitter and the one on the right is the receiver. A signal from the emitter leaves it when the emitter has a certain speed and arrives at the receiver when the receiver has a higher speed. In this freely-falling frame view, the redshift is just a Doppler shift. (To apply this analysis in the general relativistic case considered here, where we have a uniform gravitational field, we are of course also assuming the strong equivalence principle.) The only difference in case 2 is that the shape of the receiver worldline in the spacetime diagram for a freely falling observer is different from the shape of the emitter worldline, because it curves over more slowly (lower proper acceleration).

The redshift calculation for case 2 can be found in [7, Sect. 15.6] along with a critical discussion of the way semi-Euclidean coordinate systems are interpreted. Case 1 here is straightforward in the freely falling frame. The worldlines of E and R are (see Fig. 13)

$$x_{\mathrm{E}}(t) = \frac{c^2}{g}\left[\left(1 + \frac{g^2 t^2}{c^2}\right)^{1/2} - 1\right], \tag{48}$$

as in (24) on p. 69, and $x_{\mathrm{R}}(t) = x_{\mathrm{E}}(t) + \kappa$, for some constant κ. The two worldlines have the same shape, because they have the same uniform acceleration, by hypothesis 1. Then

$$v_{\mathrm{E}}(t) = \frac{gt}{(1 + g^2 t^2/c^2)^{1/2}} = v_{\mathrm{R}}(t). \tag{49}$$

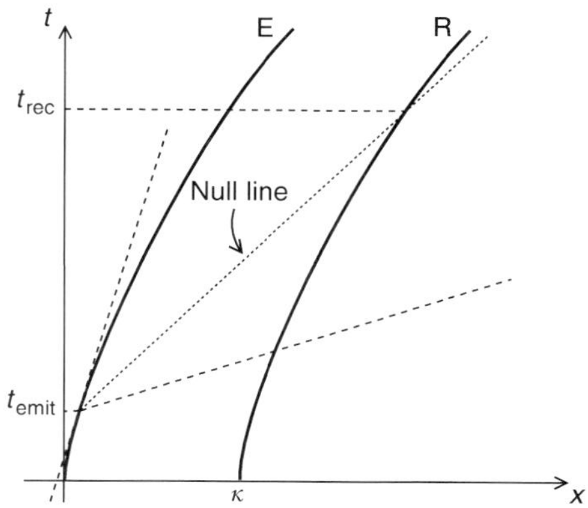

Fig. 13 Calculating the redshift in case 1. The emitter E emits a signal at time t_{emit} which is received by the receiver R at time t_{rec}. The two worldlines have the same shape. The receiver worldline is identical to the emitter worldline but shifted a distance κ along the space axis. By the time the signal reaches the receiver, the receiver worldline has curved over due to the increasing speed of the receiver

To get the redshift, imagine a light signal sent from the worldline of E at time t_{emit} and find the time of reception t_{rec} on the worldline of R. Find the apparent relative velocity

$$v := v_{\mathrm{R}}(t_{\mathrm{rec}}) - v_{\mathrm{E}}(t_{\mathrm{emit}}) = v_{\mathrm{E}}(t_{\mathrm{rec}}) - v_{\mathrm{E}}(t_{\mathrm{emit}}), \tag{50}$$

and plug it into the usual special relativistic formula for the Doppler shift. If z is the redshift, then

$$1 + z = \frac{(1 + v/c)^{1/2}}{(1 - v/c)^{1/2}}. \tag{51}$$

Interestingly, the result is much less elegant than in the standard case 2. However, it is easy to show that

$$z \approx g\kappa/c^2, \tag{52}$$

for small κ, i.e., the same result as for case 2. On the other hand, for large κ, the two results will obviously differ. What is not obvious is whether experimental accuracy could yet distinguish the two cases.

2 Rigid Motion

This section is based on the discussion of rigid motions in B.S. DeWitt's Stanford lectures on relativity [11]. These lectures (to be published by Springer) deal at some length with the problem of continuous media in the context of curved spacetime. The ideas below extend naturally to general relativity, but we consider a flat spacetime for the presentation below.

2.1 General Motion of a Continuous Medium

The component particles of the medium are labeled by three parameters ξ^i, $i = 1, 2, 3$, and the worldline of particle ξ is given by four functions $x^\mu(\xi, \tau)$, $\mu = 0, 1, 2, 3$, where τ is its proper time. In general relativity, the x^μ may be arbitrary coordinates in curved spacetime, but here we assume them to be standard coordinates of some inertial frame.

If $\xi^i + \delta\xi^i$ are the labels of a neighbouring particle, its worldline is given by the functions

$$x^\mu(\xi + \delta\xi, \tau) = x^\mu(\xi, \tau) + x^\mu{}_{,i}(\xi, \tau)\delta\xi^i,$$

where the comma followed by a Latin index denotes partial differentiation with respect to the corresponding ξ (DeWitt's notation). Note that the quantity $x^\mu{}_{,i}(\xi, \tau)\delta\xi^i$, representing the difference between the two sets of worldline functions, is formally a 4-vector, being basically an infinitesimal coordinate difference. However, it is not generally orthogonal to the worldline of ξ. In other words, it does not lie in the hyperplane of simultaneity of either particle.

To get such a vector one applies the projection tensor onto the instantaneous hyperplane of simultaneity:

$$P^{\mu\nu} = \eta^{\mu\nu} + \dot{x}^\mu \dot{x}^\nu,$$

where the dot denotes partial differentiation with respect to τ, and we note that in general relativity the projection tensor takes the form

$$P^{\mu\nu} = g^{\mu\nu} + \dot{x}^\mu \dot{x}^\nu,$$

with $g^{\mu\nu}$ the metric tensor of the curved spacetime. The result is

$$\delta x^\mu := P^\mu{}_\nu x^\nu{}_{,i}(\xi, \tau)\delta\xi^i. \tag{53}$$

One finds that application of the projection tensor corresponds to a simple proper-time shift of amount

$$\delta\tau = \eta_{\mu\nu}\dot{x}^\mu \dot{x}^\nu{}_{,i}\delta\xi^i,$$

so that

$$\delta x^\mu = x^\mu(\xi + \delta\xi, \tau + \delta\tau) - x^\mu(\xi, \tau).$$

Indeed,

$$x^\mu(\xi + \delta\xi, \tau + \delta\tau) = x^\mu(\xi, \tau) + x^\mu{}_{,i}\delta\xi^i + \dot{x}^\mu\delta\tau,$$

and feeding in the proposed expression for $\delta\tau$, we do obtain precisely δx^μ as defined above.

What can we conclude from this analysis? The two particles ξ and $\xi + \delta\xi$ appear, in the instantaneous rest frame of either, to be separated by a distance δs given by

$$(\delta s)^2 = (\delta x)^2 = \gamma_{ij}\delta\xi^i\delta\xi^j, \tag{54}$$

where

$$\gamma_{ij} = P_{\mu\nu}x^\mu{}_{,i}x^\nu{}_{,j}. \tag{55}$$

DeWitt calls the quantity γ_{ij} the proper metric of the medium.

2.2 *Rigid Motion of a Continuous Medium*

At this point, one can introduce a notion of rigidity. One says that the medium undergoes rigid motion if and only if its proper metric is independent of τ. This is therefore expressed by

$$\dot{\gamma}_{ij} = 0. \tag{56}$$

Under rigid motion the instantaneous separation distance between any pair of neighbouring particles is constant in time, as they would see it. Note that this criterion is independent of the coordinates used because γ_{ij} is a scalar.

Let us see whether this coincides with the notion of rigidity discussed earlier, i.e., whether the rigid rod of Sect. 1.2 is DeWitt rigid, or put differently, whether the rod described in Sect. 1.2 is undergoing rigid motion according to the criterion (56). DeWitt's ξ correspond to s in Sect. 1.2 (see p. 64). In a given inertial frame, particle s has motion described by $X(s,t)$, where

$$\frac{\partial X}{\partial s} = \frac{1}{\gamma}, \qquad \gamma = \gamma\big(v(s,t)\big),$$

and

$$v(s,t) = v_A(t) + \frac{\partial X}{\partial t},$$

where $v_A(t)$ is the speed of the end of the rod. Suppose we now change to a frame moving instantaneously at speed $v(s,t)$ and measure the distance between particle s and particle $s + \delta s$ as viewed in this frame. Will it be constant in this model, as required for DeWitt rigid motion? In the original frame where both particles are moving, we have separation

$$X(s+\delta s,t) - X(s,t) = \frac{\partial X}{\partial s}\delta s = \frac{\delta s}{\gamma}.$$

In the new frame moving at speed $v(s,t)$, this has length

$$\gamma \frac{\delta s}{\gamma} = \delta s = \text{constant}.$$

This is what DeWitt rigid motion requires.

2.3 Rate of Strain Tensor

The aim here is to express the rigid motion condition $\dot{\gamma}_{ij} = 0$ in terms of derivatives with respect to the coordinates x^μ by introducing the relativistic analog of the rate of strain tensor in ordinary continuum mechanics.

The non-relativistic strain tensor can be defined by

$$e_{ij} := \frac{1}{2}\left(\frac{\partial u_j}{\partial x_i} + \frac{\partial u_i}{\partial x_j}\right),$$

where $u_i(x)$ are the components of the displacement vector of the medium, describing the motion of the point originally at x when the material is deformed. One also defines the antisymmetric tensor

$$\omega_{ij} := \frac{1}{2}\left(\frac{\partial u_j}{\partial x_i} - \frac{\partial u_i}{\partial x_j}\right),$$

which describes the rotation occurring when the material is deformed. Clearly,

$$e_{ij} - \omega_{ij} = \frac{\partial u_i}{\partial x_j},$$

and hence, if all distortions are small,

$$\Delta u_i = (e_{ij} - \omega_{ij})\Delta x_j.$$

We can consider that e_{ij} describes non-rotational distortions, i.e., stretching, compression, and shear.

In the present discussion, u_i is replaced by a velocity field v_i and we have a rate of strain tensor. The non-relativistic rate of strain tensor is

$$r_{ij} = v_{i,j} + v_{j,i}, \tag{57}$$

where v_i is a 3-velocity field and the differentiation is with respect to ordinary Cartesian coordinates. Let us look for a moment at this tensor. The nonrelativistic condition for rigid motion is

$$r_{ij} = 0 \qquad \text{everywhere}.$$

This equation implies

$$0 = r_{ij,k} = v_{i,jk} + v_{j,ik}, \tag{58}$$

$$0 = r_{jk,i} = v_{j,ki} + v_{k,ji}. \tag{59}$$

Subtracting (59) from (58) and commuting the partial derivatives, we find

$$v_{i,jk} - v_{k,ji} = 0, \tag{60}$$

which, upon permutation of the indices j and k, yields also

$$v_{i,kj} - v_{j,ki} = 0. \tag{61}$$

Adding (58) and (61), we obtain

$$v_{i,jk} = 0,$$

which has the general solution

$$v_i = -\omega_{ij} x_j + \beta_i, \tag{62}$$

where ω_{ij} and β_i are functions of time only. The condition $r_{ij} = 0$ constrains ω_{ij} to be antisymmetric, i.e.,

$$\omega_{ij} = -\omega_{ji},$$

and nonrelativistic rigid motion is seen to be, at each instant, a uniform rotation with angular velocity

$$\omega_i = \frac{1}{2}\varepsilon_{ijk}\omega_{jk}$$

about the coordinate origin, superimposed upon a uniform translation with velocity β_i. Because the coordinate origin may be located arbitrarily at each instant, rigid motion may alternatively be described as one in which an arbitrary particle in the medium moves in an arbitrary way while at the same time the medium as a whole rotates about this point in an arbitrary (but uniform) way. Such a motion has six degrees of freedom.

Note that when r_{ij} is zero, we can also deduce that $v_{i,i} = 0$, i.e., $\operatorname{div} v = 0$, which is the condition for an incompressible fluid. This is evidently a weaker condition than rigidity.

Let us see how this generalises to special relativity. We return to the continuous medium in which particles are labelled by ξ^i, $i = 1, 2, 3$. Just as the coordinates x^μ are functions of the ξ^i and τ, so the ξ^i and τ can be regarded as functions of the x^μ, at least in the region of spacetime occupied by the medium. Following DeWitt [11], we write

$$u^\mu := \dot{x}^\mu, \qquad u^2 = -1, \qquad P_{\mu\nu} = \eta_{\mu\nu} + u_\mu u_\nu.$$

If f is an arbitrary function in the region occupied by the medium then

$$f_{,\mu} = f_{,i}\xi^i_{,\mu} + \dot{f}\tau_{,\mu},$$

where the comma followed by a Greek index μ denotes partial differentiation with respect to the coordinate x^μ. We also have

$$\dot{x}\cdot\ddot{x} = 0 \quad \text{or} \quad u\cdot\dot{u} = 0,$$

since $u^2 = -1$, and

$$\begin{gathered}
u_\mu u^\mu{}_{,\nu} = 0, \qquad \dot{u}_\mu = u_{\mu,\nu}u^\nu, \qquad u_\mu u^\mu{}_{,i} = 0,\\
x^\mu{}_{,i}\xi^i{}_{,\nu} + \dot{x}^\mu\tau_{,\nu} = \delta^\mu{}_\nu,\\
\xi^i{}_{,\mu}x^\mu{}_{,j} = \delta^i{}_j, \qquad \xi^i{}_{,\mu}\dot{x}^\mu = 0,\\
\tau_{,\mu}x^\mu{}_{,i} = 0, \qquad \tau_{,\mu}\dot{x}^\mu = 1,\\
P_{\mu\nu}\dot{x}^\nu{}_{,i} = P_{\mu\nu}u^\nu{}_{,i} = u_{\mu,i}\,.
\end{gathered}$$

We now define the rate of strain tensor for the medium:

$$\begin{aligned}
r_{\mu\nu} &:= \dot{\gamma}_{ij}\xi^i{}_{,\mu}\xi^j{}_{,\nu}\\
&= \Big(\dot{P}_{\sigma\tau}x^\sigma{}_{,i}x^\tau{}_{,j} + P_{\sigma\tau}\dot{x}^\sigma{}_{,i}x^\tau{}_{,j} + P_{\sigma\tau}x^\sigma{}_{,i}\dot{x}^\tau{}_{,j}\Big)\xi^i{}_{,\mu}\xi^j{}_{,\nu}\\
&= (\dot{u}_\sigma u_\tau + u_\sigma\dot{u}_\tau)(\delta^\sigma{}_\mu - u^\sigma\tau_{,\mu})(\delta^\tau{}_\nu - u^\tau\tau_{,\nu})\\
&\quad + u_{\tau,i}\xi^i{}_{,\mu}(\delta^\tau{}_\nu - u^\tau\tau_{,\nu}) + (\delta^\sigma{}_\mu - u^\sigma\tau_{,\mu})u_{\sigma,j}\xi^j{}_{,\nu}\\
&= \dot{u}_\mu u_\nu + u_\mu\dot{u}_\nu + \dot{u}_\mu\tau_{,\nu} + \tau_{,\mu}\dot{u}_\nu + u_{\nu,\mu} - \dot{u}_\nu\tau_{,\mu} + u_{\mu,\nu} - \dot{u}_\mu\tau_{,\nu}\\
&= u_{\mu,\sigma}u^\sigma u_\nu + u_\mu u^\sigma u_{\nu,\sigma} + u_{\nu,\mu} + u_{\mu,\nu}\\
&= P_\mu{}^\sigma P_\nu{}^\tau(u_{\sigma,\tau} + u_{\tau,\sigma}).
\end{aligned}$$

This is to be compared with (57) to justify calling it the rate of strain tensor. At any event x^μ, it lies entirely in the instantaneous hyperplane of simultaneity of the particle ξ^i that happens to coincide with that event.

Note in passing that this generalises to curved spacetimes. We define

$$r_{\mu\nu} := \dot{\gamma}_{ij}\xi^i{}_{,\mu}\xi^j{}_{,\nu}, \tag{63}$$

as before, noting that it is a tensor, since γ_{ij}, $\dot{\gamma}_{ij}$, ξ^i and ξ^j are scalars under change of coordinates. At any x, there are coordinates such that $g_{\mu\nu,\sigma}\big|_x = 0$, whence covariant derivatives with respect to the Levi-Civita connection are just coordinate derivatives at x, and it follows immediately that

$$r_{\mu\nu} = P^\sigma_\mu P^\tau_\nu(u_{\sigma;\tau} + u_{\tau;\sigma}), \tag{64}$$

where semi-colons denote covariant derivatives and $P^{\mu\nu}$ is given by

$$P^{\mu\nu} = g^{\mu\nu} + \dot{x}^{\mu}\dot{x}^{\nu},$$

for metric $g^{\mu\nu}$.

Returning now to the context of special relativity, the result

$$r_{\mu\nu} := \dot{\gamma}_{ij}\xi^{i}{}_{,\mu}\xi^{j}{}_{,\nu} = P^{\sigma}_{\mu}P^{\tau}_{\nu}(u_{\sigma,\tau} + u_{\tau,\sigma}) \tag{65}$$

expresses the rate of strain tensor in terms of coordinate derivatives of the four-velocity field of the medium. We now characterise relativistic rigid motion by

$$r_{\mu\nu} = 0, \qquad \dot{\gamma}_{ij} = 0. \tag{66}$$

Once again, we observe that the criterion for rigid motion, viz., $r_{\mu\nu} = 0$, is independent of the coordinates, because $r_{\mu\nu}$ is a tensor, even in a curved spacetime.

2.4 Examples of Rigid Motion

The next problem is to find some examples. We choose an arbitrary particle in the medium and let it be the origin of the labels ξ^i. The problem here is to choose these labels smoothly throughout the medium. Let the worldline $x^{\mu}(0,\tau)$ of the point $\xi^i = 0$ be arbitrary (but timelike). We now introduce a local rest frame for the particle, characterized by an orthonormal triad $n_i^{\mu}(\tau)$:

$$n_i \cdot n_j = \delta_{ij}, \qquad n_i \cdot u_0 = 0, \qquad u_0^2 = -1, \qquad u_0{}^{\mu} := \dot{x}^{\mu}(0,\tau).$$

We now assume that the worldlines of all the other particles of the medium can be given by

$$x^{\mu}(\xi,\tau) = x^{\mu}(0,\sigma) + \xi^i n_i{}^{\mu}(\sigma), \tag{67}$$

where σ is a certain function of the ξ^i and τ to be determined. On the left, τ is the proper time of the particle labelled by ξ. To achieve a relation of this type, given τ and ξ, we must find the unique proper time σ of the particle $\xi = 0$ such that the point $x^{\mu}(\xi,\tau)$ is simultaneous with the event $x^{\mu}(0,\sigma)$ in the instantaneous rest frame of the particle $\xi = 0$. Then the label ξ^i for our particle is defined by the above relation. There is indeed an assumption here, namely that these ξ^i really do label particles. That is, if we look at events with the same ξ^i but varying τ, we are assuming that we do follow a single particle. It is unlikely that all motions of the medium could be expressed like this, but we can obtain some rigid motions, as we shall discover.

To determine the function $\sigma(\xi^i, \tau)$, write

$$u^\mu = \dot{x}^\mu(\xi, \tau) = ({u_0}^\mu + \xi^i \dot{n}_i{}^\mu)\dot{\sigma},$$

all arguments being suppressed in the final expression. Here and in what follows, it is to be understood that dots over u_0 and the n_i denote differentiation with respect to σ, while the dot over σ denotes differentiation with respect to τ.

In order to proceed further, one must expand $\dot{n}_i$ in terms of the orthonormal tetrad u_0, n_i :

$$\dot{n}_i{}^\mu = a_{0i}{u_0}^\mu + \Omega_{ij}n_j{}^\mu. \tag{68}$$

The coefficients a_{0i} are determined, from the identity

$$\dot{n}_i \cdot u_0 + n_i \cdot \dot{u}_0 = 0,$$

to be just the components of the absolute acceleration of the particle $\xi = 0$ in its local rest frame [see an example in (22) on p. 69]:

$$a_{0i} = n_i \cdot \dot{u}_0, \tag{69}$$

and the identity

$$\dot{n}_i \cdot n_j + n_i \cdot \dot{n}_j = 0$$

tells us that Ω_{ij} is antisymmetric:

$$\Omega_{ij} = -\Omega_{ji}.$$

We now have

$$u^\mu = \Big[(1 + \xi^i a_{0i}){u_0}^\mu + \xi^i \Omega_{ij} n_j{}^\mu\Big]\dot{\sigma}.$$

But

$$-1 = u^2 = -\Big[(1 + \xi^i a_{0i})^2 - \xi^i \xi^j \Omega_{ik}\Omega_{jk}\Big]\dot{\sigma}^2,$$

whence

$$\dot{\sigma} = \Big[(1 + \xi^i a_{0i})^2 - \xi^i \xi^j \Omega_{ik}\Omega_{jk}\Big]^{-1/2}. \tag{70}$$

The right hand side of this equation is a function solely of σ and the ξ^i. Therefore the equation may be integrated along each worldline ξ = const., subject to the boundary condition

$$\sigma(\xi, 0) = 0.$$

We shall, in particular, have the necessary condition

$$\sigma(0,\tau) = \tau.$$

Note that the medium must be confined to regions where

$$\left(1 + \xi^i a_{0i}\right)^2 > \xi^i \Omega_{ik} \xi^j \Omega_{jk} \quad (\geq 0). \tag{71}$$

Otherwise, some of its component particles will be moving faster than light.

We can now calculate the proper metric of the medium. We have

$$\begin{aligned}
n_i \cdot u &= -\Omega_{ij}\xi^j \dot{\sigma}, \qquad (72)\\
x^\mu{}_{,i} &= n_i{}^\mu + (u_0{}^\mu + \xi^j \dot{n}_j{}^\mu)\sigma_{,i} = n_i{}^\mu + u^\mu \dot{\sigma}^{-1}\sigma_{,i},\\
u_\mu x^\mu{}_{,i} &= -\Omega_{ij}\xi^j \dot{\sigma} - \dot{\sigma}^{-1}\sigma_{,i},\\
\gamma_{ij} &= P_{\mu\nu} x^\mu{}_{,i} x^\nu{}_{,j}\\
&= \delta_{ij} - \Omega_{ik}\xi^k \sigma_{,j} - \Omega_{jk}\xi^k \sigma_{,i} - \dot{\sigma}^{-2}\sigma_{,i}\sigma_{,j}\\
&\quad + \left(\Omega_{ik}\xi^k \dot{\sigma} + \dot{\sigma}^{-1}\sigma_{,i}\right)\left(\Omega_{jl}\xi^l \dot{\sigma} + \dot{\sigma}^{-1}\sigma_{,j}\right)\\
&= \delta_{ij} + \dot{\sigma}^2 \Omega_{ik}\Omega_{jl}\xi^k \xi^l\\
&= \delta_{ij} + \frac{\Omega_{ik}\Omega_{jl}\xi^k\xi^l}{\left(1 + \xi^m a_{0m}\right)^2 - \xi^n \xi^r \Omega_{ns}\Omega_{rs}}, \qquad (73)
\end{aligned}$$

using the above expression (70) for $\dot{\sigma}$.

From this expression we see that there are two ways in which the motion of the medium can be rigid:

- All the Ω_{ij} are zero.
- All the Ω_{ij} and all the a_{0i} are constants, independent of σ.

In the second case the motion is one of a six-parameter family, with the Ω_{ij} and the a_{0i} as parameters. DeWitt refers to these special motions as superhelical motions. One example, constant rotation about a fixed axis, is discussed in Sect. 2.6. Let us first consider the case where all the Ω_{ij} are zero.

2.5 Rigid Motion Without Rotation

Saying that the Ω_{ij} are all zero amounts to saying that the triad $n_i{}^\mu$ is Fermi–Walker transported along the worldline of the particle $\xi = 0$. Let us see briefly what this means.

If $u_0(\sigma)$ is the 4-velocity of the worldline, the equation for Fermi–Walker transport of a contravector A^μ along the worldline is

$$\dot{A}^\mu = (A \cdot \dot{u}_0)u_0 - (A \cdot u_0)\dot{u}_0. \tag{74}$$

This preserves inner products, i.e., if A and B are FW transported along the worldline, then $A \cdot B$ is constant along the worldline. Furthermore, the tangent vector u_0 to the worldline is itself FW transported along the worldline, and if the worldline is a spacetime geodesic (a straight line in Minkowski coordinates), then FW transport is the same as parallel transport.

Now recall that the Ω_{ij} were defined by

$$\dot{n}_i^\mu = a_{0i}u_0^\mu + \Omega_{ij}n_j^\mu. \tag{75}$$

When $\Omega_{ij} = 0$, this becomes

$$\dot{n}_i^\mu = a_{0i}u_0^\mu. \tag{76}$$

This is indeed the Fermi–Walker transport equation for n_i^μ, found by inserting $A = n_i$ into (74), because we insist on $n_i \cdot u_0 = 0$ and we have $a_{0i} = n_i \cdot \dot{u}_0$ [see (69) on p. 94].

In fact, the orientation in spacetime of the local rest frame triad n_i^μ cannot be kept constant along a worldline unless that worldline is straight (we are referring to flat spacetimes here). Under Fermi–Walker transport, however, the triad remains as constantly oriented, or as rotationless, as possible, in the following sense: at each instant of time σ, the triad is subjected to a pure Lorentz boost without rotation in the instantaneous hyperplane of simultaneity. (On a closed orbit, this process can still lead to spatial rotation of axes upon return to the same space coordinates, an effect known as Thomas precession.) For a general non-Fermi–Walker transported triad, the Ω_{ij} are the components of the angular velocity tensor that describes the instantaneous rate of rotation of the triad in the instantaneous hyperplane of simultaneity.

Of course, given any triad n_i^μ at one point on the worldline, it is always possible to Fermi–Walker transport it to other points by solving (74). We are then saying that motions that can be given by (67), viz.,

$$x^\mu(\xi, \tau) = x^\mu(0, \sigma) + \xi^i {n_i}^\mu(\sigma), \tag{77}$$

where the ξ^i are assumed to label material particles in the medium, are rigid in the sense of the criterion given above.

Furthermore, the proper geometry of the medium given by the proper metric γ_{ij} in (55) on p. 89 is then flat, i.e.,

$$\gamma_{ij} = \delta_{ij}.$$

We also note that (σ, ξ^i) are the semi-Euclidean coordinates for an observer with worldline $x^\mu(0, \sigma)$, moving with the base particle $\xi = 0$. This generalises the construction of Sect. 1.4 to the case of a general 3D acceleration.

What we are doing here is to label the particle ξ^i by its spatial coordinates ξ^i in the semi-Euclidean system moving with the particle $\xi = 0$. Geometrically, we have the worldline of the arbitrarily chosen particle O at the origin, viz., $x^\mu(0, \sigma)$, with σ its proper time. We have another worldline $x^\mu(\xi^i, \tau)$ of a particle P labelled by ξ, with proper time τ. For given τ, we seek σ such that $x^\mu(\xi^i, \tau)$ is in the hyperplane of simultaneity of O at its proper time σ. Then (ξ^i) is the position of P in the tetrad moving with O. Indeed, $\{\xi^i\}$ are the space coordinates of P relative to O in that frame.

As attested by (72) on p. 95, we also have

$$n_i \cdot u = 0, \tag{78}$$

so that the instantaneous hyperplane of simultaneity of the particle at $\xi = 0$ is an instantaneous hyperplane of simultaneity for all the other particles of the medium as well, and the triad n_i^μ serves to define a rotationless rest frame for the whole medium. In other words, the coordinate system defined by the particle labels ξ^i may itself be regarded as being Fermi–Walker transported, and all the particles of the medium have a common designator of simultaneity in the parameter σ. In the semi-Euclidean system, σ is taken to be the time coordinate.

Put another way, (78) says that the $n_i(\sigma)$ are in fact orthogonal to the worldline of the particle labelled by ξ^i at the value of τ corresponding to σ. This happens because $u(\xi, \tau) = u_0(0, \sigma)$. In words, the 4-velocity of particle ξ at its proper time τ is the same as the 4-velocity of the base particle when it is simultaneous with the latter in the reckoning of the base particle (quite a remarkable thing).

Because σ is not generally equal to τ, however, it is not possible for the particles to have a common synchronization of standard clocks. The relation between σ and τ is given by (70) on p. 94 as

$$\dot{\sigma} = \left(1 + \xi^i a_{0i}\right)^{-1}.$$

We can thus find the absolute acceleration a_i of an arbitrary particle in terms of a_{0i} and the ξ^i :

$$\begin{aligned} a_i &= n_i \cdot \dot{u} = n_i \cdot \frac{\partial u}{\partial \sigma} \dot{\sigma} = \dot{\sigma} n_i \cdot \frac{\partial}{\partial \sigma}\left[\left(1 + \xi^j a_{0j}\right) u_0 \dot{\sigma}\right] \\ &= \dot{\sigma} n_i \cdot \dot{u}_0 \\ &= \frac{a_{0i}}{1 + \xi^j a_{0j}}. \end{aligned} \tag{79}$$

Here we have used the fact that $u = \left(1 + \xi^j a_{0j}\right) u_0 \dot{\sigma} = u_0$. We see that, although the motion is rigid and rotationless in the sense described above, not all parts of the medium are subject to the same acceleration.

It is important to note that, when we find ξ^i and σ, they constitute semi-Euclidean coordinates (adapted to $\xi = 0$) for the point $x^\mu(\xi, \tau)$ whether or not that point follows a particle for fixed ξ. What we have here are material particles that follow all these points with fixed ξ, for a whole 3D range of values of ξ.

In these coordinates, the metric tensor takes the form

$$g_{00} = \frac{\partial x^\mu}{\partial \sigma}\bigg|_\xi \frac{\partial x^\nu}{\partial \sigma}\bigg|_\xi \eta_{\mu\nu} = u^2\dot{\sigma}^{-2} = -(1+\xi^i a_{0i})^2,$$

$$g_{i0} = g_{0i} = \frac{\partial x^\mu}{\partial \xi^i}\bigg|_\sigma \frac{\partial x^\nu}{\partial \sigma}\bigg|_\xi \eta_{\mu\nu} = (n_i \cdot u)\dot{\sigma}^{-1} = 0,$$

$$g_{ij} = \frac{\partial x^\mu}{\partial \xi^i}\bigg|_\sigma \frac{\partial x^\nu}{\partial \xi^j}\bigg|_\sigma \eta_{\mu\nu} = n_i \cdot n_j = \delta_{ij},$$

which has a simple diagonal structure. We note that this metric becomes static, i.e., time-independent, with the parameter σ playing the role of time, in the special case in which the absolute acceleration of each particle is constant. This should be compared with (18) and (19) on p. 69, and also (23) on p. 69 (but note that the sign convention for the metric has been reversed).

We conclude that this rigid motion possesses only the three degrees of freedom that the particle $\xi = 0$ itself possesses. The base particle $\xi = 0$ can move any way it wants, but the rest of the medium must then follow in a well defined way.

2.6 Rigid Rotation

The simplest example of a medium undergoing rigid rotation is obtained by choosing

$$a_{0i} = 0, \qquad \Omega_{12} = \omega, \qquad \Omega_{23} = 0 = \Omega_{31}.$$

The worldline of the particle at $\xi = 0$ is then straight, but the worldlines of all the other particles are helices of constant pitch. We have

$$\dot{\sigma} = \Big\{1 - \omega^2\big[(\xi^1)^2 + (\xi^2)^2\big]\Big\}^{-1/2}$$

and the proper metric of the medium takes the form

$$(\gamma_{ij}) = \begin{pmatrix} 1+(\dot{\sigma}\omega\xi^2)^2 & -(\dot{\sigma}\omega)^2\xi^1\xi^2 & 0 \\ -(\dot{\sigma}\omega)^2\xi^1\xi^2 & 1+(\dot{\sigma}\omega\xi^1)^2 & 0 \\ 0 & 0 & 1 \end{pmatrix}.$$

Relabelling the particles by means of three new coordinates r, θ, z given by

$$\xi^1 = r\cos\theta, \qquad \xi^2 = r\sin\theta, \qquad \xi^3 = z, \tag{80}$$

the proper metric of the rotating medium takes the form

$$\mathrm{diag}\left(1, \frac{r^2}{1-\omega^2 r^2}, 1\right).$$

Indeed, we have

$$\dot{\sigma}^2 = \frac{1}{1-\omega^2 r^2},$$

whence

$$\begin{aligned}
\gamma_{rr} &= \frac{\partial \xi^i}{\partial r}\frac{\partial \xi^j}{\partial r}\gamma_{ij} \\
&= \cos^2\theta\big[1 + (\dot{\sigma}\omega r)^2 \sin^2\theta\big] - 2(\dot{\sigma}\omega r)^2 \sin^2\theta \cos^2\theta \\
&\quad + \sin^2\theta\big[1 + (\dot{\sigma}\omega r)^2 \cos^2\theta\big] \\
&= 1, \\
\gamma_{r\theta} &= \gamma_{\theta r} = \frac{\partial \xi^i}{\partial r}\frac{\partial \xi^j}{\partial \theta}\gamma_{ij} \\
&= -r\sin\theta\cos\theta\big[1 + (\dot{\sigma}\omega r)^2 \sin^2\theta\big] - r(\dot{\sigma}\omega r)^2 \sin\theta\cos^3\theta \\
&\quad + r(\dot{\sigma}\omega r)^2 \sin^3\theta\cos\theta + r\sin\theta\cos\theta\big[1 + (\dot{\sigma}\omega r)^2\cos^2\theta\big] \\
&= 0, \\
\gamma_{rz} &= \gamma_{zr} = \frac{\partial \xi^i}{\partial r}\frac{\partial \xi^j}{\partial z}\gamma_{ij} = 0, \\
\gamma_{\theta\theta} &= \frac{\partial \xi^i}{\partial \theta}\frac{\partial \xi^j}{\partial \theta}\gamma_{ij} \\
&= r^2\sin^2\theta\big[1 + (\dot{\sigma}\omega r)^2 \sin^2\theta\big] + 2r^2(\dot{\sigma}\omega r)^2 \sin^2\theta\cos^2\theta \\
&\quad + r^2\cos^2\theta\big[1 + (\dot{\sigma}\omega r)^2\cos^2\theta\big] \\
&= r^2\big[1 + (\dot{\sigma}\omega r)^2\big] = r^2\left(1 + \frac{\omega^2 r^2}{1-\omega^2 r^2}\right) = \frac{r^2}{1-\omega^2 r^2}, \\
\gamma_{\theta z} &= \gamma_{z\theta} = \frac{\partial \xi^i}{\partial \theta}\frac{\partial \xi^j}{\partial z}\gamma_{ij} = 0, \qquad \gamma_{zz} = \frac{\partial \xi^i}{\partial z}\frac{\partial \xi^j}{\partial z}\gamma_{ij} = 1.
\end{aligned}$$

In terms of these coordinates the proper distance δs between two particles separated by displacements δr, $\delta\theta$, and δz therefore takes the form

$$\delta s^2 = (\delta r)^2 + \frac{r^2}{1-\omega^2 r^2}(\delta\theta)^2 + (\delta z)^2.$$

We are merely applying (54) on p. 89 for the new particle labels. This gives the distance of one particle as reckoned in the instantaneous rest frame of the neighbouring particle. The second term on the right of this equation may be understood as arising from relativistic contraction. At first sight, it may look odd to find that, when a disc of radius r is set spinning with angular frequency ω about its axis, so that radial distances are unaffected by relativistic contraction, distances in the direction of rotation contract in such a way that the circumference of the disc gets reduced to the value $2\pi R\sqrt{1-\omega^2 R^2}$. It appears to contradict the Euclidean nature of the ordinary 3-space that the disc inhabits! DeWitt describes this as follows [11]:

> What in fact happens is that, when set in rotation, the disc must suffer a strain that arises for kinematic reasons quite apart from any strains it suffers on account of centrifugal forces. In particular, it must undergo a stretching of amount $(1-\omega^2 r^2)^{-1/2}$ in the direction of rotation, to compensate the Lorentz contraction factor $(1-\omega^2 r^2)^{1/2}$ that appears when the disc is viewed in the inertial rest frame of its axis, thereby maintaining the Euclidean nature of 3-space. It is this stretching factor that appears in the proper metric of the medium.

Let us try to put this more explicitly. Suppose A and B are two neighbouring particles at distance R from the centre and with labels θ and $\theta + \delta\theta$. When the disk is not rotating, the proper distance between them as reckoned by either in its instantaneously comoving inertial frame (ICIF) is $R\delta\theta$. When the disk is rotating, the expression for γ_{ij} tells us that the proper distance between them in the new ICIF will increase to $R\delta\theta/(1-\omega^2 R^2)^{1/2}$. Seen by an inertial observer moving with the centre of the disk, this separation will thus be $R\delta\theta$, as before, and there will be no contradiction with the edicts of Euclidean geometry. This shows that the matter between A and B is stretched in the sense of occupying a greater proper distance as judged in an ICIF moving with either A or B.

There is a direct parallel with the two accelerating rockets mentioned at the beginning of Bell's well known paper *How to Teach Special Relativity* [10]. The separation of A and B seen by an inertial observer moving with the centre of the disk is unchanged when the rotation gets under way, so their proper separation is greater, leading to a strain which DeWitt claims to be due to kinematic reasons. If the disk could somehow be made of a very fragile material already stretched to its limit in the inertial frame moving with the center of the disk, it would shatter under rotation, just as the fragile thread joining Bell's two accelerating rockets was doomed to break.

The above discussion does assume that θ labels the material particles! And this follows from the relations in (80) and the fact that ξ^1, ξ^2, ξ^3 label the particles. It would be easy to miss this point. There remains therefore the question as to whether any association of material particles could have, or is likely to have this motion.

We note that the medium must be confined to regions where $r < \omega^{-1}$ and that its motion will not be rigid if ω varies with time. There are no degrees of freedom in this kind of (superhelical) motion: once the medium gets into superhelical motion, it must remain frozen into it if it wants to stay rigid. We note also that the proper geometry of the medium is not flat, i.e., $\gamma_{ij} \neq \delta_{ij}$.

2.7 Rigid Motion in Schwarzschild Spacetime

As an example in a curved spacetime, let us show that a medium in which particles are labelled by Schwarzschild coordinates $\xi := (r, \theta, \phi)$ is in rigid motion. The metric, displayed as a matrix, is

$$(g_{\mu\nu}) = \begin{pmatrix} -c^2 B(r) & 0 & 0 & 0 \\ 0 & B(r)^{-1} & 0 & 0 \\ 0 & 0 & r^2 & 0 \\ 0 & 0 & 0 & r^2 \sin^2\theta \end{pmatrix}, \qquad B(r) := 1 - \frac{A}{r},$$

where $A := 2GM/c^2$ is the usual constant. The motion we have in mind is described by the four functions

$$x^0(\xi, \tau) = t, \quad x^1(\xi, \tau) = r, \quad x^2(\xi, \tau) = \theta, \quad x^3(\xi, \tau) = \phi,$$

following the general scheme set out at the beginning of Sect. 2.1. It is then a very simple matter indeed to show that [2]

$$\gamma_{ij} = g_{ij}, \qquad i, j \in \{1, 2, 3\},$$

and also

$$\dot{\gamma}_{ij} = 0.$$

This is all rather obvious and it is easy to see how to obtain a host of rigid motions in curved spacetimes where the metric has a static form. Alternatively, one can calculate the rate of strain tensor $r_{\mu\nu}$ of (64) on p. 92, and it is a trivial matter to show that $r_{\mu\nu} = 0$. So any medium in which the particles could be labelled by the Schwarzschild space coordinates is undergoing a rigid motion according to this criterion.

Let us think back briefly to the short quotation from a standard textbook presentation on p. 77. In fact it is interesting to see how that account continues with regard to the related question of proper distance [4]:

> Note that the [increment in the] proper radius R of a two-sphere [centered on the singularity], obtained from the spatial line element by setting θ = const., ϕ = const., is
>
> $$\mathrm{d}R = (1 - 2m/r)^{-1/2}\mathrm{d}r > \mathrm{d}r. \qquad (81)$$
>
> In other words, the proper distance between spheres of radius r and radius $r + \mathrm{d}r$ is $\mathrm{d}R > \mathrm{d}r$, and hence larger than in flat space.

It is intriguing to wonder what the last comment means. For this is not really a comparison with any spheres in flat space. The coordinate interval dr need not be at a point where the spacetime is even approximately flat. The so-called proper distance is something that is related to the coordinate r in this way. In fact, the quoted relation (81) is telling us how to understand the coordinates.

As an aside, we have the same kind of pedagogical difficulty in the following, still in the context of the Schwarzschild metric [4]:

> Let us consider proper time for a stationary observer, i.e., an observer at rest at fixed values of r, θ, ϕ. Proper time is related to coordinate time by
>
> $$d\tau = (1 - 2m/r)^{1/2} dt < dt. \tag{82}$$
>
> Thus clocks go slower in a gravitational field.

But they go slower than what? Of course, this is a neat inequality and very simple. But does it really tell us that the clock is going slower than the same clock in flat spacetime? I do not think so. dt is a coordinate change at a place where $r \neq \infty$ and spacetime is not flat. The above relation tells us how to understand the coordinate t at the relevant point, provided that we understand how to interpret proper time as given by the metric.

Now a rod permanently occupying $[r_1, r_2]$ would be undergoing rigid motion, and so would a rod permanently occupying $[r'_1, r'_2]$. But we do not yet know whether there is some motion of the points making up the rod occupying $[r_1, r_2]$ that could serve as a transition of the same rod from the unprimed to the primed state. It seems likely that one could find a DeWitt rigid motion making the transition from $[r_1, r_2]$ to $[r'_1, r'_2]$ if and only if the proper lengths (rather than the coordinate lengths) are the same, and indeed it is not difficult to give a heuristic argument. It is worth drawing the analogy with a rod in Minkowski (flat) spacetime when it is accelerated from one state of constant velocity to another, as illustrated by the 4D region shaded in Fig. 10 (see p. 81). In fact, we have a similar problem here to the one discussed in Minkowski spacetime: we may know, or assume, that the proper length of a rod will be different for a given inertial observer $\mathcal{I}$ when it moves at different constant velocities relative to $\mathcal{I}$, but we do not have a theory for what it will look like in the transition between the two velocity states.

In the usual special relativistic discussion, rigid means suitably contracted in one uniform velocity state as compared with the other, but we do not usually try to say what rigid means during the transition between the states. In Fig. 10, the rod, initially in one velocity state, then under acceleration, then in the final velocity state, is represented by the shaded 4D region. The proposal in this paper is just one proposal, i.e., we have found a possible solution for the motion during the transition, but it is not based on any microscopic theory of the atomic structure of the rod.

Rigid motion is a natural enough notion, but what of a microscopic theory? There is a clear parallel with the discussion of the acceleration of an atom in Minkowski spacetime, as discussed by Bell [10] and mentioned in relation to Fig. 10 on p. 81. When Bell's (pre-quantum) atom is accelerated slowly enough, we expect it to contract in exactly the way proposed for Fig. 10, i.e., so that it always has the same radius to the instantaneously comoving inertial observer. Slowly enough just means that many periods of the electron orbit fit in before the acceleration has changed the velocity very much.

What about a Bell atom in Schwarzschild spacetime? In fact, a version of the equivalence principle shows that, if an atom has radius r_{atom} in flat spacetime, then when held fixed at the value R of the Schwarzschild radial coordinate in such a way that the plane of the electron orbit contains the Schwarzschild radial direction, it will have Schwarzschild coordinate radius [8]

$$\left(1 - \frac{A}{R}\right)^{1/2} r_{\mathrm{atom}} \tag{83}$$

as viewed in the hyperplane of simultaneity of the Schwarzschild observer; whence its proper radius will still be r_{atom} in the hyperplane of simultaneity of the SO. This is seen as follows. One finds coordinates $\{z^{\mu}\}$ at R such that an atomic nucleus fixed at R is at the origin $z^{\mu} = 0$ and instantaneously has speed zero relative to these coordinates, and such that

$$g^{z}_{\mu\nu}\Big|_{z=0} = \eta_{\mu\nu}, \qquad \Gamma^{\mu}_{z\nu\sigma}\Big|_{z=0} = 0.$$

Assuming that the electron orbits at small enough radius and with short enough period relative to the curvature, a standard rather strong version of the equivalence principle says that it will behave relative to these coordinates like an atom in flat spacetime for a certain number of orbits, e.g., following a circular orbit with radius r_{atom} and the same period as in flat spacetime too. When we transform this orbit to the Schwarzschild coordinate description, we find the Schwarzschild coordinate radius (83).

By this kind of argument, the strong equivalence principle shows that a thing like an atom, or a rod made of a row of such atoms (without worrying about how binding forces affect it), always measures proper length in whatever hyperplane of simultaneity it is observed, provided that it is instantaneously stationary there relative to the relevant coordinates, where proper length is the quantity usually obtained from the metric, and usually just assumed without further discussion to represent the lengths of such real (if ideal) physical objects. One might say that, wherever it is, whatever it is doing, this kind of atom or rod always measures proper lengths if used correctly. The last proviso just refers to the fact that the atom must be instantaneously stationary relative to suitable coordinates.

Is there a link with rigidity as we have been describing it? Are these Bell atoms rigid? It looks as though they are. Such an atom can sit at constant Schwarzschild space coordinates and have constant coordinate radius. Moved elsewhere, if moved slowly enough, its Schwarzschild coordinate radius changes in such a way that its proper radius in the hyperplane of simultaneity of the SO is roughly constant, just like the above infinitesimal rod subjected to an approximate DeWitt rigid motion.

As an aside, the Schwarzschild coordinates arise in a purely mathematical way in many presentations of this metric, by solving Einstein's equations, and no attempt is made to associate some physical counterpart with them. Although the notation may be suggestive, the discussion after the solution should perhaps address the question: how do we now relate these coordinates to what we measure? Furthermore, one should perhaps also ask why clock readings and measuring stick readings correspond the way they do to our coordinates, bearing in mind what a measuring stick must do to lie quietly between the points r_1 and r_2 with all its atoms under different 4-accelerations. What principle of the theory are we applying?

The above idea of a rigid rod (measuring stick) in Schwarzschild spacetime is thus that we can support it in the gravitational field in such a way that the 4D region it sweeps out crosses any Schwarzschild plane of simultaneity in the fixed coordinate interval from r_1 to r_2. The term 'support' covers up for some complex continuum of different 4-forces on its various atoms. Perhaps we can just hold one end of it at r_1, say, and let the internal forces within the rod do the rest naturally. According to the above analyses, the material of such a rod would indeed be undergoing rigid motion if all particles in the rod could be labelled by constant Schwarzschild space coordinates. This would then be a rigid rod, quite analogous to the one discussed for an accelerating observer.

It seems that our measuring sticks have to be like this for the theory to have a practical application, and the principle hiding away here is (a version of) the equivalence principle.

3 Conclusion

Rigid rods are commonly referred to in the special theory of relativity. In a certain sense they hardly need to be rigid. If one is moving inertially with a rigid rod, it has length L, and if one then changes to another inertial motion, it has another length L' which is shorter than L. Nothing is required of the rod here.

One does not even have to be present in this scenario. In the paradigm provided by the general theory of relativity, one just has to adopt coordinates. It does not matter what the observer is doing, only what coordinates he or she may adopt. In this view, the new length of the rod is just an illusion, not caused by anything real. The proper length depends on the choice of spacelike hypersurface used to intersect the essentially 4D spacetime region occupied by the rod.

This is not the view described by Bell [10]. The relationship between observer and rod, or between coordinate frame and rod, in which one moves relative to the other, can be achieved in another way, namely, by accelerating the rod. The observer does not have to do anything. The rod is accelerated and one would like to say that as a consequence the particles making it up adjust their relationship to one another in such a way that the rod becomes shorter as judged by the observer. There is even a theory for this: Maxwell's theory of electromagnetism.

In this view, rigidity would just be something like the assumption that there are no transient oscillatory effects during or after acceleration, or that such effects can be neglected. It would seem that the notions of rigidity discussed here are an approximation of this kind.

Put another way, rigidity is a constraint on the motion of (the particles within) a measuring stick that allows one to say exactly what is happening to it when it makes the transition from one uniform velocity state to another (in flat spacetime). Bell was considering just such a transition in his paper [10], but using the microphysical theory provided by Maxwell, which is presumably more realistic. The aim here is to draw attention to the fact that the rigidity constraint is artificial, and show that the standard, often uncritically interpreted semi-Euclidean coordinate system adapted to the worldline of an accelerating observer (in flat spacetime) fundamentally uses this constraint, and hence remind us that we ought to be wary of non-inertial coordinate systems (see also [7]).

The distinguishing feature of special relativity, when it is considered as a special case of general relativity, is that there are preferred frames of reference adapted to observers with inertial motion. However, even in special relativity, there are no preferred frames of reference adapted to accelerating observers. If they know the theory, they may as well adopt inertial coordinates (relative to which they accelerate, of course). One may nevertheless wonder what such people would measure with a measuring stick, or with the kind of (pre-quantum, non-radiating) atom described by Bell [10]. If the acceleration is not too great, one expects the Bell atom to adjust rather quickly, whereas the rigid rod described in Sect. 1 (accelerated along its axis) adjusts immediately for any acceleration to measure proper length in the instantaneously comoving inertial frame of the observer, i.e., in the spacelike hypersurface borrowed from an instantaneously comoving inertial observer. In other words, the rigid rod satisfies what is usually known as the ruler hypothesis.

Something like the clock and ruler hypotheses are necessary to interpret the Pound–Rebka experiment. This is used as an example to illustrate the idea that one should be wary of naive interpretations of appealing coordinate systems, which often involve assumptions of this kind in a covert way.

In a curved spacetime, one expects to find a rigid motion of a rod between two states if and only if the two states correspond to the same proper length relative to suitably adapted frames. Once again, an atom of the type described by Bell would provide an approximation.

References

1. W. Rindler: *Introduction to Special Relativity*, Oxford University Press, New York (1982)
2. Full details of any missing calculations can be obtained from the author
3. M. Friedman: *Foundations of Space–Time Theories*, Princeton University Press, NJ (1983)
4. M. Blau: *Lecture Notes on General Relativity*, available on the Web
5. A.D. Yaghjian: *Relativistic Dynamics of a Charged Sphere*, Lecture Notes in Physics 686, Springer-Verlag, New York (2006)
6. E. Pierce: *The lock and key paradox and the limits of rigidity in special relativity*, Am. J. Phys. **75** (7), 610–614, 2007
7. S. Lyle: *Uniformly Accelerating Charged Particles. A Threat to the Equivalence Principle*, Fundamental Theories of Physics 158, Springer-Verlag, Berlin, Heidelberg (2008a). See in particular Chap. 2
8. S. Lyle: *Extending Bell's Approach to General Relativity*, unpublished (2008b)
9. H.R. Brown: *Physical Relativity. Spacetime Structure from a Dynamical Perspective*, Clarendon Press, Oxford (2007). For more discussion of this theme, see also: H.R. Brown:

The behaviour of rods and clocks in general relativity, and the meaning of the metric field. In: *Beyond Einstein: Essays on Geometry, Gravitation, and Cosmology*, ed. by D.E. Rowe, Einstein Studies, Vol. 12, Birkhäuser, Boston (2009)
10. J.S. Bell: *Speakable and Unspeakable in Quantum Mechanics*, 2nd edn., Cambridge University Press, Cambridge (2004)
11. B. DeWitt: *Lectures on Relativity*, Springer-Verlag, Heidelberg (to be published)

Minkowski Space and Quantum Mechanics

Paul O'Hara

1 Introduction

A paradigm shift distinguishes general relativity from classical mechanics. In general relativity the energy-momentum tensor is the effective cause of the ontological space-time curvature and vice-versa, while in classical physics, the structure of space-time is treated as an accidental cause, serving only as a backdrop against which the laws of physics unfold. This split in turn is inherited by quantum mechanics, which is usually developed by changing classical (including special relativity) Hamiltonians into quantum wave equations. For example the Dirac equation is obtained by substituting the momentum operator for the four-momentum term in the linearized relativistic Hamiltonian. Similarly, Erwin Schrodinger used the "*purely* formal procedure" [1] of replacing $\frac{\partial W}{\partial t}$ in the Hamilton-Jacobi equation with $\pm\frac{h}{2\pi i}\frac{\partial}{\partial t}$ to obtain his wave equation. In both cases, the transition to quantum mechanics relies upon additional formal assumptions associated with Hilbert Space theory, and the final form of the wave equation does not in principle depend upon the underlying geometry of Minkowski space, although in the case of special relativity, the Hamiltonian indirectly reflects the geometric structure of Minkowski space.

In this article, we try to remedy this situation by taking the metrics of general relativity as the starting point of quantum mechanics. We will associate wave equations in a natural way with those operators which are duals of differential one-forms (expressed locally as a Minkowski metric) rather than with operators derived from a Hamiltonian, thus enabling the ontological structure of space-time itself to determine in a natural and unique way the wave equations of quantum mechanics.

This too has philosophical implications for the unity of physics. First, it means the metric structure of the space determines the general form of the wave equation, and consequently implies that space-time is not merely a backdrop for the laws of mechanics but is in fact an effective and quasi-formal cause of the resulting wave equations. Secondly, because of the equivalence principle the wave equation can always be written in a form that locally reflect a wave equation in Minkowski space, although globally in the case of non-geodesic motion other factors need to

V. Petkov (ed.), *Space, Time, and Spacetime*, Fundamental Theories of Physics 167,
DOI 10.1007/978-3-642-13538-5_4, © Springer-Verlag Berlin Heidelberg 2010

be considered. Thirdly, the use of test particles to analyze the motion of a massive body within the structure of the space-time further complicates matters in that it suggests a return to the pre-relativistic mentality of considering the laws of mechanics as something operative within space-time but not contributing to its ontological structure.

With regard to this last point, it is precisely here that gauge theory plays a key role. It enables new parameters like charge to be introduced into the space-time structure by means of non-gravitational connections which are not intrinsically related to the geometry of that space-time. This becomes particularly pronounced when we analyze for example the hydrogen atom. As we shall see in Sect. 4 the understanding of the hydrogen atom is transposed into a question about an electron as a point charge-mass moving within a space-time whose geometry is determined by the Reissner-Walker metric of the proton. Consequently, in this case the electron is considered to be a test particle with both charge and mass which are determined by a gauge term, while by way of contrast the charge of the proton contributes to the energy-momentum tensor. We could say that taken together the two approaches produce a methodological unification but not an ontological one. Nevertheless, the usefulness of the approach cannot be underestimated in that it allows us to understand much about the structure of the hydrogen and other atoms.

Specifically, then the paper will be structured to reflect the three points above. Section 2 will lay out the formalism relating metrics and waves, Sect. 3 will consider the relationship between quantum mechanics and classical mechanics based on this formalism, Sect. 4 will discuss non-geodesic motion and then apply the theory to the hydrogen atom.

2 Metrics and the Dirac Equation

Before laying out the formalism proper, we need to clarify notation. Throughout the paper, $(\mathcal{M}, g)$ will denote a space-time pair, where "$\mathcal{M}$ is a connected four dimensional Hausdorff manifold" and g is a metric of signature -2 on $\mathcal{M}$ [3]. At every point p on the space-time manifold $\mathcal{M}$ we erect a local tetrad $e_0(p), e_1(p), e_2(p), e_3(p)$ such that a point x has coordinates $x = (x^0, x^1, x^2, x^3) = x^a e_a$ in this tetrad coordinate system, while the spinor ψ can be written as $\psi = \psi^i e_i(p)$, where ψ^i represent the coordinates of the spinor with respect to the tetrad at p. Also at p we can establish a tangent vector space $T_p(\mathcal{M})$, with basis $\{\partial_0, \partial_1, \partial_2, \partial_3\}$ and a dual 1-form space, denoted by T_p^* with basis $\{dx_0, dx_1, dx_2, dx_3\}$ at p, defined by

$$dx^\mu \partial_\nu \equiv \partial_\nu x^\mu = \delta^\mu_\nu. \tag{1}$$

We refer to the basis $\{dx^0, dx^1, dx^2, dx^3\}$ as "the basis of one forms dual to the basis $\{\partial_0, \partial_1, \partial_2, \partial_3\}$ of vectors at p."

2.1 General Formalism

We begin with an intuitive and non-rigorous approach to our methodology by indicating two ways in which quantum mechanical wave equations can be obtained from the metrics of general relativity, without any explicit recourse to Lagrangians or Hamiltonians. We will then combine the results of the two approaches into a mathematical theorem. Later in the next section, we will impose more rigorous constraints, which will enable us to identify the spinor formulation given here with the usual Hilbert Space formulation of quantum mechanics.

In a previous paper [12] we have shown that the quantum-mechanical wave equations can be derived as the dual of the Dirac "square-root" of the metric. In other words, if

$$ds^2 = g_{\mu\nu} dx^\mu dx^\nu = \eta_{ab} dx^a dx^b \tag{2}$$

where a and b refer to local tetrad coordinates and η to a rigid Minkowski metric of signature -2, then associated with this metric and the vector **ds** is the scalar ds and a matrix $\tilde{ds} \equiv \gamma_a dx^a$ respectively, where $\{\gamma_a, \gamma_b\} = 2\eta_{ab}$, with γ_a transforming as a covariant vector under coordinate transformations. Note also that $g_{\mu\nu}(x) = \eta_{ab} e^a_\mu(x) e^b_\nu(x)$ with $e^a_\mu(x)$ forming local tetrads at x. Moreover, since ds is an invariant scalar, and $\tilde{ds}^2 = ds^2$ we can identify the "eigenvalue" ds with the linear operator $\tilde{ds}$ by forming the spinor eigenvector equation $\tilde{ds}\xi = ds\xi$. This is equivalent to associating the metric

$$ds^2 = g_{\mu\nu} dx^\mu dx^\nu = \eta_{ab} dx^a dx^b \tag{3}$$

with the spinor equation:

$$ds\xi = \gamma_a dx^a \xi. \tag{4}$$

It follows from the general theory of eigenvectors that if ξ is a solution so also is $f(z_0, z_1, z_2, z_3)\xi$ where f is any complex scalar valued function. Indeed, there is no reason why f cannot be an L^2 function, and correspond to a quantum-mechanical wave function.

As previously mentioned, corresponding to each tangent vector $\frac{\partial}{\partial x^a}$, there exists a dual one-form dx^a. In a similar way, the $\tilde{ds}$ matrix above can be seen as the dual of the expression $\tilde{\partial}_s \equiv \gamma^a \frac{\partial}{\partial x^a}$, where γ^a is defined by the relationship $\{\gamma^a, \gamma_b\} = 2\delta^a_b$ and the dual map defined by

$$\left\langle \tilde{ds}, \tilde{\partial}_s \right\rangle \equiv \frac{1}{\mathrm{Tr}(dx^i \partial_j)} \gamma_a \gamma^b dx^a \frac{\partial}{\partial x^b} \equiv \frac{1}{\delta^i_i} \gamma_a \gamma^b \frac{\partial x^a}{\partial x^b} = 1, \tag{5}$$

remains invariant. Moreover, if we let s describe the length of a particle's trajectory along a curve $(x^0(s), x^1(s), x^2(s), x^3(s)) \in (\mathcal{M}, g)$ then s can be regarded as an independent parameter with an associated 1-form ds, which is the dual of the tangent vector ∂_s. Note that in terms of the basis vectors for $T_p(\mathcal{M})$ and $T_p^*(\mathcal{M})$ we can write $\partial_s = \frac{\partial x^a}{\partial s}\partial_a$ and $ds = \frac{\partial s}{\partial x^a}dx^a$. It also follows from this and (1) that its dual map is given by $ds.\partial_s = \frac{\partial s}{\partial x^i}\frac{\partial x^i}{\partial s} = 1$. Putting these two results together allows us to consider (4) as the dual of the equation:

$$\frac{\partial \psi}{\partial s} = \gamma^a \frac{\partial \psi}{\partial x^a}, \tag{6}$$

where $\frac{\partial}{\partial s}$ refers to differentiation along a curve parameterized by s. We will refer to (6) as a (generalized) Dirac equation and will show later on how it relates to the usual form of this equation. At times, too, we shall refer loosely to it as a "dual wave-equation."

Now consider the motion of a test particle of mass m along a timelike geodesic. Let $p^a = m(dx^a/d\tau)$, where τ is the proper time (i.e $ds = cd\tau$). Then

$$ds^2 = \eta_{ab}dx^a dx^b \qquad \text{is equivalent to} \qquad (mc)^2 = \eta_{ab}p^a p^b. \tag{7}$$

This can be expressed in spinor notation by

$$ds\xi = \gamma_a dx^a \xi \qquad \text{which is equivalent to} \qquad \gamma^a p_a \xi(p) = mc\xi(p). \tag{8}$$

Indeed, if (6) is subjected to the constraints of (8), as it should be for motion along the timelike geodesic, we find that $\psi = \psi^i(\int_{x_0}^{x(s)} p_a dx^a)e_i$ is a solution of (6), provided the integration is taken along the curve $s = c\tau$ and $\xi(p) = \frac{d\psi^i(p)}{d\tau}e_i$ (cf Theorem 1 below). It is also worth noting that if all of ψ^i are equal then $\psi = f(x)u$ where u is a spinor independent of x, and $f(x)$ is a function. In this particular case the Dirac equation takes on the form

$$(\tilde{\partial}_s f)u = \frac{\partial f}{\partial s}u. \tag{9}$$

Moreover, in terms of a 4-dimensional (complex) Euclidean space E^4, this equation can be directly related to the expression [4]

$$df = \mathbf{ds}.\nabla f = ds\frac{\partial f}{\partial s}, \tag{10}$$

by observing that $\tilde{\partial}_s f$ is the matrix form of the vector ∇f. Also, if $\mathbf{ds}.\nabla f$ is invariant with respect to both rotations and reflections, then the associated spinor equation can be immediately written in a covariant manner in a natural way. It is sufficient to note that for the Lorentz spinor transformation $D(\Lambda(x))$ applied to the Dirac equation (9), we get

$$D(\Lambda(x))(\tilde{\partial}_s f)u = D(\Lambda(x))\frac{\partial f}{\partial s}u \tag{11}$$

$$\Rightarrow \quad D(\Lambda(x))(\tilde{\partial}_s f)D^{-1}(\Lambda(x))D(\Lambda(x))u = \frac{\partial f(s))}{\partial s}D(\Lambda(x))u \tag{12}$$

$$\Rightarrow \quad \tilde{\partial f}_{s'}u'(x') = \frac{\partial f(x)}{\partial s}u'(x'), \tag{13}$$

which expresses the covariance. Note $\tilde{\partial}_{s'} \equiv \gamma^a \frac{\partial}{\partial x'_a}$.

2.2 A Gauge Approach to Mass

Another approach to the above formalism is to introduce test particles by means of a gauge term. Specifically, let $(\mathcal{M}, g)$ be a pseudo-Riemannian manifold, with metric tensor g determined by Einstein's field equations, and introduce a massless test particle into the field. By the principle of equivalence we can choose a local tetrad $\{dx_0, dx_1, dx_2, dx_3\}$ such that the massless test particle travels along a null geodesic given by $0 = dx^a dx_a$, which in terms of a spinor basis can be written as

$$\gamma^a dx_a \xi = 0. \tag{14}$$

Next define the wave equation corresponding to the metric by taking the dual of the 1-form space:

$$\gamma^a \partial_a \psi = 0, \tag{15}$$

which can be interpreted as the wave equation of a massless particle.

We now introduce a test particle of mass m by means of a minimal principle [2], by adopting the same technique that is usually used to introduce test charges into a field. In other words, let

$$0 = \gamma^a(\partial_a - p_a)\psi, \tag{16}$$

which in turn gives the fundamental wave equation

$$\gamma^a \partial_a \psi = \gamma^a p_a \psi. \tag{17}$$

This immediately suggests the particular solution

$$\psi = e^{\int^x p^a dy_a}\xi(p_0). \tag{18}$$

Moreover, if the gauge term describes a test particle of mass m moving along a timelike geodesic as defined in (8), then

$$\gamma^a \partial_a \psi = mc\psi. \tag{19}$$

Once again, we have obtained a Dirac equation.

However, there is a difference between the two. In the first approach we strictly work with the dual of a metric to obtain the desired equation. In the gauge approach we not only use the dual of a null metric to obtain an answer but it implicitly restricts us to eigenfunction solutions, in contrast to the other method which allows for an entire family of solutions, including eigenfunctions. Another advantage of the dual approach is that mass is intrinsically linked to the structure of space-time itself through the metric, in contrast to the use of a gauge where mass is introduced as extrinsic to the space.

2.3 A Theorem

The above results can be summarized in the following theorem:

Theorem 1. *Let* $\xi(p) = [\frac{d}{ds}\psi^i(s)]e_i$, *where* $mcs = \int^x p^a dx_a$ *along a timelike geodesic then* $\gamma^a p_a \xi = mc\xi$ *iff* $\psi(p) = \psi^i(\int^x p^a dx_a)e_i$ *is a solution of*

$$\frac{\partial}{\partial s}\psi(p) = \gamma^a \partial_a \psi(p),$$

where $\int^x p^a dx_a$ *is Lorentz invariant, and integration is taken along the curve with tangent vector* $p^a = m\frac{dx^a}{d\tau}$, *where* τ *is proper time.*

Proof. Noting that $\psi(p) = \psi^i(\int^x p^a dx_a)e_i$ and assuming $\gamma^a p_a \xi = mc\xi$ then

$$\gamma^a \frac{\partial \psi}{\partial x^a} = \gamma^a p_a \frac{\partial \psi^i}{\partial s} e_i \tag{20}$$

$$= \gamma^a p_a \xi \tag{21}$$

$$= mc\xi \qquad \text{given} \tag{22}$$

$$= \frac{\partial \psi}{\partial s} \tag{23}$$

To prove the converse it is sufficient to substitute $\psi(p) = \psi^i(\int^x p^a dx_a)e_i$ into $\tilde{\partial}_s \psi = \partial_s \psi$ to get answer. □

Corollary 1. *In the case of* $\psi^i(\int^x p^a dx_a) = e^{\int^x p^a dx_a}$ *then* $\gamma^a \partial_a \psi = mc\psi$.

Proof. Clearly $\frac{\partial \psi}{\partial s} = mc\psi$. □

Corollary 2. *If $\psi(\int^x p^a dx_a) = f(\int^x p^a dx_a)u$ where u is a spinor independent of x^a then the equation*

$$\tilde{\partial}_s f u = \frac{\partial f}{\partial s} u. \tag{24}$$

has the same solutions as

$$\tilde{\partial}_s \psi = \frac{\partial}{\partial s} \psi. \tag{25}$$

Proof. Substitute. □

2.4 Covariance

Theorem 1 also enables us to write down a covariant form for the generalized Dirac equation which depends directly upon the covariance of its dual metric equation. We begin by showing that the equation $\tilde{ds}\xi = ds\xi$ is covariant under Lorentz transformations. Specifically, if $dx^a = \frac{\partial x^a}{\partial x'^b} dx'^b = \Lambda^a_b dx'^b$ then $\tilde{ds}\xi = ds\xi$ transforms under Lorentz transformations $D(\Lambda(x)) = D(x)$ into

$$D(x)\tilde{ds}\xi(x) = dsD(x)\xi(x). \tag{26}$$

Now the left hand side can be rewritten as

$$\begin{aligned} D(x)\tilde{ds}\xi &= D(x)\tilde{ds}D^{-1}(x)D(x)\xi(x) & (27)\\ &= \tilde{ds}'D(x)\xi(x), \qquad \text{where } \tilde{ds}' \equiv \gamma_a dx'^a & (28)\\ &= \tilde{ds}'\xi'(x'). & (29) \end{aligned}$$

Equating the two equations (26) and (29) then gives

$$\tilde{ds}'\xi'(x') = ds\xi'(x'), \tag{30}$$

which establishes the covariance.

The covariance of the Dirac equation follows if we can re-write (6) in the form

$$(\tilde{\partial}_s \psi^i)e_i = \frac{\partial \psi^i}{\partial s} e_i. \tag{31}$$

This is always possible to do in Minkowski space by choosing a single fixed tetrad for the entire space or along geodesics in more general spaces by Fermi transporting the tetrad along the curve, which is the case in this article. From Theorem 1, we already know that equation (31) is equivalent to the covariant metric $\tilde{ds}\xi = ds\xi$ provided $\xi = (d\psi^i/ds)e_i$, where s indicates differentiation along the timelike geodesic parameterized by s (recall $mcs = \int^x p_a dx^a$). It follows that $\tilde{ds}\xi = ds\xi$

is covariant iff $\gamma_a p^a \xi = mc\xi$ is covariant iff $\gamma_a p^a (d\psi^i/ds)e_i = mc(d\psi^i/ds)e_i$ is covariant iff (31) is covariant with respect to the Lorentz transformation $D(x)$, along the geodesic. Moreover, this latter restriction of motion along a geodesic, may actually be relaxed and the following more general theorem can be proven:

Theorem 2. *The Dirac equation defined over the manifold* $(\mathcal{M}, g)$ *is Lorentz covariant under the transformation* $D(x)$ *defined with respect to a tetrad* e_i, *provided the equation is written in the form*

$$\gamma^a \frac{\partial \psi^i}{\partial x^a} e_i = \frac{\partial \psi^i}{\partial s} e_i .$$

Proof. Let $D(x)$ be a local Lorentz transformation at x then:

$$\begin{aligned} D(x)\tilde{\partial_s}\psi(s) &= D(x)(\tilde{\partial_s}\psi^i)e_i && (32)\\ &= (D(x)\tilde{\partial_s}\psi^i)D^{-1}(x)D(x)e_i && (33)\\ &= (\tilde{\partial_{s'}}\psi^i)e_i' && (34)\\ &= \left(\frac{\partial \psi^i}{\partial s}\right)e_i' && (35)\end{aligned}$$

□

Remark 1. Although this equation is covariant, it will turn out that in the case of non-geodesic motion (6) and (8) cannot both hold at the same time and a more complicated formula is involved (see Sect. 4). Also, in regular Minkowski space, the covariant form of the generalized Dirac equation can be reduced to the form of (6) in any inertial frame.

3 Quantum and Classical Mechanics

At this stage the reader may be wondering how the usual formulation of quantum mechanics emerges. Indeed, the wave equations above seem to express the wave equations of classical mechanics more than quantum mechanics, in that there is no expression for Planck's constant h, nor does the expression $i = \sqrt{-1}$ appear with the operators. With regard to the latter point, we note that i could be seen as absorbed into the γ matrices, but we postpone a full discussion of this until later in this section. First, we analyze the solutions of the wave equation for a massless particle from three perspectives to help us better grasp the formal difference between classical and quantum mechanics. Later on, we will formulate the axioms of quantum mechanics as suggested by our analysis.

3.1 Wave Equation of a Massless Particle

The linearized metric for a massless particle is given by

$$0 = \gamma^0 c dt - \gamma^1 dx_1 - \gamma^2 dx_2 - \gamma^3 dx_3 \tag{36}$$

from which it follows by the canonical correspondence established above that the associated wave equation for the particle is given by:

$$0 = \gamma_0 \frac{\partial \psi}{c \partial t} - \gamma_1 \frac{\partial \psi}{\partial x_1} - \gamma_2 \frac{\partial \psi}{\partial x_2} - \gamma_3 \frac{\partial \psi}{\partial x_3}. \tag{37}$$

This is the Dirac equation for a massless particle. Squaring this out we get an equation analogous to the Klein-Gordan equation for a massless particle:

$$\frac{1}{c^2} \frac{\partial^2 \psi}{\partial t^2} = \sum_{i=1}^{3} \frac{\partial^2 \psi}{\partial x_i^2}, \tag{38}$$

which is also the classical wave equation. Also note that eigenfunction solutions of the Klein-Gordan equation are not necessarily eigenfunctions of the generalized Dirac equation. In this regard, it should also be noted that the Klein-Gordan equation simply prescinds from any discussion of spin or equivalently, it may be considered as the equation for a spin 0 particle. In contrast the Dirac equation has non-zero spin value solutions.[1]

Since the wave equation emerges from the structure of space-time itself, the question arises as to how to distinguish classical mechanics from quantum mechanics. We investigate this by analyzing the motion of a massless particle in a Minkowski space, subject to different sets of boundary conditions.

Case 1:

Consider the motion of a classical massless particle moving on the x-axis with uniform velocity c, but constrained by two mirrors placed at $x = 0$ and $x = \xi$ to move

[1] This also raises the question of quantum statistics. It has been noted in a previous paper [13] that Fermi-Dirac statistics is a consequence of indistinguishable particles forming spin-singlet states, while Bose-Einstein statistics follows as a consequence of breaking the rotational invariance associated with the singlet states. Moreover, the easiest way for this breaking to occur is for the spin states of the particles to be statistically independent. It follows as a trivial consequence of the above theory that bosons cannot be second quantized as fermions and fermions cannot be second quantized as bosons, in that particles which are forming spin-singlet states with probability one cannot be considered statistically independent. It also follows that spin 0 particles must obey Bose-Einstein statistics. For example, if S and T represent the spin observables of two particles such that $P(S = 0) = P(T = 0) = 1$ then $P(S = 0, T = 0) = P(S = 0)P(T = 0) = 1.1 = 1$, and hence the spin observables S and T are statistically independent.

uniformly on the interval $[0, \xi]$. We will assume that perfect reflection takes place at the mirrors and that no energy is exchanged. In this case, the equation of motion for a strictly classical particle with position $x = 0$ at $t = 0$ is given by:

$$x = \begin{cases} ct - 2n\xi, & \text{for } t \in [\frac{2n\xi}{c}, \frac{(2n+1)\xi}{c}] \\ 2(n+1)\xi - ct & \text{for } t \in [\frac{(2n+1)\xi}{c}, \frac{(2n+2)\xi}{c}], \end{cases}$$

and its wave function $\psi(x, t)$ takes on the form

$$\psi(x,t) = \begin{cases} \delta[k(x - ct)] & \text{for} \quad x - ct = -2n\xi \\ \delta[k(x + ct)] & \text{for} \quad x + ct = 2(n+1)\xi \\ 0 & \text{otherwise.} \end{cases}$$

The wave function in this case pinpoints the position of the particle with probability 1. Moreover, there is no restriction on the energy (implicit in the term k) in this case. Theoretically, it may have values ranging from 0 to ∞.

Case 2:

The classical particle is an idealized situation. In reality, the position of a massless particle constrained to move on the line is unknown and any attempt to know its exact position will be subject to Heisenberg's uncertainty relations, which we will formulate in the next section. In other words, its exact position can not be known in principle, because any attempt to pinpoint it will scuttle the position and defeat the whole purpose of the experiment. The best we can do is to describe the position by means of a uniform probability density $f(x - ct) = 1/\xi$ for $x \in [0, \xi]$ which also suggests writing $\psi(x, t) = e^{\pm ik(x-ct)}/\sqrt{\xi}$ to preserve both the boundedness and the periodic motion of the particle, as described by the above wave equation. This does not mean that causality is violated nor that the particle does not have an exact position, at least in the above case. It simply affirms that our initial conditions have to be defined statistically, and also in such a way as to reflect the periodic motion of the particle. As a consequence the future evolution of the wave function of the system is best interpreted in a statistical way. Finally, note that in this model the energy of the particle can once again vary from 0 to ∞ in a continuous manner.

Case 3:

The particle may be constrained to move in a potential well in such a way that the wave function is continuous ($= 0$) at the boundaries. In the case of the above problem, this means that the wave function has harmonic solutions of the form $\psi(\xi, t) = Ae^{i\nu' t}\sin(kx)$, where A is a constant. Substituting, we will find that $k = \frac{n\pi}{\xi}$ and the photon energy becomes quantized and of the form $E \equiv kc = \nu'$. It should be noted that this solution corresponds to the motion of a harmonic oscillator, and is the key to the quantization process associated with quantum field theory in general [5]. Indeed, if we were to re-scale our units of energy by defining $\nu' = h\nu$, then $E = h\nu$, with h having units $J.s$, and the standard wavelength becomes $\lambda = \frac{ch}{E}$. In the next section, h will be introduced in a more formal way.

The purpose of the above three examples is to highlight the importance of the boundary conditions when distinguishing between a classical type problem and a quantum mechanical problem, a point also stressed by Lindsey and Margenau [6]. Classical and quantum laws are not in opposition to each other. There is not one set of laws on the microscopic level and another on the macroscopic. On the contrary, classical and statistical methodologies are complimentary to each other and are in principle, applicable at all levels. However, on the microscopic level, statistical fluctuations will be more pronounced and consequently in practice (and in principle) the effects associated with quantum physics will become more apparent.

3.2 *Quantum Mechanics and Hilbert Space*

The above analysis permits us to better understand something of the difference between quantum mechanics and classical mechanics from the perspective of general relativity. As we have noted, it suggests the difference is to be found in the boundary conditions, which in the case of quantum mechanics is subjected to statistical conditions. With this in mind, we formulate a few axioms which not only respect the manifold structure of general relativity, but also enable us to distinguish quantum mechanics from classical physics, in a formal way.

Essentially what we have noted is that the metric of general relativity forces (real) eigenvalue solutions for the free particle of the form

$$\frac{\partial \psi(\int p^a dx_a)}{\partial x_a} = p^a \psi(\int p^a dx_a).$$

However, since the choice of eigenfunction associated with the specific eigenvalue in this case is not unique, we restrict ourselves for the purpose of quantum mechanics to those eigenfunctions $\psi(t, \mathbf{x})$ such that for each t, $\psi(t, \mathbf{x}) \in L^2(E^3) \times H$, where H is a 4-dimensional Hilbert space. Also, we associate the dual of the 1-form dx with the self-adjoint partial differential operator $i\hbar\partial/\partial x$, where $\hbar$ is a constant. Consequently by defining the dual in this way, we not only find that

$$dx(i\hbar\partial_x) \equiv i\hbar\partial_x x = i\hbar, \tag{39}$$

But it can also be linked to the uncertainty principle (see below). At first, this may seem artificial but actually if we look more closely at (6) we will find that to associate the operator $-i\partial_x$ with a real valued momentum eigenvalue is already implicit in this equation, and indeed is a consequence of the signature of the metric tensor $\eta_{ab} = \{\gamma_a, \gamma_b\}$. In particular, if we let $\gamma_a = -i\alpha_a$ for $a = 1, 2, 3$, and set $\gamma_0 = \alpha_0$, then the α_a's are the generators of the Dirac Algebra $SL(2, C)$. Also

$\gamma_a p^a = -\alpha_a(ip^a)$, for $a = 1, 2, 3$ and the linearized metric (3) can be written in the explicit form

$$\tilde{ds} = \alpha_0 dx^0 - i\alpha_1 dx^1 - i\alpha_2 dx^2 - i\alpha_3 dx^3, \tag{40}$$

which in order to maintain the invariant relationship $\left\langle \tilde{ds}, \tilde{\partial}_s \right\rangle = 1$ (cf (5)), gives

$$\tilde{\partial}_s = \alpha^0 \partial_0 - i\alpha^1 \partial_1 - i\alpha^2 \partial_2 - i\alpha^3 \partial_3. \tag{41}$$

In other words, if we let α_a obey the Dirac Algebra, then we can associate the momentum operator with $-i\partial_a$ in a natural way. Finally, let $\hbar = h/2\pi$ where h is Plancks constant and re-scale the momentum operator by writing $-i\hbar\partial_a$ in place of $-i\partial_a$ to obtain the usual form of quantum mechanics.

This too may seem artificial, but in reality we are free to choose any scale we wish. This being the case, we choose $\hbar$ because it seems to be the scaling constant, which nature uses. Moreover if we multiply across by $-i\hbar\alpha^0$ and note that $\hat{\alpha}_a \equiv -i\alpha_0\alpha_a$ obeys the same Dirac Algebra as α_a, then the Dirac equation (6) can be rewritten as:

$$-i\hbar\alpha^0 \frac{\partial\psi}{\partial s} = (-i\hbar\partial_0 - i\hbar\hat{\alpha}^1\partial_1 - i\hbar\hat{\alpha}^2\partial_2 - i\hbar\hat{\alpha}^3\partial_3)\psi. \tag{42}$$

In particular, the eigenvalue of $-i\hbar\partial_0$ associated with the eigenfunction $\exp((i/\hbar) \int^x p_a dx^a)$ is given by p_0 which we denote by E/c, where E has the units of energy. It now follows from Corollary 1 that

$$\alpha^0 mc^2\psi = (E - ic\hbar\hat{\alpha}^1\partial_1 - ic\hbar\hat{\alpha}^2\partial_2 - ic\hbar\hat{\alpha}^3\partial_3)\psi \tag{43}$$

which gives us back the usual form of the Dirac equation except for a change in the sign of i.

Finally based on the above discussion, we formulate a few axioms which not only respect the manifold structure of general relativity, but also enable us to distinguish quantum mechanics from classical physics, in a formal way.

Definition 1. Space-time is a four dimensional manifold $(\mathcal{M}, g)$.

Definition 2. At every point p of $\mathcal{M}$ there is a tangent vector space $T_p(\mathcal{M})$ with tetrad basis $\{-i\hbar\partial_0, -i\hbar\partial_1, -i\hbar\partial_2, -i\hbar\partial_3\}$ and a dual 1-form space, denoted by T_p^* with basis $\{dx_0, dx_1, dx_2, dx_3\}$ at p, defined by $dx^a(i\hbar\partial_b) = i\hbar\frac{\partial x^a}{\partial x^b} = i\hbar\delta^a_b$.

Definition 3. Quantum mechanical operators are elements of $SL(2, C)$ Dirac Algebra, which can be viewed as a representation of the vector spaces T_p and T_p^* on $\mathcal{M}$.

Definition 4. Each element of $SL(2, C)$ algebra acts on the Hilbert Space $L^2(E^4) \times H$, where H is a 4-dimensional Hilbert Space.The elements $\psi \in L^2(E^4) \times H$ are called the states of the system.

Remark 2. It follows from the definition of the Hilbert Space that if $\psi \in L^2(E^4) \times H$ then for each t, $\psi(t, \mathbf{x}) \equiv \psi_t(\mathbf{x}) = \psi_t^i e_i \in L^2(E^3) \times H$, where each $e_i \in H$ and an inner product exists such that $\langle \psi_t, \psi_t \rangle = \int (\psi^*)^i \psi_i d^3x$. Moreover, if $\psi_t(\mathbf{x})$ is normalized for each t then $\psi^* \psi$ can be interpreted as a probability density function for position. Likewise, for each $\mathbf{x}$ we can define $\psi(t, \mathbf{x}) \equiv \psi_{\mathbf{x}}(\mathbf{t}) = \psi_{\mathbf{x}}^i e_i \in L^2(E^1) \times H$ and seek quantum effects in time.

Lemma 1. *Let $\tilde{df}$ and $\tilde{\partial}_x$ be the $SL(2, \mathcal{C})$ representation of $df \in T^*$ and $\partial_x \in T$ respectively, then $\left\langle \tilde{df}, \tilde{\partial}_x \right\rangle \psi = [\tilde{\partial}_x, \tilde{f}]\psi$, where $\psi \in L^2(E^4)$.*

Proof. Note $\tilde{df} = \frac{\partial f}{\partial x^a} d\tilde{x}^a$ and $\tilde{f} = fI$, where I is an identity matrix. Therefore

$$\left\langle \tilde{df}, \tilde{\partial}_a \right\rangle \psi = \frac{\partial f}{\partial x^a} \gamma^a \psi = [\tilde{\partial}_a, \tilde{f}]\psi.$$

The Lemma has been proven. □

Lemma 2. *(The uncertainty relationships) Let $\tilde{X} = \int_0^{x(s)} d\tilde{X}$ and $\tilde{P} = i\hbar\tilde{\partial}_s$ be the $SL(2, \mathcal{C})$ representations of position and momentum respectively, defined along a curve of length s. Also let $\bar{X} \equiv \int \psi^* \tilde{X} \psi ds$, $\bar{P} \equiv \int \psi^* \tilde{P} \psi ds$, $\Delta^2 \tilde{X} \equiv \int \psi^* (\tilde{X} - \bar{X})^2 \psi ds$ and $\Delta^2 \tilde{P} \equiv \int \psi^* (\tilde{P} - \bar{P})^2 \psi ds$ then $\Delta \tilde{X} \Delta \tilde{P} \geq \frac{\hbar}{2} \left| \int^s \psi^* \frac{1}{i\hbar} (\tilde{P}\tilde{X} - \tilde{Q}\tilde{X}) \psi ds \right|$. In particular, in the case of the components $\tilde{X}^a$, $\tilde{P}^a$ we get $\Delta \tilde{X}^a \Delta \tilde{P}^a \geq \frac{\hbar}{2}$.*

Proof. Usual proof using Cauchy-Schwartz inequality. □

3.3 Classical Mechanics

The above formulation lays the ground work for distinguishing classical from quantum mechanics. Indeed, we have seen that quantum theory is highly dependent upon $\hbar > 0$ and states $\psi \in L^2(E^4)$. Moreover, this suggests that classical mechanics can be obtained by relaxing one of these two conditions either by letting $\hbar \to 0$ or by choosing $\psi \notin L^2(E^4)$ or both. In principle what distinguishes a quantum particle from a classical one is that in contrast to quantum mechanics, the position and momentum of a classical particle can be fully pinpointed and localized as in Case 1 above. On the other hand, in the case of a particle moving uniformly between two mirrors, where the initial position is unknown, then as has already been noted in Sect. 2.1 (Case 2), the wave function is given by $exp(ikx)/\sqrt{\xi}$. However, if we take the limit as $x \to 0$ or $\hbar \to 0$ ($\hbar$ is contained in k) then this would constrain both the position and momentum to become increasingly localized and allow them to be measured more exactly, as the distance between the mirrors ξ shrinks to 0. Consequently, the resulting wave function at any time t would be of the form $\psi(0) = \sum_n a_n \delta^{(n)}$. This last equation follows from a well known result in distribution theory [7], which states

Theorem 3. *A distribution T which has a support of one point (i.e., is equal to zero except at one point) is a finite linear combination of the Dirac function and its derivatives:* $T = \sum_n a_n \delta^{(n)}$.

Based on this we can formally state that

Definition 5. A classical particle is a particle whose position operator T at any time t has support of one point.

It follows from this definition that the momentum operator $\tilde{P}$ has a support of one point. It also follows from the definition and Theorem 1 that for a classical particle with constant momentum situated at (x_0) on a geodesic, the wave function is given by $\delta^4(p^a(x-x_0)_a)$. Also the set of operators T in definition 5 clearly form a subspace of the solution set of the Dirac equation. Indeed, the set $\{\delta, \ldots, \delta^{(k)} \ldots\}$ is a spanning set for this subspace. Hence, the uncertainty in observing its value is 0 everywhere except at the single point, and the standard deviation $\Delta\tilde{X}$ and $\Delta\tilde{P}$ are also zero. In other words, the uncertainty principle fails for a classical particle. Moreover, in order for nature to circumvent classical solutions, it is sufficient that there be a fundamental unit of wavelength given by $\lambda = \frac{ch}{E} = \frac{2\pi c\hbar}{E}$, such that $\lambda > 0$ whenever $\hbar > 0$, which is another way of saying that if $\hbar > 0$ then classical solutions need not exist. It also strongly suggests that the process of localizing a particle for measurement is equivalent to confining (at least during the measuring process) the particle to a box. This, in turn, hints that in terms of wave-particle duality, particle properties emerge when we attempt to experimentally localize and isolate the wave, causing a discontinuity in the quantum solutions, closely approximated by delta type functions. We have seen an example of this above, when we considered a particle moving uniformly between two mirrors with wave-function $e^{\pm ik(x-ct)}/\sqrt{\xi}$.

In concluding this section, we note that classical solutions to the generalized Dirac Equation describing the motion of a particle, can be reduced to distributions corresponding to point masses as described in Theorem 3 above, and live on a larger space than the L^2 functions associated with quantum mechanics. To better understand this point, it might be useful to recall the definition of L^p spaces, and some of their properties [8]. Consider a fixed measure space $(X, \mathcal{M}, \mu)$. Let f be a measurable function on X such that

$$\|f\|_p = \left(\int |f|^p d\mu\right)^{\frac{1}{p}} \tag{44}$$

then we define

Definition 6. $L^p(X, \mathcal{M}, \mu) = \{f : X \to \mathcal{C} : \|f\|_p < \infty\}$.

Also, in general, we can define a bounded linear functional on L^p by $\phi_g(f) = \int fg$, such that $g \in L^q$ where $1/p + 1/q = 1$, and $\phi_g \in (L^p)^*$, the dual space of L^p. In the case of $p = 2$, the L^2 space is also a Hilbert space and therefore, its own dual. In turn this allows us to formulate quantum mechanics in a very elegant and simple manner. However, it the case of classical solutions, as described above, the Dirac δ functional is usually interpreted as a functional on the set of continuous differential functions of compact support denoted by $C_c^\infty(X)$, which in turn are dense in L^p, where $1 \leq p < \infty$. Moreover, in the case of a finite (probability) measure, $L^p \subset L^1$, for $p \geq 1$. With this distinction in mind, it now follows from our formulation that general relativity while permitting a natural unification of both quantum and classical mechanics by means of the generalized Dirac Equation, also permits a distinction by means of L^2 functions and distributions which are duals of L^1 functions.

4 Non-Geodesic Motion

In the introduction we referred to the philosophical implications of our approach for the unity of physics, in that the metric structure of the space determines the general form of the wave equation, and consequently implies that space-time is not merely a backdrop for the laws of mechanics but is in fact an effective and quasi-formal cause of the resulting wave equations. At the same time because of the equivalence principle it was noted that the general form of the wave equation is determined only locally and not globally, especially when we consider motion along a non-geodesic. In addition, the use of test particles to analyze the motion of a massive body within the structure of the space-time by its very nature seems to reflect a return to the pre-relativistic mentality of considering the laws of mechanics as something operative within space-time but not contributing to its ontological structure.

In fact, I would argue that the use of test particles highlights the mathematical difficulty of dealing with a many-body problem in General Relativity, and offers a methodological unification to the laws of physics but not necessarily an ontological one. It does not contradict our basic premise that the structure of space-time is determined by the mass-energy tensor and that the wave equations of quantum mechanics is the dual of the metric. Ideally, in the case of the hydrogen atom it would be preferable to write down a single metric for the joint proton–electron system. Unfortunately, the two body problem in general relativity is a formidable task and has not been solved. If it were then the motion of the electron would be best described by tracing the path of a vector corresponding to the center of mass (or center of charge) of the electron, and considering the dual of that path to be the wave equation proper for describing the quantum mechanical properties of the electron. Instead, as we shall see below, we circumvent the difficulty by taking the electron as a test-particle. This is easier to handle mathematically, and is probably a very good approximation to describing the actual system. Also in this approach we avoid the problem associated with our lack of understanding of the mass-charge relationship.

The test particle also brings to the fore another question. If the two-body problem were to be solved then how would non-geodesic motion be interpreted? For example in a Schwartzschild space consisting of a single point mass how would the motion of the particle be perceived if defined relative to a moving coordinate system? It would clearly give rise to a non-geodesic motion and suggest that a force is causing the motion. But where does such a force come from, and does such a frame of reference have any ontological or physical meaning? Or again, if a particle is fixed at 0 in a Cartesian frame in Minkowski space, how does one interpret its motion from the perspective of a frame in polar coordinates in circular motion around the origin? Whether the motion is virtual or real some reference frames by their very nature presuppose non-geodesic motion and the presence of forces. Consequently we can never fully avoid non-geodesic motion even if it should turn out that in a grand unification scheme all forces can be intrinsically related to space curvature.

The validity then of our methodology lies in its accuracy to make verifiable statements about nature. To know is to know something about the real, as Aristotle noted in his metaphysics. Equally, a correct method must incorporate physical observations into its system and be able to explain the data in a satisfying and simple way. It means to hit the nail on the head, as did Newton in his understanding of planetary motion. It means that a coherent methodology should be possible based on the assumption that the real world is intelligible, coherent and void of contradiction. It means that we should be able to find a solution to the current division between general relativity and quantum field theory, at least from a methodological perspective without in any way claiming to have unified the forces. One approach might be to consider space-time as an explanatory cause of the physical forces, and expect both quantum mechanics and all the forces of nature to be unified into a single unified field theory founded on the structure of space-time itself. This does not mean that the gravitational field needs to quantized. Indeed, in the previous section, where the equations of quantum mechanics are the dual of a linearized metric, the quantization takes place within the structure of space-time, but the metric (and consequently gravity) is not quantized. Others such as Fred Cooperstock also hold that gravity should not be quantized [11].

4.1 Methodological Unification

A first step in a possible methodological unification will be to attempt a description of wave-particle motion along arbitrary curves defined on manifolds with metric g_{ij} and not just on geodesics. The key to this are equations:

$$\frac{\tilde{ds}}{ds}.\tilde{\partial}_s\psi = \frac{1}{2}\left\{\frac{\tilde{ds}}{ds}, \tilde{\partial}_s\psi\right\} + \frac{1}{2}\left[\frac{\tilde{ds}}{ds}, \tilde{\partial}_s\psi\right] \tag{45}$$

$$= \frac{\mathbf{ds}}{ds}\cdot\frac{\partial\psi}{\partial s} + \frac{\mathbf{ds}}{ds}\wedge\frac{\partial\psi}{\partial s}, \tag{46}$$

which because of the Principle of Equivalence are valid locally at any point in the space. Note that in (45) the first expression on the right hand side is equivalent to a dot product of a tangent vectors with the gradient of the wave function $\mathbf{ds}\cdot\frac{\partial\psi}{\partial s}$ and the second term is equivalent to a cross product $\mathbf{ds}\wedge\frac{\partial\psi}{\partial s}$ of the same two terms. Equation (46) can be seen as defining a wave equation along an arbitrary curve $(x^0(s), x^1(s), x^2(s), x^3(s))$. Also note that curvature plays a role in the particular choice of tetrad components. For example, in the case of a Schwarzschild coordinate system a local tetrad basis at some point X can be defined by $dx_0 = (1 - \frac{2MG}{r})dt$, $dx_1 = (1 - \frac{2MG}{r})^{-1}dr$, $dx_2 = rd\theta$, $dx_3 = r\sin\theta d\phi$. However, it also highlights the fact that our formulation is always local and not global, and would work equally well on any manifold where a spinor basis can be defined at any point. Indeed, the very fact that the Dirac equation is a good predictor of the statistical behavior of an electron in the hydrogen atom, on a planet which is rotating both on its axis and around the sun indicates how useful the tetrad formulation is for local results. Finally, it is worth noting the similarity between (46) and the famous equation for Lorentz Force for a charge of size e moving in an electric field $\mathbf{E}$:

$$\mathbf{F} = e[\mathbf{E} + (\mathbf{v} \times \mathbf{B})]. \tag{47}$$

The question arises can we decompose (45) into a metric equation and a wave equation comparable to (8) and (6) above, especially since there would seem to be an infinite amount of possibilities governed by both the physics and the choice of non-inertial frames related to the acceleration introduced into the system. With this in mind we limit ourselves to two cases which may help us to better understand the issues.

In the first case we seek a spinor equation of the form

$$ds\xi = \gamma_a dx^a \xi \tag{48}$$

for the metric with a corresponding wave equation of the form

$$\lambda(s)\frac{\partial\psi}{\partial s} = \gamma^a \frac{\partial\psi}{\partial x^a}. \tag{49}$$

Once again, $\frac{\partial}{\partial s}$ refers to differentiation along a curve parameterized by s, and this equation is a generalization of (6) in that it reduces to (6) if $\lambda(s) = 1$. Indeed, this seems to be the simplest form that such an equation can take when motion is not along a geodesic.

It is important that the multiplication of these two equations be consistent with (45) in that

$$(\gamma^a dx_a)(\gamma^b \frac{\partial\psi}{\partial x^b}) = ds\frac{d\psi}{ds} + \gamma^a\gamma^b\left(dx_a\frac{\partial\psi}{\partial x^b} - dx_b\frac{\partial\psi}{\partial x^a}\right) \tag{50}$$

$$= ds\left[\frac{d\psi}{ds} + \gamma^a\gamma^b\left(\frac{dx_a}{ds}\frac{\partial\psi}{\partial x^b} - \frac{dx_b}{ds}\frac{\partial\psi}{\partial x^a}\right)\right], \tag{51}$$

should be compatible with the multiplication of (48) and (49). This requires that

$$\xi = \frac{d\psi}{ds} + \gamma^a\gamma^b\left(\frac{dx_a}{ds}\frac{\partial\psi}{\partial x^b} - \frac{dx_b}{ds}\frac{\partial\psi}{\partial x^a}\right) \tag{52}$$

$$= \lambda(s)\frac{d\psi}{ds}. \tag{53}$$

Another consequence of this identification is that $\frac{\xi^i}{\lambda}ds$ is an exact differential and suggests that λ is related to entropy.

The second approach is similar to the first and reflects the duality between curves and waves. In this case we switch the role of $\lambda(s)$ in (48) and (49) to get:

$$\lambda(s)ds\xi = \gamma_a dx^a\xi \tag{54}$$

for the metric with a corresponding wave equation of the form

$$\frac{\partial\psi}{\partial s} = \gamma^a\frac{\partial\psi}{\partial x^a}. \tag{55}$$

Once again these equations are compatible with (51) provided

$$\lambda\xi = \frac{d\psi}{ds} + \gamma^a\gamma^b\left(\frac{dx_a}{ds}\frac{\partial\psi}{\partial x^b} - \frac{dx_b}{ds}\frac{\partial\psi}{\partial x^a}\right) \tag{56}$$

$$= \lambda(s)\frac{d\psi}{ds}. \tag{57}$$

In addition this system of equations seems to be comparable to the work of Stueckelberg and Horwitz [14]. In their approach if ds is considered to be an independent variable then starting with Hamilton's equations one can derive the metric relation $ds^2 = \frac{m^2}{M^2}(cd\tau)^2$, with M an independent parameter with the units of mass. Clearly, if this equation is linearized in spinor form, it will result in (54) above provided $\lambda(s) = \frac{m}{M(s)}$. Moreover, if $M(s)$ is such that ds is an exact differential with respect to τ then $ds = \frac{m}{M}cd\tau$ is an exact differential and can be related to entropy. Indeed, it was precisely the introduction of such exact differentials by means of Pfaffian equations that permitted Caratheodory and Born [10] to put the second law of thermodynamics on a rigorous basis. Parallel mathematical conditions seems to exist in the non-commutative terms of the above product and consequently suggest a connection to entropy.

Finally, we note that both of these approaches can be used to introduce gauge transformations into the system which can be associated with external interactions and induce non-geodesic motion into the predefined space. In terms of the Stueckelberg this means that $m^2 \neq M^2$ and that the so called "mass shell" condition is violated.

4.2 The Hydrogen Atom

As an application of the work above we analyze the motion of an electron within the hydrogen atom. We do this by considering the electron as a test particle lying within the Reissner-Nordstrom metric of the proton of mass m_p, and invoke the minimal principle to incorporate charge and the vector potential in a gauge invariant way. This principle in effect allows us to incorporate a first order interaction compatible with the approach for non-geodesic motion.

Linearizing the metric we get:

$$ds\xi = [-i\alpha_1 R^{-1}dr - ir(\alpha_2 d\theta + \alpha_3 \sin\theta d\phi) + \alpha_0 Rcdt]\xi. \tag{58}$$

where $R(r) = \left(1 - \frac{2Gm_p}{c^2 r} + \frac{Ge^2}{c^4 r^2}\right)^{\frac{1}{2}}$. The corresponding wave equation for motion along geodesics becomes:

$$\frac{\partial}{\partial s}\psi' = \left\{-i\alpha_1 R\frac{\partial}{\partial r} - \frac{i}{r}\left(\alpha_2\frac{\partial}{\partial\theta} + \alpha_3\frac{1}{\sin\theta}\frac{\partial}{\partial\phi}\right) + \alpha_0 R^{-1}\frac{1}{c}\frac{\partial}{\partial t}\right\}\psi'. \tag{59}$$

Furthermore, if gauge terms for the electromagnetic field are now introduced into the wave function by means of the minimal principle then this requires that we seek solutions of the form $\psi' = \psi'(\int^x p'_a dx^a)$, where $p'_a = p_a + \frac{e}{c}A_a$ and A_a represents the 4-vector potential associated with the electromagnetic field of the proton. It also requires (in accordance with approach two above) that the metric (58) take on the modified form:

$$\lambda(s)ds\xi = [-i\alpha_1 R^{-1}dr - ir(\alpha_2 d\theta + \alpha_3 \sin\theta d\phi) + \alpha_0 Rcdt]\xi, \tag{60}$$

for non-geodesic motion, where $\xi = \frac{\partial\psi'}{\partial s}$. In particular, if we seek solutions of the type $\psi' = \exp(\frac{ie}{c\hbar}\int A_a dx^a)\psi(\frac{i}{\hbar}\int p_a dx^a)$ in tetrad coordinates then this can be reduced to the equation

$$\left[c\sum_{a=1}^{3}\hat{\alpha}_a P'_a + \alpha_0(mc^2 + eV)\right]\psi = E\psi, \tag{61}$$

where $\hat{\alpha}_a = -i\alpha_0\alpha_a$, $P'_a = -i\hbar\partial_a + \frac{e}{c}A_a$, m is the mass of the electron and V is the electrostatic potential in the rest frame of the proton. Note that this can be written in the usual form of the equation for the hydrogen atom $H\psi = E\psi$, provided $H \equiv \left[c\sum_{a=1}^{3}\hat{\alpha}_a P'_a + \alpha_0(mc^2 + eV)\right]$. However, there is also an important difference. In this formulation there is the presence of a $\alpha_0 eV$, instead of the usual eV, but as we will see below this has advantages.

If we let $\mathcal{G}_a = \frac{\partial V}{\partial x^a}$, assume $A_a = 0$ for a =1, 2 and 3, then the "square" of this equation, reduces to

$$(-c^2\hbar^2 \sum_1^3 \partial_a^2 + (mc^2 + eV)^2 + i\hbar ec\hat{\alpha}_a\alpha_0\mathcal{G}_a)\psi = E^2\psi. \tag{62}$$

Noting that $-i\hbar\hat{\alpha}_a\alpha_0 = \alpha_a$ which can also be seen as generators of the Dirac Algebra, this can be rewritten as:

$$(-c^2\hbar^2 \sum_1^3 \partial_a^2 + (mc^2 + eV)^2 - \hbar ec\alpha_a\mathcal{G}_a)\psi = E^2\psi. \tag{63}$$

Contrast this with conventional relativistic quantum mechanics where we would have found

$$(-c^2\hbar^2 \sum_1^3 \partial_a^2 + m^2c^4 - i\hbar ec\alpha_a\mathcal{G}_a)\psi = (E + eV)^2\psi. \tag{64}$$

Notice that the complex number coefficient associated with the electric moment $\mathcal{G}$ in this latter equation has been replaced with a real coefficient in (63). In other words, the spin electric moment $\frac{\hbar e}{2imc}$, which is a complex number, can be replaced with a real spin term $\frac{\hbar}{2mc}$ thus removing the ambiguity normally associated with electron spin in the hydrogen atom. For example, Lindsey and Margenau [9] warn us "not to take the electron spin to literally", because of the presence of the imaginary term. Happily, we can say that with the above approach the difficulty is resolved.

5 Conclusion

In this article we have attempted to unify general relativity and quantum mechanics by viewing any metric as a dual of a wave equation. We have noted that the resulting wave equation contains the usual Dirac equation of quantum mechanics as a special case. We have also noted that the difference between quantum and classical mechanics seems to lie in boundary conditions, with quantization (as distinct from quantum theory) emerging when the wave function is confined to a finite domain with continuous boundary conditions, and classical mechanics being the result of delta-function type solutions for the wave equation.

Overall, our approach was able to duplicate the standard results of quantum mechanics but in addition, we were able to remove the anomaly of an imaginary electric moment, when solving the hydrogen atom problem. This result in itself, should be sufficient to encourage further development. Finally, an analysis of the relationship between non-geodesic curves and wave equations was investigated.

Here the problems are more open ended and many possible wave equations are possible compatible with the structure of the metric. However, even in this case, the approach shows itself to be compatible with the work of Stueckelberg and Horwitz and also with gauge theory, and suggests further investigation. It also suggests that entropy is a key factor in any analysis of wave-particles which violate the so called "mass shell" condition.

References

1. O'Raifeartaigh L (1997) The Dawning of Gauge Theory. Princeton University Press, Princeton, New Jersey: 92.
2. O'Raifeartaigh L (1997) The Dawning of Gauge Theory. Princeton University Press, Princeton, New Jersey: 16.
3. Hawkin S & Ellis G (1995) The large scale structure of space-time. Cambridge University Press, Cambridge: 56.
4. Cartan E (1996) The Theory of Spinors. Dover, New York: 43.
5. Milonni P (1994) The Quantum Vacuum. Academic, New York.
6. Lindsey R & Margenau H (1957) Foundations of Physics. Dover, New York: 46–55.
7. Duff G & Naylor D (1966) Differential Equations of Applied Mathematics. Wiley, New York: 71.
8. Folland G (1984) Real Analysis. Wiley, New York: 170–210.
9. Lindsey R & Margenau H (1957) Foundations of Physics. Dover, New York: 511.
10. Born M (1949) Natural Philosophy of Cause and Chance. Oxford Press, Oxford: 38–45.
11. Cooperstack, F (2003) The Case for Unquantized Gravity. In: www.gravityresearchfoundation.org/pdf/abstracts/2003abstracts.pdf
12. O'Hara P (2005) Found. Phys. 35:1563–1584.
13. O'Hara P (2003) Found. Phys. 33:1349–1358.
14. Horwitz L (1992) Found. Phys. 22: 421–450.

Relativity and Quantum Field Theory

Jonathan Bain

Abstract Relativistic quantum field theories (RQFTs) are invariant under the action of the Poincaré group, the symmetry group of Minkowski spacetime. Non-relativistic quantum field theories (NQFTs) are invariant under the action of the symmetry group of a classical spacetime; i.e., a spacetime that minimally admits absolute spatial and temporal metrics. This essay is concerned with cashing out two implications of this basic difference. First, under a Received View, RQFTs do not admit particle interpretations. I will argue that the concept of particle that informs this view is motivated by non-relativistic intuitions associated with the structure of classical spacetimes, and hence should be abandoned. Second, the relations between RQFTs and NQFTs also suggest that routes to quantum gravity are more varied than is typically acknowledged. The second half of this essay is concerned with mapping out some of this conceptual space.

1 Introduction

The comparison of Minkowski spacetime with classical (i.e., non-relativistic) spacetimes has been fruitful in contemporary philosophy of spacetime in debates over the ontological nature of space and time (see e.g., Earman 1989, Chap. 2). In this essay, I extend this type of analysis to debates in the philosophy of quantum field theory. In particular, the distinction between Minkowski spacetime and classical spacetimes allows one to make a corresponding distinction between relativistic quantum field theories (RQFTs) and non-relativistic quantum field theories (NQFTs). This latter distinction is subsequently helpful, or so I shall argue, in clarifying the debate over whether or not RQFTs admit particle interpretations, and in investigating the conceptual space of possible extentions of RQFTs to include gravity.

Section 2 distinguishes between RQFTs and NQFTs in terms of the distinction between Minkowski spacetime and classical spacetimes. This distinction is then applied to an on-going debate over the ontology of QFTs. According to a Received View in this debate, RQFTs do not admit particle interpretations (Arageorgis et al. 2003; Fraser 2008; Halvorson and Clifton 2002). This view takes the existence of

V. Petkov (ed.), *Space, Time, and Spacetime*, Fundamental Theories of Physics 167,
DOI 10.1007/978-3-642-13538-5_5,

local number operators and a unique total number operator in the formulation of a QFT as necessary conditions for a particle interpretation of the theory. Given that formulations of RQFTs do not admit such objects, the Received View concludes that RQFTs cannot be given particle interpretations. I will argue that the existence of local and unique total number operators in a QFT requires the absolute temporal structure of a classical spacetime. Thus the Received View's concept of particle appears to be motivated by a non-relativistic concept of absolute time. The moral I draw is that the Received View's concept of particle is inappropriate for RQFTs.

No RQFT currently exists that consistently incorporates gravity. Section 3 reviews an example of an NQFT that does: Christian's (1997) Newtonian Quantum Gravity (NQG). NQG is an NQFT in (a version of) Newton-Cartan spacetime, the latter being an example of a curved classical spacetime. Part of the spacetime structure of NQG is dynamic and quantized, and its symmetry group is an extension of the non-relativistic Maxwell group. The latter entails that NQG is not plagued by the family of conceptual problems associated with unitarily inequivalent representations of the canonical (anti-) commutation relations, as are QFTs in curved Lorentzian spacetimes (Ruetsche 2002). In particular both local number operators and a unique total number operator are present in NQG, again due to the absolute temporal structure of classical spacetimes.

Using NQG as motivation, Sect. 4 undertakes the task of relating NQFTs, both in the presence and the absence of gravity, to RQFTs and to other theories, both of particles and fields, classical and quantum, in the presence and the absence of gravity. What emerges is a tentative map of the relations between some of the fundamental theories in physics, including the as-yet-to-be formulated, fully relativistic quantum theory of gravity (QG).

2 NQFTs and Particles

By an RQFT I will mean a quantum field theory invariant under the actions of the Poincaré group, the symmetry group of Minkowski spacetime. By an NQFT, I will mean a quantum field theory invariant under the actions of the symmetry group of a classical spacetime. Section 2.1 reviews the distinction between classical spacetimes and Minkowski spacetime. Section 2.2 indicates the significance this distinction has for the debate over particle interpretations of QFTs.

2.1 Classical Spacetimes vs. Minkowski Spacetime

Minkowski spacetime can be represented by a pair (M, η_{ab}), where M is a smooth 4-dim differentiable manifold and η_{ab} is a $(-1, 1, 1, 1)$ symmetric tensor field on M, the Minkowski metric, satisfying the compatibility condition $\nabla_a \eta_{ab} = 0$, for the derivative operator ∇_a associated with the connection on M. This condition determines a unique curvature tensor $R^a{}_{bcd}$, which vanishes, encoding spatiotemporal flatness. The isometry group of Minkowski spacetime, the Poincaré group, is generated by vector fields that Lie annihilate the Minkowski metric. Symbolically, we require $\pounds_x \eta^{ab} = 0$, where $\pounds_x$ is the Lie derivative associated with x^a. Intuitively, this means that the transformations between reference frames defined by the integral curves of the vector field x^a preserve the structure of the Minkowski metric. This structure famously entails that there is no unique way to separate time from space in Minkowski spacetime: any two observers moving inertially with respect to each other will disagree on the time interval between any two events, and on the spatial interval between any two events. In coordinate form, elements of the Poincaré group may be represented by transformations

$$x^{\mu} \rightarrow x^{\mu'} = \Lambda^{\mu}{}_{\nu} x^{\nu} + d^{\mu} \; (Poincaré) \tag{1}$$

where $\Lambda^{\mu}_{\nu} \in SL(2, \mathbb{C})$ is a pure Lorentz boost and $d^{\mu} \in \mathbb{R}^4$ is a spacetime translation.

In comparison, a classical spacetime is a spacetime that minimally admits absolute spatial and temporal metrics. More precisely, a classical spacetime may be represented by a tuple $(M, h^{ab}, t_a, \nabla_a)$, where M is a differentiable manifold, h^{ab} is a (0, 1, 1, 1) symmetric tensor field on M identified as a spatial metric, t_a is a covariant vector field on M which induces a (1, 0, 0, 0) temporal metric $t_{ab} = t_a t_b$, and ∇_a is a derivative operator associated with a (non-unique) connection on M and compatible with the metrics in the sense $\nabla_c h^{ab} = \nabla_a t_b = 0$. The spatial and temporal metrics are also required to be orthogonal in the sense $h^{ab} t_b = 0$. These conditions allow M to be decomposed into instantaneous three-dimensional spacelike hypersurfaces parameterized by a global time function. The most general classical spacetime symmetry group is generated by vector fields x^a that Lie annihilate h^{ab} and t_a. Symbolically, we require $\pounds_x h^{ab} = \pounds_x t_a = 0$, and again, this means that the transformations between reference frames defined by the integral curves of the vector fields x^a preserve the structure of the absolute spatial and temporal metrics. This entails that in any classical spacetime, there is always a unique way to separate time from space: any two observers moving inertially with respect to each other will always agree on the time interval between any two events, and on the spatial interval between any two simultaneous events. In this sense, space and time are *absolute* in a classical spacetime.

On the other hand, the compatibility conditions in a classical spacetime do not determine a unique curvature tensor. Additional constraints on the curvature may be imposed, and such constraints define different types of classical spacetimes. Two examples include Neo-Newtonian spacetime, characterized by $R^a_{bcd} = 0$, encoding spatiotemporal flatness; and Maxwellian spacetime, characterized by $R^{ab}_{cd} = 0$, encoding a rotation standard (Bain 2004, pp. 348–352). The symmetries of Neo-Newtonian spacetime form the 10-parameter Galilei group (*Gal*) generated by vector fields x^a that Lie annihilate the spatial and temporal metrics, and the connection. Symbolically, $\pounds_x h^{ab} = \pounds_x t_a = \pounds_x \Gamma^a_{bc} = 0$ (where $\Gamma^a{}_{bc}$ is the connection defined by ∇_a), and in coordinate form,

$$\begin{aligned} \mathbf{x} \to \mathbf{x}' &= R\mathbf{x} + \mathbf{v}t + \mathbf{a} \quad (Gal) \\ t \to t' &= t + b \end{aligned} \tag{2}$$

where R is a constant orthogonal rotation matrix, $\mathbf{v}$, $\mathbf{a} \in \mathbb{R}^3$ are velocity boost and spatial translation vectors, and $b \in \mathbb{R}$ is a time translation. The symmetries of Maxwellian spacetime are given by the infinite dimensional Maxwell group (*Max*) generated by vector fields x^a that Lie annihilate the spatial and temporal metrics and the rotational part of the connection. Symbolically, $\pounds_x h^{ab} = \pounds_x t_a = \pounds_x \Gamma_c^{ab} = 0$ (where $\Gamma_c^{ab} = h^{bd}\Gamma_{bc}^a$). In coordinate form,

$$\begin{aligned} \mathbf{x} \to \mathbf{x}' &= R\mathbf{x} + \mathrm{c}(t) \quad (Max) \\ t \to t' &= t + b \end{aligned} \tag{3}$$

where R is a constant orthogonal rotation matrix, $\mathbf{c}(t) \in \mathbb{R}^3$ is a time-dependent spatial boost vector, and $b \in \mathbb{R}$ is a time translation. A quick and dirty distinction between Neo-Newtonian and Maxwellian spacetime can be given in terms of the way the absolute spatial slices are "rigged": In Neo-Newtonian spacetime, the rigging consists of "straight" trajectories, whereas in Maxwellian spacetime, it consists of "straight" and "curved" trajectories. More precisely, a Neo-Newtonian connection can distinguish between a straight and a curved trajectory, whereas a Maxwellian connection cannot. Both connections can, however, distinguish between straight and curved trajectories on the one hand, and "corkscrew" trajectories on the other; i.e., in both spacetimes, there is an absolute standard of rotation.

Now, just as there can be different types of classical spacetimes, there can be different types of NQFTs. A GQFT (Galilei-invariant Quantum Field Theory), for instance, is an NQFT invariant under *Gal* (Lévy-Leblond 1967), while an MQFT (Maxwell-invariant Quantum Field Theory) is an NQFT invariant under *Max*. A slight variant of the latter is Christian's (1997) Newtonian quantum gravity reviewed in Sect. 3 below.

2.2 Particle Interpretations

According to a Received View (Arageorgis et al. 2003; Fraser 2008; Halvorson and Clifton 2002), in order to admit a particle interpretation, a QFT must satisfy the following two conditions.

(a) The QFT must admit a Fock space formulation in which local number operators appear that can be interpreted as acting on a state of the system associated with a bounded region of spacetime and returning the number of particles in that region.
(b) The QFT must admit a unique Fock space formulation in which a total number operator appears that can be interpreted as acting on a state of the system and returning the total number of particles in that state.

Condition (a) is supposed to encode the essential particle characteristic of *localizability*: For a system of particles distributed over various regions of space, an adequate theory must be able to identify the number of particles located in each region.[1] Condition (b) is supposed to encode the essential particle characteristic of *countability*: For a system of particles distributed over various regions of space, an adequate theory must be able to identify a *unique* value for the total number of particles, counted over *all* regions. (Schematically, one would hope that a unique total number operator could be defined as the sum over all regions of spacetime of local number operators.)

One can now demonstrate that Conditions (a) and (b) fail in RQFTs. The Received View concludes that RQFTs do not admit particle interpretations. However, it can also be shown that Conditions (a) and (b) hold in NQFTs precisely because of the existence of an absolute temporal metric in classical spacetimes. The moral I draw is that Conditions (a) and (b) are motivated by a non-relativistic notion of time, and hence are inappropriate in the relativistic context. What should be offered in their place as conditions of adequacy for particle interpretations in the relativistic context is best left to another essay. The remainder of this section attempts to substantiate the moral.

2.2.1 Particles in RQFTs?

It is a fairly simple matter to demonstrate that RQFTs fail to satisfy Conditions (a) and (b). In general, Condition (b) is made problematic by the existence of unitarily inequivalent Fock space representations of the canonical (anti-) commutation relations (CCRs) of an RQFT.[2] To the extent that unitary equivalence is necessary for physical equivalence, this suggests that any given RQFT admits (uncountably) many different ways to parse particle talk, one for every unitary equivalence class of Fock space representations and their attendant total number operator. One is faced with a problem of which representation to privilege (Ruetsche 2002, pg. 359). This Problem of Privilege may appear to be solved in Minkowski spacetime by appeal to the time-like isometry subgroup of the Poincaré group. Intuitively, the time-like symmetries of Minkowski spacetime provide one with a way to "split" the frequencies of solutions to relativistic field equations, and thereby construct a one-particle state space on which a Fock space representation can then be built. One can then show that this method of constructing a Fock space representation is unique up to unitary equivalence.

This is made rigorous by a result due to Kay (1979). Let (S, σ, D_t) be a classical phase space, where S is the space of (well-behaved) solutions to a field equation, σ

[1] This follows the intuitions of Halvorson and Clifton (2002, pp. 17–18). This aspect of the Received View should thus be made distinct from concepts of localized particles that require the existence of position operators and/or localized states.

[2] This is due to the failure of the Stone-von Neumann theorem for theories with infinite degrees of freedom.

is a symplectic form on S, and D_t: $S \rightarrow S$ is a one-parameter group of linear maps that preserves σ and represents the evolution of the classical system in time. A one-particle structure over $(S,\ \sigma,\ D_t)$ is a pair $(\mathcal{H},\ U_t)$, where $\mathcal{H}$ is a Hilbert space and U_t is a weakly continuous one-parameter group of unitary operators on $\mathcal{H}$ with positive energy[3], such that there is a 1-1 real linear map $K : S \rightarrow \mathcal{H}$ with the following properties: (a) The (complex) range of K is dense in $\mathcal{H}$; (b) $2\mathrm{Im}\langle Kf,\ Kg\rangle = \sigma(f,\ g)$ for all $f, g\ S$, where $\langle , \rangle$ is the inner product on $\mathcal{H}$; and (c) $D_t K = K U_t$. Kay (1979) proves that a one-particle structure associated with the classical Klein-Gordon field is unique up to unitary equivalence (similar results hold for the Dirac field). Thus, as Halvorson (2001, pg. 114) states, "... the choice of time evolution in the classical phase space suffices to determine uniquely the (first) quantization of the classical system."

However, if there is more than one choice of classical time evolution, there will be more than one choice of one-particle structure, and hence more than one unitary equivalence class of Fock space representations. Indeed, this occurs for classical fields defined over a portion of Minkowski spacetime referred to as the right Rindler wedge. The time-like isometry subgroup of the Poincaré group restricted to this portion admits two distinct time-like Killing vector fields, one associated with inertial reference frames and the other with accelerated frames. This gives rise to two unitarily inequivalent Fock space representations, the standard Minkowski representation, and the Rindler representation. This has suggested to some authors that inertial and accelerating observers will disagree over the particle content of an RQFT in Minkowski spacetime (see e.g., Wald 1994, Chap. 5). To such authors, then, the Problem of Privilege is not solved simply by appealing to Minkowski spacetime structure.[4]

Now suppose the Problem of Privilege could be solved to the satisfaction of all for non-interacting RQFTs in Minkowski spacetime. Haag's Theorem indicates that this would provide cold comfort for particle physicists engaged in experiments with what they take to be interacting particles. Under a reasonable assumption, Haag's Theorem entails that representations of the CCRs for both a non-interacting and an interacting RQFT cannot be constructed so that they are unitarily equivalent at a given time.[5] Provided, again, that unitary equivalence is a necessary condition for physical equivalence, this suggests that an interacting RQFT cannot be interpreted as consisting of a system of initially non-interacting particles that interact

[3] Such a U_t can be written $U_t = e^{itH}$ for H a positive operator.

[4] Arageorgis et al. (2003, pp. 180–181) argue that the Rindler representation is unphysical and hence, implicitly, that there is no Problem of Privilege for *physical* Fock space representations, appropriately construed, in Minkowski spacetime. They effectively argue that the time-like Killing vector field associated with accelerated frames in the right Rindler wedge should not count as a global way to "split the frequencies", in so far as it is not extendible to Minkowski spacetime as a whole.

[5] See, e.g., Earman and Fraser (2006, pg. 313). The reasonable assumption is that the representations admit unique Euclidean-invariant vacuum states. This assumption can be dropped by inserting a cut-off into the interacting RQFT and renormalizing the fields, but such tactics open up the host of conceptual problems afflicting renormalized field theories.

over a finite period of time, and then separate back into non-interacting states; a typical scenario for scattering experiments. More precisely, Haag's Theorem suggests that a Fock space representation of the CCRs of a non-interacting RQFT cannot be used to represent particle states in an interacting RQFT. One might then wonder if particle states might be represented more directly in an interacting RQFT by constructing an explicit Fock space representation of its CCRs, as opposed to piggy-backing on non-interacting representations. However, it is unclear if such a Fock space representation of the CCRs for an interacting RQFT is constructible (Fraser 2008).[6]

Thus is Condition (b) foiled in RQFTs, both non-interacting and interacting. Condition (a) is foiled in RQFTs by the consequences of the Reeh-Schlieder theorem. Briefly, the Reeh-Schlieder theorem entails that the vacuum state is *separating* for any local algebra of operators defined by an RQFT (Streater and Wightman 2000, pg. 138). This means that, given any bounded region of Minkowski spacetime, and any operator associated with that region (in the sense of being an element of the corresponding local operator algebra), if the operator annihilates the vacuum state, then it is identically zero. Now the annihilation operators that appear in Fock space formulations of QFTs are defined to annihilate the vacuum state and act non-trivially on other states. Thus separability of the vacuum state of an RQFT entails that there can be *no* annihilation operator associated with a bounded region of Minkowski spacetime; hence there can be no number operator associated with a bounded region of Minkowski spacetime. Thus "local" number operators in the sense of Condition (a) do not exist in RQFTs.

2.2.2 Particles in NQFTs?

In NQFTs, both free and interacting, Conditions (a) and (b) are satisfied, and one can argue that this is due to the presence of an absolute temporal metric in classical spacetimes. Consider Condition (b) first. What would guarantee uniqueness of a Fock space representation of the CCRs for a QFT is the presence of a unique global time function on the associated spacetime. This would provide a unique (up to unitary equivalence) means to construct a one-particle structure over the classical phase space. And such a unique global time function is only guaranteed in those spacetimes that admit an absolute temporal metric. To see this, note that the compatibility condition, $\nabla_a t_b = 0$, on the temporal metric of a classical spacetime entails t_a is closed, and thus locally exact. If M is topologically well-behaved (if, for instance, it is simply connected), then t_a is globally exact, and there exists a unique globally defined time function $t\colon M \to \mathbb{R}$ satisfying $t_a = \nabla_a t$. On the other hand, suppose there exists a unique global time function $t : M \to \mathbb{R}$. Then a temporal metric t_{ab} compatible with a connection ∇_a can be defined by $t_{ab} = (\nabla_a t)(\nabla_b t)$.

[6] Some authors have taken the moral of Haag's Theorem to be that (irreducible) representations of the CCRs are inappropriate for interacting RQFTs (Streater and Wightman 2000, pg. 101).

Thus there is no Problem of Privilege for non-interacting NQFTs. One can further demonstrate that Haag's theorem does not make trouble for interacting NQFTs, either. Haag's theorem entails the following necessary condition for the existence of an interacting quantum field unitarily equivalent to a free field: *Either the interaction polarizes the vacuum*[7] *or Poincaré-invariance does not hold* (Bain preprint). For non-relativistic quantum fields, the presence of an absolute temporal metric guarantees *both* the failure of Poincaré invariance *and* the failure of vacuum polarization. To see the latter, consider an interacting Hamiltonian $H = H_{free} + H_{int}$. Any representation of the symmetry group of a classical spacetime in which the time-translation generator is encoded in H will be unitarily equivalent (in the sense of satisfying the same commutation relations) to a representation in which the time-translation generator is encoded in H_{free}, provided that H_{int} is invariant under the group action. Thus if H_{free} annihilates the vacuum state, so will H. This does not hold true for the Lorentz group.[8]

An absolute temporal metric is also sufficient for Condition (a). While a version of the Reeh-Schlieder theorem can be proven in the NQFT context (Requardt 1982), it does not entail that the NQFT vacuum state is separating. Briefly, separability of the vacuum state for a local algebra $\mathfrak{R}(\mathcal{O})$ of operators associated with a region $\mathcal{O}$ of Minkowski spacetime is derived under the assumptions of vacuum cyclicity for $\mathfrak{R}(\mathcal{O})$ (guaranteed by the Reeh–Schlieder theorem), relativistic local commutativity, and the existence of a non-trivial causal complement of $\mathcal{O}$.[9] To extend this result to NQFTs, one must first replace relativistic local commutativity with its non-relativistic analogue.[10] This entails keeping track of the distinction between local algebras defined on *spatial* regions of spacetime, and those defined on *spatiotemporal* regions. Requardt's (1982) non-relativistic Reeh–Schlieder theorem only holds for the latter; but, due to the presence of an absolute temporal metric, spatiotemporal regions of classical spacetimes have trivial causal complements, and hence is separability denied.[11] On the other hand, the presence of a temporal metric also

[7] Vacuum polarization occurs when an interacting Hamiltonian fails to annihilate the vacuum state of the free field.

[8] Lévy-Leblond (1967, pp. 160–161) makes this comparison explicit for the particular case of the Galilei group. Due to the presence of an absolute temporal metric in Neo-Newtonian spacetime (and classical spacetimes in general), the commutation relations that define the Galilei Lie algebra (and the Lie algebra of any classical spacetime symmetry group in general) are such that the generator of time-translations is independent of the other generators. In the commutation relations that define the Lorentz Lie algebra, the time-translation generator is mixed up with the other generators.

[9] Streater and Wightman (2000, pg. 139). Vacuum cyclicity for $\mathfrak{R}(\mathcal{O})$ requires that for any operator $A \in \mathfrak{R}(\mathcal{O})$, $A\Omega$ is dense in $\mathcal{H}$, where Ω is the vacuum state. Relativistic local commutativity requires that local fields ϕ, ψ commute, $[\phi(f), \psi(g)] = 0$, when the supports of the test functions f, g are spacelike separated. The causal complement of a region $\mathcal{O}$ of Minkowski spacetime consists of all points spacelike separated from points in $\mathcal{O}$.

[10] Namely, $[\phi(f), \psi(g)] = 0$, when the supports of the test functions f, g have zero temporal and non-zero spatial separation (Lévy-Leblond 1967, pg. 164).

[11] The causal complement of a spatiotemporal region of a classical spacetime may be identified with the set of all points with zero temporal separation and non-zero spatial separation from points

guarantees that the domain of dependence for an open spatial region $\mathcal{S}$ of a classical spacetimes is just $\mathcal{S}$, and this ensures that the differential operators that appear in the parabolic PDEs of NQFTs are not anti-local for such spatial regions.[12] This has the consequence that the vacuum is not cylic for algebras associated with spatial regions, and thus is separability denied in this case, too.

3 Newtonian Quantum Gravity

While no RQFT currently exists that consistently incorporates gravity, Christian (1997) has constructed an NQFT that does. Not only is it an explicit example of an interacting NQFT that satisfies Conditions (a) and (b) of the Received View's concept of particle, it also is an instance of an NQFT in a curved classical spacetime. As such, it can be compared with QFTs in curved Lorentzian (i.e., relativistic) spacetimes.[13] This comparison will suggest, in Sect. 4, ways of extending RQFTs to incorporate gravity. This section first reviews the distinction between two particular theories of Newtonian gravity in flat and curved classical spacetimes, and then considers how Christian quantizes a particular version of the latter.

The standard way the theory of classical Newtonian gravity is formulated is as a field theory set against the backdrop of flat Neo-Newtonian spacetime. Models in this formulation may be given by a 6-tuple $(M,\ h^{ab},\ t_a,\ \nabla_a,\ \phi,\ \rho)$, where $(M,\ h^{ab},\ t_a,\ \nabla_a)$ represents classical Neo-Newtonian spacetime, and ϕ and ρ are scalar fields on M that represent a Newtonian potential field and a mass density, respectively. These latter objects are required to satisfy the Poisson equation, and an equation of motion:

$$h^{ab}\ \nabla_a\ \nabla_b\phi = 4\pi G\rho \quad (\textit{Poisson equation}) \tag{4}$$

$$\xi^a\ \nabla_a\xi^b = -h_{ab}\ \nabla_a\phi \quad (\textit{equation of motion}) \tag{5}$$

where G is the Newtonian gravitational constant, and ξ^a is a tangent vector field for a timelike particle trajectory worldline that encodes its four-velocity.

---

in the region. This assumes a prohibition on infinite causal propagations, but allows that finite causal propagations have no upper bound.

[12] The domain of dependence $D(\mathcal{O})$ of a region $\mathcal{O}$ of spacetime consists of points p for which any inextendible causal worldline through p intersects $\mathcal{O}$. A differential operator is said to be *anti-local* for a given region of spacetime just when a function and its transform under the operator can vanish in that region only if the function is identically zero. In classical spacetimes, for any open spatial region $\mathcal{S}$, $D(\mathcal{S})$ has no temporal extent. Thus if a solution ϕ to a well-posed PDE vanishes on $\mathcal{S}$, it vanishes on $D(\mathcal{S})$, but this does not guarantee that it vanishes on an open set in time. This blocks an inference to anti-locality by means of the Edge of the Wedge Theorem. One can further demonstrate that anti-locality of a differential operator entails cyclicity of the associated vacuum state. Segal and Goodman (1965) demonstrated this for the case of the Klein-Gordon operator, and subsequent authors have extended their results to cover operators associated with other relativistic field equations.

[13] A Lorentzian spacetime is a pair $(M,\ g_{\mu\nu})$, where M is a differentiable manifold and $g_{\mu\nu}$ is a metric defined on M with signature $(-1, 1, 1, 1)$.

One can also formulate Newtonian gravity by incorporating the gravitational potential field into the spacetime connection, and such theories are referred to as theories of Newton-Cartan gravity (NCG). Models of NCG eliminate the Newtonian gravitational potential, and may be given by $(M,\ h^{ab},\ t_a,\ \nabla_a,\ \rho)$. Here the objects $(M,\ h^{ab},\ t_a,\ \nabla_a)$ still represent a classical spacetime; in particular, the spatial and temporal metrics still satisfy orthogonality and compatibility constraints, and additional constraints may still be imposed on the curvature tensor defined by the derivative operator $_a$. But the Poisson equation (4) is now replaced with a generalized Poisson equation, and the equation of motion (5) is replaced with the geodesic equation:

$$R_{ab} = 4\pi G\rho t_a t_b \quad \textit{(generalized Poisson equation)} \tag{6}$$

$$\xi^a\ \nabla_a \xi^b = 0 \quad \textit{(equation of motion)} \tag{7}$$

where R_{ab} is the Ricci tensor defined, ultimately, by the derivative operator ∇_a. These changes enforce the principle of equivalence in NCG. Intuitively, the Newton-Cartan connection defined by (6) and (7) cannot distinguish "straight" inertial trajectories from "curved" gravitationally accelerated trajectories. In this sense, gravity is geometricized in NCG. Now there are different ways this geometrization procedure can be carried out, depending on additional constraints one might impose on the curvature tensor. Christian (1997) considers the following two constraints:

$$R^{[a\ c]}_{[b\ d]} = 0 \tag{8}$$

$$R^{ab}_{cd} = 0 \tag{9}$$

Let "strong NCG" refer to the theory of NCG that, in addition to the compatibility and orthogonality constraints of classical spacetimes, satisfies (6), (7), (8), (9), and call the classical spacetime associated with it strong Newton-Cartan spacetime. In strong Newton-Cartan spacetime, as in all classical spacetimes, there is a global time function that may be associated with absolute time, and there are globally defined spatial slices that may be interpreted as absolute space at an instant. And as with other examples of curved classical spacetimes, what is "curved" is the way these spatial slices are rigged together by the connection. Recall in Maxwellian spacetime, the rigging is determined by condition (9) above and consists of either "straight" or "curved" trajectories (a Maxwellian connection cannot tell these apart), but not "cork-screw" trajectories (there still is a standard of rotation). In strong Newton-Cartan spacetime, "curved" rigging is restricted to gravitationally accelerated trajectories, subject to the additional condition (8). More precisely, whereas the symmetries of Maxwellian spacetime are characterized by the Maxwell group (3), those of strong Newton-Cartan spacetime are characterized by an extension of the Maxwell group, and thus are slightly more constrained.[14]

[14] See, e.g., Bain (2004, pg. 372).

Christian (1997) demonstrates that Conditions (8) and (9) are sufficient to recast strong NCG as a constrained Hamiltonian system, and thus to quantize it. The reduced phase space (Christian 1997, pg. 4867) consists of variables encoding the matter degrees of freedom, and variables that encode the dynamical degrees of freedom of the strong NCG connection, which are identified as gravitational degrees of freedom. The matter variables are solutions to the Schrödinger equation in strong Newton-Cartan spacetime.[15] The connection variables take the form of extended Maxwell frames; i.e., rigid, non-rotating, gravitationally accelerating frames. This phase space has a nondegenerate symplectic structure, and a unique one-parameter family of time evolution maps (due to the absolute temporal metric of strong Newton-Cartan spacetime). Hence it admits a unique one-particle structure, and thus a unique Fock space representation of the CCRs. The result is Christian's Newtonian Quantum Theory of Gravity (NQG, hereafter), an interacting (extended) Maxwell-invariant QFT set in strong Newton Cartan spacetime.

NQG is a concrete example of an interacting NQFT that satisfies the Received View's necessary Conditions (a) and (b) for a particle interpretation. It is also an interacting NQFT that successfully incorporates gravity; in particular, the gravitational degrees of freedom in NQG are *both* fully dynamical and fully quantized. This is in stark contrast with attempts to incorporate gravity into RQFTs. For instance, the fact that the NQG gravitational degrees of freedom are fully *dynamical* distinguishes NQG from the program of QFTs in curved Lorentzian spacetimes. This program attempts to construct RQFTS that incorporate gravity by treating it classically as a manifestation of the curvature of spacetime. This is done by breaking the dynamical link between spacetime and matter forged in general relativity. The curved Lorentzian spacetime in such an RQFT is absolute in the sense that it has no dynamical degrees of freedom. In NQG, on the other hand, strong Newton-Cartan spacetime has quantized *dynamical* degrees of freedom; namely, those associated with the quantized strong Newton-Cartan connection. Intuitively, these quantized degrees of freedom are associated with the dynamical "rigging" of the absolute spatial slices. Moreover, as indicated above, NQG does not face the Problem of Privilege in determining a Fock space representation of the CCRs: the absolute temporal metric of strong Newton-Cartan spacetime decides the matter uniquely up to unitary equivalence. This is in contrast to QFTs in curved (Lorentzian) spacetimes in which

[15] Christian (1997, pg. 4855) refers to this as the Schrödinger-Kuchar equation after Kuchar (1980), who demonstrated that it can be quantized to produce a non-interacting Galilei-invariant NQFT in strong Newton-Cartan spacetime. Christian's NQG is an extention of Kuchar's non-interacting theory to one in which the quantized Schrödinger field interacts with a quantized strong Newton-Cartan connection field (thus Christian's NQG is a fully interacting NQFT). The key to this extention is Christian's construction of a Lagrangian density that produces not just the Schrödinger-Kuchar equation, but also the field equations of Strong NCG. In particular, all Lagrangian densities associated with NCG prior to Christian (1997) failed to recover the generalized Poisson equation (6).

there is not even a guarantee that the spacetime will admit time-like isometries in the first place.[16]

Finally, note that the fact that the NQG gravitational degrees of freedom are fully *quantized* distinguishes NQG from semi-classical approaches to incorporating gravity into RQFTs. These approaches attempt to include dynamical degrees of freedom associated with the gravitational field into an RQFT by replacing the stress-energy tensor in the Einstein equations with its expectation value with respect to quantized matter fields. In such approaches, one treats gravity classically (the metric is not quantized), but one quantizes the matter fields.

4 Intertheoretic Relations

NQG has suggested to Christian (1997, 2001) a novel route to formulating a fully relativistic quantum theory of gravity (QG, hereafter); namely, by relativizing NQG. This section reviews this strategy and expands on Christian's picture of intertheoretic relations associated with it. In particular, the existence of NQFTs suggests modifications to Christian's picture, which open up additional routes to QG. Since a full investigation of all such additional routes is beyond the scope of the current essay, this section will content itself with an initial explorative expedition.

To begin, Christian (1997, pg. 4847; 2001, pg. 307) views NQG as a means to fill a void in the "great dimensional monolith of physics". This is a diagrammatic representation of the relations between fundamental theories in physics. It takes the form of a cube with axes representing the Newtonian gravitational constant G, Planck's constant h, and the inverse speed of light $1/c$ (see Fig. 1).

The vertices of Christian's cube are meant to represent the following theories: classical mechanics (CM), special relativity (SR), general relativity (GR), Newton-Cartan gravity (NCG), Newtonian quantum gravity (NQG), Galilei-invariant quantum mechanics (GQM), relativistic quantum field theory (RQFT), and fully-relativistic quantum gravity (QG). Schematically, these theories can be described by their coordinates $(G, h, 1/c)$ in monolith space. GR, for instance, may be given the coordinates $(1, 0, 1)$, indicating that G and $1/c$ are "turned on", whereas h is "turned off". The cube thus entails that there are three distinct approaches to constructing QG: quantizing GR (epitomized in "background independent" approaches like loop quantum gravity); "turning on" gravity in an RQFT (epitomized in "background dependent" approaches like string theory); and the approach, novel to Christian (1997), of "relativizing" NQG.

[16] As Ruetsche (2002, pg. 361) notes, one way practitioners have attempted to address this problem is by becoming "algebraic imperialists" and elevating the status of the underlying abstract C^*-algebra over concrete Hilbert space realizations of it. (Doing so provides one access to notions of "physical equivalence" weaker than unitary equivalence.) This strategy is adopted by Christian (1997, pg. 4870) as a way of interpreting NQG, but this seems unnecessary, given that NQG does not face the problem of privilege in the first place.

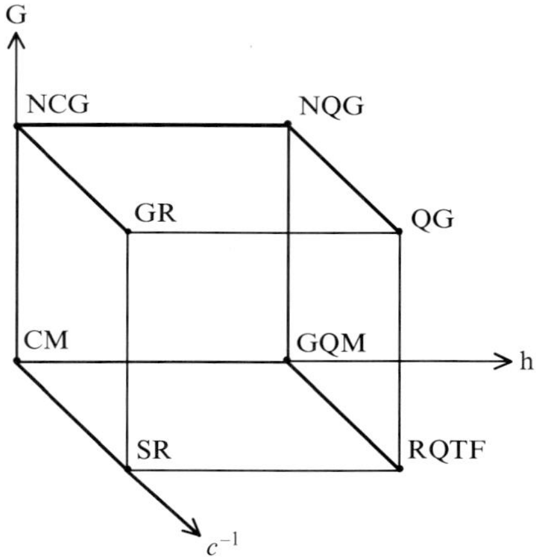

Fig. 1 Christian's (1997) dimensional monolith

To better understand Christian's monolith, and ways of extending it, requires understanding the nature of the limits that define the links in Fig. 1. Under closer inspection, multiple problems arise.

(a) First, the $1/c \rightarrow 0$ limit that "turns off" relativity might initially be thought of as a contraction of the Poincaré group to obtain the Galilei group (see e.g., Bacry and Lévy-Leblond 1968). However, more than one such limit can be taken for a given relativistic theory. Such limits depend in particular on the form of the dynamical equations of the theory. For instance, there are two distinct non-relativistic limits of the Maxwell equations (Holland and Brown 2003). Moreover, the $1/c \rightarrow 0$ link between GR and NCG cannot be described by a group contraction. On the one hand, the Poincaré group is not the symmetry group associated with GR (under one interpretation, the latter is Diff(M)). On the other hand, as Sect. 3 indicates, there is more than one version of NCG, depending on how the geometrization procedure is carried out. One of these versions can indeed be shown to be the $1/c \rightarrow 0$ limit of GR, but this version does not have the Galilei group as its symmetry group.[17]
(b) The $G \rightarrow 0$ limit might be associated simply with setting G to zero in the relevant dynamical equation (thus "turning off" gravity). But this would make the link between GR and SR problematic. Setting G to zero in the Einstein equations results in a Ricci-flat ($R_{ab} = 0$) Lorentzian spacetime, whereas Minkowski spacetime is spatiotemporally flat ($R^a{}_{bcd} = 0$). (Note that Ricci-flatness only entails spatiotemporal flatness in conformally flat (4-dim)

[17] This version can be referred to as "weak NCG" (Bain 2004, pg. 346). It differs from strong NCG by dropping Condition (8). Bain (2004, pg. 365) identifies the symmetry group of weak NCG with an extention of the Leibniz group, another classical spacetime symmetry group.

spacetimes, in which the Weyl tensor vanishes.) This problematizes the other $G \rightarrow 0$ links as well, in so far as there can be Ricci-flat classical spacetimes other than Neo-Newtonian spacetime, which, presumably, is the spacetime of CM and GQM.

(c) Finally, one might describe the $h \rightarrow 0$ limit as the inverse of quantization. But just how the quantization procedure should be characterized is far from settled. For instance, the quantization procedure that represents the link between SR and RQFT is not unique: For a theory of a classical relativistic field with infinite degrees of freedom, the failure of the Stone-von Neuman theorem entails that there are uncountably many unitarily inequivalent representations of the CCRs of the corresponding QFT. Furthermore, inequivalent quantizations are not only associated with systems with infinite degrees of freedom; they also arise for finite systems with topologically non-trivial state spaces.[18] This problematizes the link between CM and GQM, as well as the link between NCG and NQG (in the latter case, for topologically trivial gravitational fields, appeal to the unique global time function in classical spacetimes solves the Problem of Unique Quantization (viz., Privilege), as explained in Sect. 3).

In addition to these issues with the extant links in Christian's diagram, there also seems to be a deeper, structural problem. This problem manifests itself explicitly in the links between NQG and GQM, and RQFT and GQM:

1. First, Christian's NQG is an NQFT that incorporates gravity. Thus, one might expect that turning off gravity would result in an NQFT *sans* gravity. One might then wonder about the referent of "GQM": Is it meant to include infinite-dimensional non-relativistic quantum theories (viz, NQFTs) as well as finite-dimensional non-relativistic quantum theories (viz, non-relativistic quantum particle dynamics)? And moreover, it is not immediately clear that it should refer to a Galieli-invariant theory.
2. A second related concern involves the link between RQFT and GQM. The $1/c \rightarrow 0$ limit of an RQFT might be characterized by a contraction of the Poincaré group to yield the Galilei group, with the qualifications mentioned above. But this maneuver by itself does not take us from an RQFT to a theory of GQM, if we allow that the latter includes theories with finite degrees of freedom.

These concerns stem from the fact that NQFTs are missing from Christian's diagram. NQFTs may be thought of as appropriately qualified $1/c \rightarrow 0$ limits of RQFTs. Now suppose we relabel Christian's GQM as NQM (Non-relativistic Quantum Mechanics) and restrict its referent to finite-dimensional non-relativistic quantum theories of particle dynamics (i.e., finite theories of quantum particles invariant under the symmetry group of a classical spacetime). Then, for $N =$

[18] An example of such a system is a charged particle moving in a region external to an operating solenoid. Quantization of this system produces the Aharonov-Bohm effect (see e.g., Belot 1998, pg. 546).

degrees of freedom, NQMs may be thought of, schematically, as the "inverse thermodynamic" limit $N \to 0$ of NQFTs. This limit is intended to be applicable to quantum theories independently of classical theories, and vise-versa (i.e., it is intended to be "orthogonal" to the $h \to 0$ limit). So, for instance, it should also hold between a classical theory with an infinite number of degrees of freedom (a non-relativistic classical field theory, for instance), and a classical theory with finite degrees of freedom (a non-relativistic classical theory of particle dynamics, for instance). Whether such a limit can be precisely defined is a matter for another essay.[19] What it informally suggests is that Christian's cube should be replaced by a 4-dim hypercube with an additional axis representing degrees of freedom N. Suppressing the G-dimension, we then have the diagram in Fig. 2.

The vertices in Fig. 2 represent the following theories: non-relativistic classical particle mechanics (NCM), relativistic classical particle mechanics (RCM), non-relativistic classical field theory (NCFT), relativistic classical field theory (RCFT), non-relativistic quantum particle mechanics (NQM), relativistic quantum particle mechanics (RQM), non-relativistic quantum field theory (NQFT), and relativistic quantum field theory (RQFT). The distinctions here are between theories (classical and quantum, relativistic and non-relativistic) with infinite degrees of freedom, and theories (classical and quantum, relativistic and non-relativistic) with finite degrees of freedom.[20]

Theories in hypermonolith space are coodinatized by 4-tuples $(G, h, 1/c, N)$. There are now four distinct approaches to constructing relativistic QG: quantizing the classical field theory of GR, with coordinates (1, 0, 1, 1); "turning on" gravity in an RQFT with coordinates (0, 1, 1, 1); "relativizing" a non-relativistic QFT of gravity (such as Christian's NQG) with coordinates (1, 1, 0, 1); or "taking the thermodynamic limit" of a relativistic quantum particle theory of gravity with coordinates (1, 1, 1, 0). Just what the latter might involve requires further analysis.

As an example of how this investigation might proceed, consider how the eight $G \to 0$ links in the hypercube could be fleshed out (these all end in the vertices/theories that appear in Fig. 2). They may be divided into links in which gravity is turned off in a field theory, and links in which gravity is turned off in a particle theory.

1. (Non-relativistic classical field theory of gravity (1, 0, 0, 1)) $\to$ NCFT.
   An example of a theory with coordinates (1, 0, 0, 1) that produces an NCFT in the $G \to 0$ limit is asymptotically flat weak NCG. This is a version of

[19] Landsman (2007) discusses a rigorous way of defining an $N \to \infty$ limit that holds between a quantum system with N degrees of freedom and a classical system. The definition makes use of the C^*-algebra formulation of quantum and classical systems. This formalism also admits a rigorous definition of an $h \to 0$ limit, and Landsman notes that the former limit is a special case of the latter.

[20] For simplicity's sake, the former are identified as field theories and the latter as particle theories. This ignores field-theoretic systems on lattices (with finite degrees of freedom), as well as particle systems with infinitely many particles; and it also glosses over conceptual issues concerning the nature of a particle vis-à-vis a field; but nothing in the following hangs on this simplifying means of expediency.

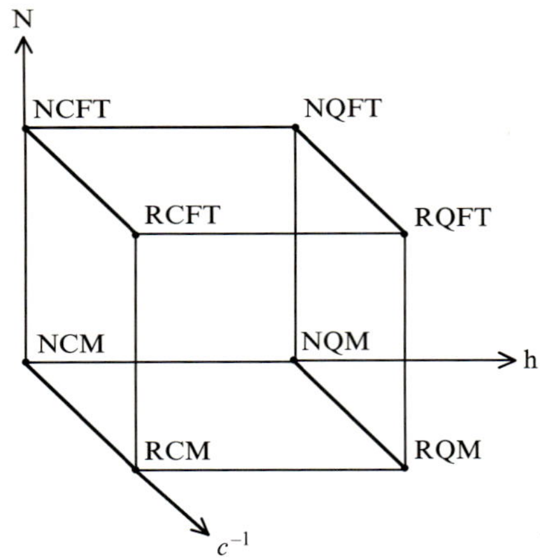

Fig. 2 Relations between theories in the absence of gravity

NCG that drops condition (9) in Sect. 3 above, and imposes asymptotic spatial flatness to enforce Galilei-invariance. Bain (2004, pg. 358) indicates that it is empirically equivalent to a version of (non-geometricized) Newtonian gravity in Neo-Newtonian spacetime in which an "island universe" boundary condition is imposed (namely, $\phi \rightarrow 0$ at spatial infinity). Hence turning off gravity in asymptotically flat weak NCG is equivalent to turning off gravity in (non-geometricized) Newtonian gravity in Neo-Newtonian spacetime under the island universe assumption, and this evidently yields a Galilei-invariant classical field theory in Neo-Newtonian spacetime.

2. (RCFT of gravity (1, 0, 1, 1)) → RCFT.
   GR is a theory with coordinates (1, 0, 1, 1). Turning off gravity in GR results in a field theory in a Ricci-flat Lorentzian spacetime (providing non-gravitational fields are present). This does not by itself guarantee the theory is Poincaré-invariant. To assure coherence here, one might additionally impose the requirement of conformal flatness (although whether this can be motivated on physical or other grounds remains to be seen). Alternatively, one might simply expand one's concept of a relativistic theory to include theories invariant under the symmetries of Lorentzian spacetimes in general.
3. (NQFT of gravity (1, 1, 0, 1)) → NQFT.
   NQG is an NQFT of gravity in strong Newton-Cartan spacetime. Evidently, turning off gravity yields an NQFT in a Ricci-flat classical spacetime satisfying conditions (8) and (9).
4. (RQFT of gravity (1, 1, 1, 1)) → RQFT.
   The expectation here is that the full-blown relativistic theory of quantum gravity will reproduce a relativistic quantum field theory in the limit of no gravity (just as it should produce GR in the classical limit).

The remaining four links involve turning off gravity in a particle theory:

5. (Non-relativistic classical particle theory of gravity (1, 0, 0, 0)) → NCM.
6. (Non-relativistic quantum particle theory of gravity (1, 1, 0, 0)) → RCM.
7. (Relativistic classical particle theory of gravity (1, 0, 1, 0)) → NQM.
8. (Relativistic quantum particle theory of gravity (1, 1, 1, 0)) → RQM.

Whether examples of all the theories on the left hand side in links 5–8 can be identified is best left to another essay, with particular interest directed at an example of Link 8. Such an example, together with an appropriately formulated thermodynamic limit that links field theories with particle theories, would open up a fourth route to the elusive fully relativistic theory of quantum gravity.

5 Conclusion

This essay has used the distinction between Minkowski spacetime and classical spacetimes as a tool to probe two contemporary issues in philosophy of quantum field theory; namely, the debate over particle interpretations of RQFTs, and the status of approaches to a fully relativistic quantum theory of gravity. First, the distinction between Minkowski spacetime and classical spacetimes suggested a distinction between RQFTs and NQFTs which in turn suggested that the concept of particle that a Received View adopts in arguing against particle interpretations of RQFTs is motivated by a non-relativistic notion of absolute time. Second, the existence of NQFTs, and in particular, consistent NQFTs of gravity, also suggested that routes to fully relativistic quantum gravity are more varied than the current literature suggests.

Finally, a general moral can be drawn. The existence of NQFTs suggests that the distinction between relativistic and non-relativistic theories should not be couched in terms of Poincaré-invariance vs. Galilei-invariance. On the one hand, as is already evident in GR, a relativistic theory need not be Poincaré-invariant. On the other hand, as is evident in NQFTs, a non-relativistic theory need not be Galilei-invariant. The discussion in Sect. 4 of this essay suggests that a more appropriate distinction should be based on theories that are invariant under the symmetries of a Lorentzian spacetime vs. theories that are invariant under the symmetries of a classical spacetime.

References

Arageorgis, A., J. Earman, and L. Ruetsche (2003) 'Fulling Non-uniquenes and the Unruh Effect: A Primer on Some Aspects of Quantum Field Theory', *Philosophy of Science, 70*, 164–202.

Bain, J. (2010) 'Quantum Field Theories in Classical Spacetimes and Particles', forthcoming in *Studies in History and Philosophy of Modern Physics*.

Bain, J. (2004) 'Theories of Newtonian Gravity and Empirical Indistinguishability', *Studies in History and Philosophy of Modern Physics, 35*, 345–376.

Bacry, H. and J.-M. Lévy-Leblond (1968) 'Possible Kinematics', *Journal of Mathematical Physics, 9*, 1605–1614.

Belot, G. (1998) 'Understanding Electromagnetism', *British Journal for the Philosophy of Science, 49*, 531–555.

Christian, J. (2001) 'Why the Quantum Must Yield to Gravity'. In C. Callender and N. Huggett (eds.), *Physics Meets Philosophy at the Planck Scale* (pp. 204–338). Cambridge: Cambridge University Press.

Christian, J. (1997) 'Exactly Soluble Sector of Quantum Gravity, *Physical Review D, 56*, 4844–4877.

Earman, J. (1989) *World Enough and Spacetime*, Cambridge: MIT.

Earman, J. and D. Fraser (2006) 'Haag's Theorem and its Implications for the Foundations of Quantum Field Theory', *Erkenntnis, 64*, 305–344.

Fraser, D. (2008) 'The Fate of "Particles" in Quantum Field Theories with Interactions', *Studies in History and Philosophy of Modern Physics, 39*, 841–859.

Halvorson, H. (2001) 'Reeh-Schlieder Defeats Newton-Wigner: On Alternative Localization Schemes in Relativistic Quantum Field Theory', *Philosophy of Science, 68*, 111–133.

Halvorson, H. and R. Clifton (2002) 'No Place for Particles in Relativistic Quantum Theories?', *Philosophy of Science, 69*, 1–28.

Holland, P. and H. Brown (2003) 'The Non-Relativistic Limits of the Maxwell and Dirac Equations: The Role of Galilean and Gauge Invariance', *Studies in History and Philosophy of Modern Physics, 34*, 161–187.

Kay, B. (1979) 'A Uniqueness Result in the Segal-Weinless Approach to Linear Bose Fields', *Journal of Mathematical Physics, 20*, 1712–1713.

Kuchar, K. (1980) 'Gravitation, Geometry, and Nonrelativistic Quantum Theory', *Physical Review D*, 22, 1285–1299.

Landsman, N. (2007) 'Between Classical and Quantum'. In J. Butterfield and J. Earman (eds.) *Handbook of the Philosophy of Physics*, Amsterdam: North-Holland, 417–554.

Lévy-Leblond, J.-M. (1967) 'Galilean Quantum Field Theories and a Ghostless Lee Model', *Communications in Mathematical Physics, 4*, 157–176.

Requardt, M. (1982) 'Spectrum Condition, Analyticity, Reeh-Schlieder and Cluster Properties in Non-Relativistic Galilei-Invariant Quantum Theory', *Journal of Physics A, 15*, 3715–3723.

Ruetsche, L. (2002) 'Interpreting Quantum Field Theory', *Philosophy of Science, 69*, 348–378.

Segal, I. and R. Goodman (1965) 'Anti-locality of Certain Lorentz-Invariant Operators', *Journal of Mathematics and Mechanics, 14*, 629–638.

Streater, R. and A. Wightman (2000) *PCT, Spin and Statistics, and All That*, Princeton: Princeton University Press.

Wald, R. (1994) *Quantum Field Theory in Curved Spacetimes and Black Hole Thermodynamics*, Chicago: Chicago University Press.

Ether, the Theory of Relativity and Quantum Physics

Eduardo V. Flores

Abstract In this paper we revisit some of the reasons given by Einstein that resulted in his change of mind about the ether from denying to defending its existence. The ether proposed by Einstein we call Einstein's new ether. We consider the potential use of Einstein's new ether in quantum mechanics. The standard model of elementary particles reveals the existence of at least one component of Einstein's new ether. In this work we explore additional properties of Einstein's new ether. In particular, we consider a recent experiment known as the Afshar experiment due to its implications for the wave particle duality paradox. The Afshar experiment is perhaps the first experiment that provides clear evidence that wave and particle aspects of the photon have some sort of physical reality beyond the limits imposed by complementarity. We propose that the physical reality of the wave aspect of the photon has its origin in Einstein's new ether. Here, we report on consequences of the Afshar experiment for Einstein's new ether.

1 Introduction

"Ether and the Theory of Relativity" was the title of an Address delivered on May 5th, 1920, in the University of Leyden by Albert Einstein [1]. The central point of the address was the need to retain some kind of physical ether. Einstein's new proposal would have been unthinkable some 15 years earlier when he wrote his famous paper on special relativity. In 1905 Einstein was the first to realize that physicists should abandon the fruitless and misleading concept of the ether. In essence, he accepted the apparent fact that light propagates through vacuum, and that vacuum is really empty [2]. This was due to Einstein's study of the nineteenth century physics of the theory of ether which got him to a series of contradictions and difficulties with the ether that led him to deny its existence. But his experience dealing with general relativity and his philosophy of natural phenomena led him to state at the address [1]: "More careful reflection teaches us, however, that the special theory of relativity does not compel us to deny ether." Later he adds: "We shall see later that this point of view is justified by the results of the general theory of relativity." He concluded

V. Petkov (ed.), *Space, Time, and Spacetime*, Fundamental Theories of Physics 167,
DOI 10.1007/978-3-642-13538-5_6,

that "To deny the ether is ultimately to assume that empty space has no physical qualities whatever."

The ether proposed by early researchers fails to describe experimental observation when it comes to detecting its presence through relative motion. The problem seems to stem not from the physical reality of the ether but from the erroneous physical properties ascribed to it. This point is made clear in a recent article Barceló and Jannes who study condensed matter systems that play the role of a medium for its excited states [3]. When the collective excitations of a condensed matter system are described by fields satisfying equations of motion formally indistinguishable from those of relativistic field theory then relativistic effects, similar to those of the special theory of relativity, arise naturally. In their article they claim that "By proposing a thought experiment based on the construction of a Michelson-Morley interferometer made of quasi-particles, we show that a real Lorentz-FitzGerald contraction takes place, so that internal observers are unable to find out anything about their 'absolute' state of motion. Therefore, we also show that an effective but perfectly defined relativistic world can emerge in a fishbowl world situated inside a Newtonian (laboratory) system...". It is remarkable that even in the background of a Newtonian world observers may not be able to identify their state of uniform motion with respect to the medium if the collective excitations in this medium can be described by relativistic-like equations of motion. We do not propose that the ether is made of regular matter but the example from the condensed matter field shows that relativistic-like mediums are possible even for regular matter.

Einstein's view of the physical ether was that it could not be made of regular matter since he supposed it was more fundamental than regular matter. We propose here that Einstein's ether is a new physical object. We define the components of Einstein's new ether as whatever is left in a given region of spacetime after all regular matter and radiation has been removed. In his address Einstein pointed out the first property of the ether [1]: "We know that it determines the metrical relations in the space-time continuum." Thus, we propose that the first fundamental property of Einstein's new ether is to determine the metrical relations of spacetime.

There is physical evidence that Einstein's new ether exists. Astronomical observations reveal of the presence of a non-zero cosmological constant or dark energy [4]. Einstein's equation of general relativity shows that the cosmological constant originates not from regular matter but from empty space or vacuum energy [5]. The standard model of elementary particles uncovers two important sources of vacuum energy: the Higgs mechanism for mass generation and the zero point energy of every field [6]. The zero point energy of the vacuum due to the electromagnetic field has a measurable physical effect known as Casimir's effect [7]. Thus, the existence of at least one component of Einstein's new ether, the non-zero vacuum energy, is practically undeniable. Therefore, it is clear that Einstein's new ether as defined here exists.

The study of Einstein's new ether would be far more relevant if it could shed light on aspects of quantum theory that need resolution. An important aspect of quantum mechanics that needs clarification is the wave-particle duality paradox. The wave-particle paradox as embodied in Bohr's principle of complementarity has been a cornerstone in the interpretation of quantum mechanics since its inception [8]. Einstein was an advocate of physical reality and was troubled by the status and interpretation of quantum mechanics in his days. Fortunately, there have been interesting developments on the complementarity issue. For instance, it is clear now that in some cases complementarity is bound to the uncertainty principle [9], in other cases complementarity is independent of the uncertainty principle but it is related to quantum entanglements between the detectors and the particle [10], and in other cases it appears that complementarity could be circumvented [11, 12]. The last possibility is the case of a new experiment known as the Afshar experiment [12, 13]. Wave and particle aspects appear to coexist for the photon beyond the limitations of complementarity. One of the implications of the Afshar experiment is the high level of physical reality of the wave and particle aspects of the photon. It is not easy to describe the physical reality of a wave associated with a particle, but some properties may be derived from experimental constraints. From the Afshar experiment and other quantum mechanical experiments we know that the wave aspect does not originate from regular matter. However, as for any wave propagation with physical reality a medium or ether is required. We propose that Einstein's new ether provides the medium for the wave phenomenon as revealed by the Afshar experiment.

The major contribution of this work is the discovery of some properties of Einstein's new ether from the Afshar experiment. Thus, in Sect. 2 we give an extensive summary of the Afshar experiment. In Sect. 3 we report on the consequences for the photon from the results of the Afshar experiment. In Sect. 4 we propose that the origin of the physical reality of the wave aspect of the photons is to be found in Einstein's new ether and we enumerate properties of the wave aspect derived from experimental observation. In Sect. 5 we give an overall description of Einstein's new ether and make some final remarks.

2 The Afshar Experiment

Afshar et al. reported experimental work that provides evidence of simultaneous particle and wave aspects of the photon beyond the limits set by complementarity [12]. The paradoxical results of the Afshar experiment have produced some debate on the validity and interpretation of the experiment [14–24]. These paradoxical findings have been fully explained only recently [28]. However, the Afshar experiment result that particle and wave coexistence is possible in quantum mechanics still stands. We note that particle and wave coexistence is a principle of Bohmian mechanics [29]. Bohmian mechanics has been claimed to have an output equivalent to standard quantum mechanics at the non-relativistic level. The relevant outcome of the Afshar experiment for our purpose is the evidence of some sort of physical reality associated with quantum waves beyond the limitations imposed by complementarity.

Since the results of the Afshar experiment are crucial to this work, we present a summary of the experiment. It turns out that a version known as the modified

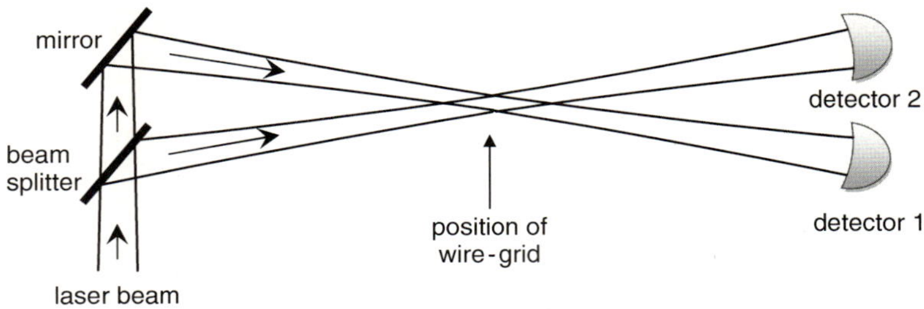

Fig. 1 Modified Afshar experiment

Afshar experiment is a simpler and more transparent version of the Afshar experiment for calculation and analysis purposes [16]. A laser beam impinges on a 50:50 beam splitter and produces two spatially separated coherent beams of equal intensity (Fig. 1). The beams overlap at some distance. Beyond the region of overlap the two beams fully separate again. There, two detectors are positioned such that detector 1 detects only the photons originating from the mirror, and detector 2 detects only photons originating from the beam splitter. Where the beams overlap the waves interfere forming a pattern of bright and dark fringes. At the center of the dark fringes thin wires may be placed.

When the wires are not present, photons are free all the way from the mirror or beam splitter to its corresponding detector. Thus, the application of momentum conservation is straight forward. When a detector clicks, momentum conservation allows us to identify the particular path the photon took. We associate knowledge of the path of the photon with the which-way information parameter K. Thus, once a detector clicks we have full which-way information, $K = 1$, about the path of the photon from the time it enters the interferometer until it hits the detector. The wave aspect is gauged by the presence of interference fringes. When interference fringes are formed we measure them using the visibility parameter V. If we do not verify wave interference by a measurement we may not assume the presence of interference fringes and we write, $V = 0$. In our particular case we get $V^2+K^2 = 1$ which is in agreement with the Greenberger-YaSin inequality, $V^2 + K^2 \leq 1$, a modern version of the principle of complementarity [25].

However, Afshar [13] proposed a way to test whether or not an interference pattern is present when the beams cross. The idea is simple: place thin wires at the center of the presumed dark fringes. If the wires scatter a considerable number of photons then there is no interference but if the readings at the detectors hardly change then there is evidence of destructive interference at the location of the wires. Experimental evidence shows that there are dark fringes. From the data it is fairly easy to set a lower limit for the visibility of the interference pattern, $V \geq 0.968$ [12, 20, 23]. Thus, there is evidence of a sharp interference pattern at the location of the wires. Interference is a reliable indicator of the wave aspect of the photon. Since the wires are so small and do not seem to interact significantly with the photons it is expected that the which-way information has not been altered significantly, $K \approx 1$. Therefore, we obtain the interesting result, $V^2 + K^2 \geq 1.93$. However, it is still possible that the wires have dramatically changed the which-way information of the photons by randomly sending photons to either detector. This is a possibility that has to be checked with a calculation.

We have performed a calculation that shows the change in photon momentum due to thin wires located at the center of dark fringes [23]. In our calculation we simulate the conditions of the experiment performed by Afshar et al. so as to be able to compare results. Thus, we use light with $\lambda = 638$ nm. We use six wires each with a thickness of 32 μm, the center to center wire separation is 319 μm and the beam width is 3.22 mm. Two laser beams propagate symmetrically on a plane and cross each other at an angle of 0.002 radians. The wires are placed at the center of the dark fringes where the beams cross. Our purpose is to obtain the intensity of light diffracted by the wires as a function of the angle θ on the plane formed by the beams and measured from the axis of symmetry of the incoming beams.

We notice that the resulting diffraction pattern from the calculation has been experimentally confirmed using the Afshar experiment setup [24]. The outcome of the calculation is that the decrease in photon count at either detector is 0.25%. Afshar et al. reported that the percent decrease in photon count at one of their detectors was 0.31% [2]. Our calculation is in reasonable agreement with this measurement. An interesting result from our calculation is that this 0.25% decrease in photon count can be analyzed.

To understand the results, consider 100,000 photons that come from the mirror or beam splitter towards its corresponding detector one at a time. The calculation shows that the detector will only collect 99,750 since 250 photons have been lost. The losses can be easily accounted. The wires stop 126 photons. The total number of diffracted photons away from the detector is 125. So far the losses are 251 photons, however, the calculation also shows that one photon is diffracted towards the detector. The net losses are $251 - 1$, thus, we have accounted for the 250 decrease in photon count at the detector. We notice that diffracted light does not have which-way information as its origin could be either the mirror or the beam splitter, we do not know. Fortunately, most of the diffracted light, 125 photons, do not reach the detectors. Only one diffracted photon reaches the detector. The remaining 99,749 photons that reach the detector are not the result of diffraction. Thus, these photons have full which-way information and the value of K is indeed close to 1. Therefore, the calculation confirms that the complementarity relation, $V^2 + K^2 \leq 1$, has been violated. We note that our analysis of the Afshar experiment is based on the assumption that the conservation laws in quantum mechanics apply just like they apply in classical mechanics [28].

3 Consequences of the Afshar Experiment

The most important result of the Afshar experiment for our work is the realization that the wave aspect of the photon has some sort of physical reality. Any level of physical reality ascribed to the wave aspect of the photons has physical consequences. Thus, some properties of the photon that can be deduced directly from the results of the Afshar experiment are reported here:

1. Sharp particle-like and sharp wave-like aspects of the photon coexist, $K^2+V^2 \approx 2$, beyond the limitations of complementarity, $K^2+V^2 \leq 1$. Thus, it appears that particle and wave both have simultaneous physical reality.
2. The *particle* component becomes explicit when the photon hits a detector. The photon path is identified through momentum conservation. A single momentum and a unique path are particle properties.
3. The presence of the *wave* component is evident by the small decrease in photon count when the two beam paths are open. The small decrease in photon count occurs only if the wires are located at the center of dark fringes. To get dark fringes we must have wave interference.
4. In each trial there is only one particle in the system, thus, the particle must be located in one of the two beams. We expect that the particle would influence a beam proportional to its proximity. Thus, the two beams cannot be influenced identically as the particle would affect the closer beam more. However, the experimental results are consistent with the presence of two identical beams. Thus, the particle does not influence the wave (beam). The wave (beam) must be shaped by boundary conditions of the experiment.
5. The wave component shows that there is external influence on the particle. This is clear by what takes place at the wire grid. When only one beam is on, the beam is uniform since there is no interference and a large number of photons hit the wires. In this case the decrease in photon count due to the wire grid is large, 14.55%. However, when both beams are on, destructive interference takes place at the location of the wires and the decrease in photon count is only 0.25%. This means that the particle, under external influence, is somehow steered away from regions of destructive interference.
6. In the experiment, at any given time, there is only one particle and two beams. We observe that only one detector clicks when energy and momentum are deposited there. Assuming that the particle is the carrier of energy and momentum, we conclude that the detector that clicks is the one that gets hit by the particle. We can also conclude that the detectors do not respond to the presence of the empty beam (wave).
7. The location of the particle in one of the two beams appears to be a random event.

4 Properties of the Wave Aspect of a Photon

Coexistence of particle and wave aspects of a photon implies the need for a more accurate interpretation of quantum mechanics. An immediate consequence is a new level of physical reality for both wave and particle. In our case we are interested in the wave aspect. We propose that the wave aspect originates in Einstein's new ether. The results of the Afshar experiment together with quantum theory imply the following properties and limitations for the wave component of a photon:

1. Physical reality of the wave component of a photon implies the existence of a medium or ether.
2. The observation that the particle is randomly located in one of the two beams (waves) could be a basis for the statistical interpretation of the wave component.
3. A physical origin for the wave component of a photon does not preclude a statistical interpretation of the wave component in relation to the particle.
4. If the wave component is limited to statistical interpretation it may mean that the wave component can only provide an overall description of the particle interactions with the ether which lead to wave-like effects.
5. The wave component shows where the particle can or cannot go. The particle does not go to regions of destructive interference. Thus, the wave component appears to describe the presence of force fields that "steer" the particle.
6. The only candidate that could provide a force field on a free particle is the ether or structure of spacetime. Thus, it is likely that the wave component represents an overall description of the interaction of a particle with the ether. In particular, the wave component could be related to the state of the ether as experienced by the particle.
7. Probability theory shows that if the wave component of a particle has probabilistic interpretation then the net wave due to a set of non-interacting particles should be the product of wave components associated with individual particles. This observation can lead to a quantum field theory approach to deal with many particles.
8. The relativistic Lorentz invariant treatment of the N-body problem is successfully handled with field theory tools. A physical origin for the wave component of particles does not preclude the application of field theory techniques to deal with the N-body problem.

5 Einstein's New Ether and Final Remarks

There is evidence for the existence of a physical vacuum or Einstein's new ether. However, there are important constraints that need to be considered. First of all, Einstein's new ether is the structure of spacetime, thus, it should be four-dimensional. Second, this structure is physical but not made of regular matter as Einstein pointed out [1]. Third, it provides an environment or stage for regular matter to exist. Fourth, it provides the medium for the propagation of the wave component of particles.

The Afshar experiment provides strong evidence of the physical reality of the wave component of the photon. A physical reality for the wave component of the photon is problematic as it implies the existence of a medium for wave propagation. In an otherwise perfect vacuum, totally void of regular matter, the only candidate for a medium is the structure of spacetime itself, as there is nothing left to be considered. We call the structure of spacetime Einstein's new ether. Thus, we propose that the physical origin of the wave component of the photon is to be found in Einstein's new ether.

We have deduced from the results of the Afshar experiment the role played by the wave component of the photon. The wave component of the photon shows a direct correlation between the position of the particle and the value of the wave at a given place, thus, the standard statistical interpretation is well suited. In view of Einstein's new ether, the wave component shows how a particle responds to changes that occur in its vicinity. For a free particle its vicinity is the ether. The state of the ether at the location of the particle would have the greatest influence on the motion of an otherwise free particle. Thus, the most likely scenario is that the wave component shows the state of the ether as experienced by the particle. Research on the details of the connection between the value of the wave component and the state of the ether as experienced by the particle should be conducted. The possibility that the ether could be involved in the resolution of the wave-particle duality paradox would have delighted Einstein.

To provide "the metrical relations in the space-time continuum" [1] Einstein's ether would have to be described mathematically by the metric tensor, $g_{\mu\nu}$. If the ether has an overall energy density Λ, its energy momentum tensor would be $\Lambda g_{\mu\nu}$. For the particular case when Λ is constant, the term $\Lambda g_{\mu\nu}$ in Einstein's equation plays the role of an energy momentum tensor not associated with regular matter but with empty space [5]. Thus, we may think of $\Lambda g_{\mu\nu}$ as the energy momentum tensor of the structure of spacetime or ether. Note that in local geodesic coordinates the above term becomes $\Lambda \eta_{\mu\nu}$, where $\eta_{\mu\nu}$ is the Minkowski metric tensor of flat spacetime. Thus the overall properties of the ether are homogeneous, isotropic and Lorentz invariant when viewed in these coordinates [26].

The overall description of Einstein's new ether, $\Lambda g_{\mu\nu}$, is similar to the cosmological constant term in Einstein's equation of general relativity [5]. Researchers have studied some of the consequences of a non-zero cosmological constant and have come to the conclusion that it has important physical effects [4]. An important conclusion is that the cosmological constant represents a different type of energy density; they have renamed it dark energy [4]. Dark energy is present wherever space is; if space expands dark energy expands with it without being diluted. It is as if dark energy would be an intrinsic property of spacetime. The origin of dark energy is one of the open problems in physics since the theory predicts a very large value originating from the zero point energy of fields while the observed value is very small [3]. The possibility that the ether or structure of spacetime is involved in the near cancellation of the dark energy has been considered before [27]. Thus, the small value of the cosmological constant or net dark energy could be another piece of evidence for the existence of a physical structure of spacetime or Einstein's new ether.

References

1. A. Einstein, Sidelights on Relativity, Dover Publications, New York, (1983)
2. R. Eisberg and R. Resnick, Quantum Physics of Atoms, Molecules, Wiley, NY, (1974)
3. C. Barceló and G. Jannes, A real Lorentz-FitzGerald contraction, Found Phys 38:191–199 (2008)

4. P.J.E. Peebles and Bharat Ratra, The cosmological constant and dark energy, Rev Mod Phys 75:559–606 (2003)
5. L.D. Landau and E.M. Lifshitz, The Theory of Classical Fields, Vol. 2, Pergamon Press, London, (1975)
6. S. Weinberg, Quantum Theory of Fields, Cambridge University Press, Cambridge, (1996)
7. H.B.G. Casimir, "On the Attraction Between Two Perfectly Conducting Plates," Proc. Kon. Ned. Akad. Wetenschap 51:793 (1948)
8. N. Bohr, The quantum postulate and the recent development of atomic theory, Nature 121, 580 (1928)
9. S. Dürr, G. Rempe, Can wave–particle duality be based on the uncertainty relation? Am J Phys 68:1021 (2000)
10. B.-G. Englert, Fringe visibility and which-way information: An inequality, Phys Rev Lett 77:2154 (1996)
11. M. Kolar, T. Opatrny, N. Bar-Gill, N. Erez, and G. Kurizki, New J Phys 9:129 (2007), http://www.njp.org/
12. S.S. Afshar, E. Flores, K.F. McDonald, and E. Knoesel, Paradox in wave-particle duality for non-perturbative measurements, Found Phys 37:295 (2007), arXiv:quant-ph/0702188
13. S.S. Afshar, Violation of the principle of complementarity, and its implications Proc. SPIE 5866:229–244 (2005), arXiv:quant-ph/0701027v1
14. J.G. Cramer, A farewell to Copenhagen? (2004) *E-print* www.analogsf.com/0410/altview2.shtml
15. R.E. Kastner, Why the Afshar experiment does not refute complementarity, Stud His Philos M P 36:649–658 (2005), (arXiv:quant-ph/0502021v3)
16. E. Flores and E. Knoesel, Why Kastner analysis does not apply to a modified Afshar experiment E, Proc SPIE 6664 66640O (2007), (arXiv:quant-ph/0702210v1)
17. A. Drezet, Complementarity and Afshar's experiment (2005) *Preprint* arXiv:quant-ph/0508091v3
18. P. O'Hara, Entanglement and quantum interference (2006) *Preprint* http://arxiv.org/abs/quant-ph/0608202arXiv:quant-ph/0608202v1
19. O. Steuernagel, Found Phys 37:1370–1385 (2007), (arXiv:quant-ph/0512123v2)
20. E. Flores, Reply to comments of Steuernagel on the Afshar's experiment, Found Phys 38: 295 (2008), arXiv:0802.0245v1
21. R.E. Kastner, On the visibility in the Afshar two-slit experiment (2008) *Preprint* arXiv:0801.4757v2
22. D.D. Georgiev, Single photon experiments and quantum complementarity, Prog Phys 2:97–103 (2007)
23. E. Flores, "Modified Afshar experiment: Calculations," Proc. SPIE Vol. 7421, 74210W (2009) arXiv:0803.2192v2
24. R. Buonpastore, E. Flores and E. Knoesel, "Diffraction of Coherent Light with Sinusoidal Amplitude by a Thin-Slit Grid." Optics 121:1009–1012 (2010)
25. D. Greenberger and A. Yasin, Simultaneous wave and particle knowledge in a neutron interferometer, Phys Lett A 128: 391 (1988)
26. H. Ohanian and R. Ruffini, Gravitation and Spacetime, 2nd edn, W. W. Northon, New York, 1994
27. E. Flores, "Physical implications of the cosmological constant," International Journal of Theoretical Physics, 32, 8 (1993)
28. E. Flores and J. De Tata, "Complementarity Paradox Solved: Surprising Consequences." Foundations of Physics, DOI 10.1007/s10701-010-9477-4, June 09, 2010
29. D. Durr, S. Goldstein, R Tumulka, and N Zangh, "Bohmian Mechanics," Compendium of Quantum Physics: Concepts, Experiments, History and Philosophy, edited by D. Greenberger, K. Hentschel, and F. Weinert, (Springer, 2009)

Part II
Implications of Minkowski's Ideas for the Philosophy of Space and Time

Minkowski's Proper Time and the Status of the Clock Hypothesis[1]

Richard T.W. Arthur

Abstract In this chapter I argue that the concept of *proper time* must be regarded as one of Minkowski's enduring contributions to physics. I examine some confusions that still interfere with an appreciation of this, including a conflation of proper time with the co-ordinate time of the inertial frame of a system at rest, and the related mistaken notion that Special Relativity cannot be applied to accelerating systems. This sets the stage for a treatment of the so-called *clock hypothesis*, according to which the instantaneous rate of a clock depends only on its instantaneous speed. I argue that this does not have the status of an independent hypothesis, but is simply a description of the behaviour of an ideal clock as predicted by (classical, special and general) relativity theory. The question whether this hypothesis holds, moreover, must be distinguished from the question of whether the restorative acceleration of the mechanism within any real system acting as a clock is sufficiently great (relative to the acceleration undergone by the system) that the system will be able to approximate such an ideal clock. The failure of the clock hypothesis would entail the falsity of relativity theory in the form proposed by Einstein, as Weyl had sought to demonstrate with his unified theory of gravity and electromagnetism in 1918. I argue that it is the Strong Equivalence Principle in General Relativity that preserves the chronometric significance that the metric had in Special Relativity, and thereby preserves the relation of inertia to time assumed classically.

Discussion of Hermann Minkowski's mathematical reformulation of Einstein's Special Theory of Relativity as a four-dimensional theory usually centres on the ontology of spacetime as a whole, on whether his *hypothesis of the absolute world* is an original contribution showing that spacetime is the fundamental entity, or whether his whole reformulation is a mere mathematical *compendium loquendi*. I shall not be adding to that debate here. Instead what I wish to contend is that Minkowski's most profound and original contribution in his classic paper of 100 years ago lies

[1] I am grateful to Stephen Lyle, Vesselin Petkov, and Graham Nerlich for their comments on the penultimate draft; any remaining infelicities or confusions are mine alone.

V. Petkov (ed.), *Space, Time, and Spacetime*, Fundamental Theories of Physics 167, DOI 10.1007/978-3-642-13538-5_7,

in his introduction or discovery of the notion of *proper time*.[2] This, I argue, is a physical quantity that neither Einstein nor anyone else before him had anticipated, and whose significance and novelty, extending beyond the confines of the special theory, has become appreciated only gradually and incompletely. A sign of this is the persistence of several confusions surrounding the concept, especially in matters relating to *acceleration*. In this paper I attempt to untangle these confusions and clarify the importance of Minkowski's profound contribution to the ontology of modern physics. I shall be looking at three such matters in this paper:

1. The conflation of proper time with the time co-ordinate as measured in a system's own rest frame (*proper frame*), and the analogy with *proper length*.
2. Misconceptions that Special Relativity (SR) applies only to objects in inertial motion, and *not to accelerated systems*, and that therefore one must introduce General Relativity to solve Langevin's Twin Paradox.
3. Misconceptions surrounding the status of the so-called *clock hypothesis* (CH), according to which the instantaneous rate of a clock depends only on its instantaneous speed, and not on its acceleration. I shall argue that the CH is a criterion for ideal clocks that is implicit in SR, and does not have the status of an independent assumption; and that it also performs this role in GR as a consequence of the strong equivalence principle.

1 Proper Time, Local Time and Proper Length

The first symptom of this under-appreciation of the novelty of proper time I wish to discuss is that it is often taken to be the time co-ordinate measured by a clock at rest in an inertial frame: in its own frame, therefore, *proper*, as opposed to *local time*. 'Local time', of course, was originally Lorentz's term for the transformed co-ordinate time t' of an inertial frame of reference moving with velocity v with respect to the stationary frame. By Lorentz's own admission, the chief cause of his failure to discover Special Relativity "was my clinging to the idea that only the variable t can be considered as the true time, and that my local time t' must be regarded as no more than an auxiliary mathematical quantity".[3] Einstein's correction lay in seeing that all of these time co-ordinates or local times are on a par and equally entitled to be regarded as the true time. It is perhaps for these reasons that the illusion arose that proper time is simply the local time of the system's own rest frame, the time co-ordinate as measured in an inertial frame in which the system is at rest. And this

[2] Here I concur with Roberto Torretti, who rates Minkowski's introduction of proper time as "probably his most important contribution to physics" (1983, 96).

[3] Quoted from (Torretti 1983, p. 85).

in turn could explain why it is not uncommon to find the introduction of proper time attributed to Einstein, even by authors with a keen sense of the history of physics.[4]

The conflation of proper time with co-ordinate time in a system's own rest frame is also perhaps fostered by the numerical equivalence of the value of the proper time elapsed for a body moving along an inertial path with the value measured by the time-co-ordinate in its rest frame. Thus it is often said that proper time is simply time measured in a body's "proper frame", *as if a body keeps its own inertial frame while accelerating!* There are two confusions here: first, the idea that a body "has" an inertial frame, when a reference frame is just a point of view for representing the body's motion, and (according to the principle of relativity) one can represent this motion equivalently from *any* inertial frame; and second, of course, the idea that the body could stay in the same inertial frame even though it is accelerating, and therefore moving *non-inertially*. At any rate, this is a confused idea of proper time, which is *not* a time co-ordinate and was not introduced by Einstein, but by Minkowski, in his famous paper of 1908 (Lorentz et al. 1923, 73–91). Introducing the concept, he asks us to imagine at any point P (x, y, z, t) in spacetime a worldline running through that point, so that the magnitude corresponding to the timelike vector dx, dy, dx, dt laid off along the line is

$$d\tau = \sqrt{(c^2dt^2 - dx^2 - dy^2 - dz^2)/c} \tag{1}$$

Proper time is now defined as the integral of this quantity along the world line in question: "The integral $\tau = \int d\tau$ of this quantity, taken along the worldline from any fixed starting point P_0 to the variable endpoint P, we call the *proper time* of the substantial point at P." (85) Thus for Minkowski spacetime, the proper time is:

$$\tau = \int_P d\tau = \int_P \sqrt{(c^2dt^2 - dx^2 - dy^2 - dz^2)/c} \tag{2}$$

or, equivalently,

$$\tau = \int_{P0}^{P} \{1 - 1/c^2[(dx/dt)^2 + (dy/dt)^2 + (dz/dt)^2]\}^{1/2} dt \tag{3}$$

As defined, the proper time cannot be evaluated without adopting a system of co-ordinates. But because of the signature of the Minkowski metric, the interval $d\tau$ and its integral τ are both *invariants*, so that their values are independent of what system of co-ordinates is adopted. Thus for the time elapsed along any worldline, τ gives a measure that is *independent of the co-ordinates*, even if a particular frame must be adopted in order to calculate its value. The whole content of relativity theory

[4] See for example J.-P. Provost, C. Bracco and B. Raffaelli 2007, 498: "If one starts with SR, ...the first important notion, as Einstein told us in 1905, is the proper time τ".

can now be framed in terms of such invariants, so that co-ordinates are no longer regarded as primitive, as they had been in Einstein's way of conceiving SR. For instance, as Minkowski proceeded to explain, x, y, z and t – the components of the vector OP, where O is the origin – are considered as functions of the proper time τ, and the first derivative of the components of this vector with respect to the proper time, $dx/d\tau$, $dy/d\tau$, $dz/d\tau$ and $dt/d\tau$, are those of the *velocity vector* at P, which is also a four-dimensional invariant. As is well known, the resulting four-dimensional co-ordinate-free rendering of special relativity is of immense utility for the further development of relativity theory, even if Einstein did not at first appreciate its significance.

The misidentification of proper time as the time co-ordinate in its rest frame is also encouraged by an analogy with *proper length*, which is the length of a body in its rest frame. The analogy, it is often claimed, is *perfect*, and the invariance of proper time is no objection. For just as the length of a path joining two events in timelike separation is invariant under change of frame, so is the length of a curve joining two events in spacelike separation.[5] This can be seen by comparing the expressions for proper time and proper length written in the tensor form necessary for general relativistic spacetimes. Here the appropriate generalization of the expression given by Minkowski for his flat spacetime to curved spacetimes is[6]

$$\tau = \int_P d\tau = \int_P (-g_{\alpha\beta} dx^\alpha dx^\beta)^{1/2} \tag{4}$$

The flat-space expression for the proper length should be generalized so that it is the exact analogue of proper time, a line integral along a curve joining two spacelike separated events:

$$L = \int_P ds = c \int_P (g_{\alpha\beta} dx^\alpha dx^\beta)^{1/2} \tag{5}$$

There is no doubt that this defines a proper distance, that between the endpoints of a path in spacetime along which no process can travel, a spacelike curve. An arbitrary curve joining two spacelike separated events, however, is *not generally the length of an object*. It can only be the length of an object if all the points on the curve are simultaneous in some given reference frame; for an object, such as a body or wave front, is a three dimensional object existing at a given time. And while the path integral along such an arbitrary curve is indeed the length of a path, and is independent

---

[5] Cf. the article on proper length in Wikipedia (http://en.wikipedia.org/wiki/Proper_length: March 24, 2009). The author suggests a generalization of proper length so that (in either Special or General Relativity) it is given by the line integral $L = c \int_P \sqrt{[-g_{\mu\nu} dx^\mu dx^\nu]}$, where $g_{\mu\nu}$ is the metric tensor for the spacetime with $+{-}{-}{-}$ signature, normalized to return a time, and P is the spacelike path. The author also notes that "Proper length has also been used in a more restricted sense to help with discussions of length contraction by textbooks, where it is defined as the length of an object when measured by someone at rest relative to that object."

[6] Cf. Misner et al. (1973, 393).

of the choice of reference frame, it has no particular physical significance. Proper length is correctly defined as the path integral, *not along an arbitrary curve* joining the endpoints of the path at the same time, but *along the shortest curve*, which is a straight line joining them in the frame at which they are at rest. If (elapsed) proper time were the strict analogue of this, it would be the longest time between two timelike separated events, which would be the time in a frame of reference at rest, i.e. the co-ordinate time in a body's rest frame. It is precisely this interpretation that I am contesting: proper time, according to Minkowski's definition above, *is not a co-ordinate time*; and it is not defined for only the shortest path, i.e. only within the body's rest frame, but is defined *for any timelike curve in spacetime*. Because proper length is the interval between two events *at the same co-ordinate time*, it is specific to a particular reference frame. Proper time, as defined by Minkowski, is not.

Thus *proper time* has a fundamentally different character from proper length.[7] Although both are invariant under change of frame, proper length is the length of an object in its own rest frame, whereas proper time is independent of frame. In this respect proper length is analogous to proper mass. (It differs from the latter, however, in that proper mass seems to be an essential characteristic of an elementary body (such as an electron), whereas proper length is a contingent one.) At any rate, there is a fundamental dissymmetry between duration and length in Special Relativity, somewhat obscured by talk of their embodiments in observers' clocks and rods. For whereas an observer's clock measures proper time elapsed along a path, a dynamical variable specifiable independently of reference frame, the proper length of the observer's measuring rod is specific to the inertial frame in which the observer is at rest. Thus we see that, ironically, there is a sense in which Minkowski's introduction of proper time undermines his famous pronouncement at the beginning of his paper about the demise of time:

> Henceforth space by itself, and time by itself, are doomed to fade away into mere shadows, and only a kind of union of the two will preserve an independent reality. (Minkowski 1908, 75)[8]

2 Acceleration and Appeals to General Relativity

Relatedly, it is often stated that Special Relativity (SR) applies only to objects in inertial motion, and not to accelerated systems, for which an appeal to General Relativity (GR) is necessary. This one finds especially in treatments of the Twin Paradox (a.k.a. the Clock Paradox), where it is claimed that a proper resolution of the paradox

---

[7] For a complementary analysis that comes to the same conclusion, namely that there is a profound difference between proper time and proper length, see Petkov (2009, 89–95).

[8] This famous pronouncement of Minkowski's is echoed by Einstein in his essay "The Problem of Space, Ether and the Field in Physics": "Hitherto it had been silently assumed that the four-dimensional continuum of events could be split up into time and space in an objective manner... With the discovery of the relativity of simultaneity, space and time were merged in a single continuum..." (1954, 281–282).

must therefore involve General Relativity.[9] In this connection one finds references to Einstein's "Dialogue on Objections to the Theory of Relativity" (1918), which is interpreted as having shown how to solve the paradox using General Relativity, through an application of the Equivalence Principle (EP). On this misreading it is thought that since the acceleration of the travelling twin is responsible for the difference in the twins' ages (i.e. in the proper times of their journeys), and accelerations fall outside the scope of the special theory, and since (by the EP) this acceleration will be equivalent to a *gravitational time dilation*, the paradox must receive its explanation in GR.

On the contrary, the paradox receives a complete explanation within SR, which is perfectly applicable to accelerated motions.[10] This was already made clear by Arnold Sommerfeld in a very succinct note on Minkowski's paper of 1908, published in (Lorentz et al. 1913), with an English translation appearing in 1923 and again in 1952:

> [T]he element of proper time $d\tau$ is not a complete differential. Thus if we connect two world-points O and P by two different world-lines 1 and 2, then
>
> $$\int_1 d\tau \neq \int_2 d\tau$$
>
> If 1 runs parallel to the t-axis, so that the first transition in the chosen system of reference signifies rest, it is evident that
>
> $$\int_1 d\tau = t, \int_2 d\tau < t$$
>
> On this depends the retardation of the moving clock compared with the clock at rest. (Lorentz et al. 1913, 71, and 1952, 94)

As for Einstein's 1918 paper, there Einstein is concerned to defend the consistency of the explanation of time dilation given in SR, and to show its compatibility with an equivalent explanation in GR. He argues that the differences in the time of the two journeys is an "inevitable result" of the special theory of relativity. In his version, there are two identical clocks, U_1 and U_2, and U_2 is accelerated until it reaches a velocity $-v$ relative to U_1, travels with this velocity for a while, and is then decelerated until its motion is reversed, moving with a velocity v back to rejoin U_1. U_2 will then be retarded with respect to U_1 by an amount $-\Delta t$. To the objection that all

[9] Cf. this analysis on the *Encyclopedia Britannica* internet site: "The answer is that the paradox is only apparent, for the situation is not appropriately treated by special relativity. To return to Earth, the spacecraft must change direction, which violates the condition of steady straight-line motion central to special relativity. A full treatment requires general relativity, which shows that there would be an asymmetrical change in time between the two sisters. Thus, the "paradox" does not cast doubt on how special relativity describes time, which has been confirmed by numerous experiments." http://qa.britannica.com/eb/article-252886. The number of similar confusions on individual physicists' web sites seems to have decreased sharply of late; but there is no excuse for anyone who has read Misner, Thorne and Wheeler (1973): they have a separate section (6.1) titled "Accelerated Observers Can be Analyzed Using Special Relativity" (p. 163).

[10] See my thorough treatment of the twin paradox in SR in my (2008).

motion is relative, he rejoins that in SR this is so only for systems in mutual relative unaccelerated motion. But since here the clock U_2 is accelerated, no contradiction to SR is forthcoming. The clock U_2 will be running behind U_1: as in Sommerfeld's argument above, $\int_2 d\tau = \int_1 d\tau - \Delta t < \int_1 d\tau$.

But his imaginary opponent objects that in General Relativity one is not constrained to take only reference frames in inertial motion: one could adopt a coordinate frame co-moving with the accelerated clock. Einstein responds by applying the Equivalence Principle to show how things would appear from that frame. Now a static homogeneous gravitational field appears, and clock U_1 undergoes acceleration in free fall up to a velocity v (at which point the field disappears), while the clock U_2 is prevented from moving by the action of an external force. U_1 then travels with this velocity for a while, until a static homogeneous gravitational field appears, directed in the opposite direction to before, in which it decelerates in free fall until its motion is reversed, moving with a velocity $-v$ back to rejoin U_2, which had remained at rest. Einstein explains that although the clock will be retarded while undergoing the inertial legs of its journey, resulting in U_1 being retarded by the same amount $-\Delta t$ as was U_2 in the previous scenario, when it is free-falling in the third leg of its journey it is located at a higher gravitational potential than U_2, and, as a consequence of general relativistic time dilation, it will be speeded up by exactly $2\Delta t$, thus completely disposing of the paradox.

The moral of the story is that time dilation in SR arises from the fact the twins *trace different paths through spacetime* because of the different accelerations they undergo, they but it is *not a direct effect of the accelerating motion itself.* In the idealized scenario of the twin paradox, the time dilation due to the acceleration of the travelling twin is not normally considered, but it could be made arbitrarily small compared to that produced by the inertial motions. The time dilation in the second scenario considered by Einstein is produced not by the acceleration, but by the difference in gravitational potential at two points in the field. Thus in the SR case, it is the difference in the paths that results in a time dilation for the accelerated twin; and analogously in the GR case, the compensating gravitational time dilation is due to the difference in gravitational potential at two points in the field rather than being an effect of the accelerating motion itself. This is what Einstein showed in 1918.[11]

3 Acceleration and the Clock Hypothesis

A much more subtle set of issues surrounds the "clock hypothesis", for which the previous two sections set the scene. Wolfgang Rindler renders it as follows:

[11] An updated treatment of Einstein's solution is given by Jones and Wanex (2006), who demonstrate that the SR and GR paradox solutions are identical for finite accelerations (as well as the infinite ones shown by Møller (1955)), by using the destination distance as the key observable parameter.

> If an *ideal* clock moves non-uniformly through an inertial frame, we shall *assume* that acceleration as such has no effect on the rate of the clock, i.e. its instantaneous rate depends only on its instantaneous speed v according to the above rule. This we call the *clock hypothesis*. It can also be regarded as the definition of an "ideal" clock. (Rindler 1977, 43).

What Rindler refers to here as "the above rule" is the standard formula for time dilation in Special Relativity, representing the time interval T recorded by a clock in an inertial frame S in terms of the time interval T_0 recorded by a clock at rest in a frame S moving inertially with speed v:

$$T = \gamma T_0 = T_0/(1 - v^2/c^2)^{1/2} \tag{6}$$

This, however, is not a good definition of the clock hypothesis, since here T and T_0 are both coordinate times, so that strictly speaking formula (6) only applies to inertial motions, not accelerated ones. Thus a better definition is to say that an ideal clock is one that measures proper time as given by formula (2) above:

$$\tau = \int_P d\tau = \int_P \sqrt{(c^2dt^2 - dx^2 - dy^2 - dz^2)}/c \tag{2}$$

or in the generalized form also appropriate to General Relativity that we gave above,

$$\tau = \int_P d\tau = \int_P (-g_{\alpha\beta}dx^\alpha dx^\beta)^{1/2} \tag{4}$$

On the necessity and status of this hypothesis, opinion is divided. Rindler claims it is an assumption that it is necessary to make in order to get from "purely kinematic laws about acceleration" to the dynamics of really accelerated systems (1966, 28) and Harvey Brown claims something similar in his recent book (Brown 2005, 9). Brown adds that it is the clock hypothesis that "allows for the identification of the integration of the metric along an arbitrary time-like curve – not just a geodesic – with the proper time. This hypothesis is no less required in general relativity than it is in the special theory." (Brown 2005, 9).

On the other hand, Jim Hartle holds that Minkowski's formula for the proper time holds "even for accelerating clocks, i.e., when the velocity is dependent on the time" (Hartle 2003, 62), and he makes no use of the clock hypothesis in his textbook. Roberto Torretti allows that the clock hypothesis "may be viewed as a conventional definition of what we mean by clock accuracy, and hence by physical time" (1996, 96), but argues that "Special Relativity would doubtless have been rejected or, at any rate, deeply modified, if the clock hypothesis were not fulfilled – to a satisfactory approximation – by the timepieces actually used in physical laboratories." And according to Misner, Thorne and Wheeler, "one *defines an 'ideal'* ... *clock* to be one which measures ... proper time as given by $(-g_{\alpha\beta}\, dx^\alpha\, dx^\beta)^{1/2}$ ..." (1973, 393).

On these latter views, provided a given process approximates well enough an ideal clock, the clock hypothesis seems to amount to little more than the desideratum that, with the metric locally Minkowskian, the predictions of SR should agree with experimental fact. So the question is, why should it be necessary to state it as an independent hypothesis? Two main sets of considerations have been adduced. As we have seen, one kind of justification has been that, since many natural clocks are subject to accelerations which result in their failing to satisfy the clock hypothesis, we need to appeal to the hypothesis in passing from the kinematics of acceleration of ideal clocks to the dynamics of really moving clocks. A second kind of justification has to do with the different status of Special Relativity within the General Relativistic context, where, as is shown by the example of Weyl's unified theory of electromagnetism and gravitation, the metric could be locally Minkowskian and yet the rate of clocks could be path-dependent in such a way that the instantaneous rate would depend on the way the clock had been accelerated hitherto, contrary to the clock hypothesis. This, it is argued, proves the independence of the clock hypothesis from the assumption that spacetime is locally Minkowskian.

Let's look at the question of ideal clocks first. Rindler stresses that it is not the case that any natural process serving as a clock will meet the condition stated in the clock hypothesis. He gives the example of "a spring-driven pendulum clock whose bob is connected by two coiled springs to the sides of the case (so that it works without gravity)" (43), pointing out that it "will clearly increase its rate as it is accelerated upward". Rindler allows that "certain natural clocks (vibrating atoms, decaying muons,) conform very accurately to the clock hypothesis", and observes that in general "this will happen if the clock's internal driving forces greatly exceed the accelerating force" (43). Similarly, Harvey Brown states that the "key issue is the comparison of the magnitude of the external force producing the acceleration and that of the forces at work in the internal mechanism of the clock." (Brown 2005, 95) As he points out, "an important part of the history of time has been the search for accurate clocks which withstand buffeting" (94). Exemplary in this regard were the wonderful timepieces constructed by John Harrison in the eighteenth century in an effort to win the Admiralty Prize for a clock accurate enough for use in determining longitude on board ship (see Sobel 1995 for an engaging account). But Brown's discussion of this issue is given in the course of an inquiry into what happens when clocks are no longer moving inertially (94). He concludes that the justification of the clock hypothesis "rests on accelerative forces being small in the appropriate sense", i.e. if they are "small in relation to the internal restorative forces of the clock" (95). The "effect of motion on the clock depends accumulatively only on its instantaneous speed, not its acceleration" (95), provided the accelerative force is small in comparison with the restorative forces. Similarly, Brown and Pooley write

> The claim that the length of a specified segment of an arbitrary time-like curve in Minkowski spacetime – obtained by integrating the Minkowski line-element *ds* along the segment – is related to proper time rests on the assumption (now commonly dubbed the 'clock hypothesis') that the performance of the clock in question is unaffected by the acceleration it may be undergoing. It is widely appreciated that this assumption is not a consequence of Einstein's 1905 postulates. Its justification rests on the contingent dynamical requirement that the external forces accelerating the clock are small in relation to the internal 'restoring' forces at work inside the clock. (Brown and Pooley 2001, 264–265)

But it seems to me that this way of describing the situation conflates two distinct issues: the clock hypothesis as a criterion for an ideal clock, an ideal clock being a clock that will keep proper time; and the separate problem of whether the restorative acceleration of the mechanism within any real system acting as a clock is sufficiently great (relative to the acceleration undergone by the system) that the system will be able to approximate such an ideal clock.[12] As Misner, Thorne and Wheeler write (2001, 393), once one has defined an ideal clock, "one must then determine the accuracy to which a given ... clock is ideal by using the laws of physics to analyze its behavior". So, I maintain (*contra* Brown), it is not the justification of the clock hypothesis that depends on the accelerative force being small in comparison with the restorative forces, but the justification of whether a given naturally occurring periodic process approximates sufficiently well the behaviour of an ideal clock.

As an example, consider the case discussed by Misner, Thorne and Wheeler (393) of an atomic clock accelerated to $2g$ in an airliner (with the airliner not in free fall, and accelerating to avoid a mid-air collision). As they explain, in order to determine empirically whether the clock "will still measure proper time $d\tau = (-g_{\alpha\beta}\, dx^{\alpha}\, dx^{\beta})^{1/2}$ along its world line to nearly the same accuracy as if it were freely falling" – i.e. to determine whether it approximates an ideal clock – "one can analyze the clock in its own 'proper reference frame' [§13.6 (pp. 327–332)], with Fermi-Walker transported basis vectors, using the standard local Lorentz laws of quantum mechanics as adapted to accelerated frames (local Lorentz laws plus an 'inertial force', which can be treated as due to a potential with a uniform gradient)" (393). In the same vein they give a proof that a pendulum clock at rest on the Earth's surface is ideal (394–395).

This distinction between an ideal clock's being implicitly determined by theory, and a real world time-piece's time-keeping qualities being determined by how well it will be able to conform to such an ideal clock, is not something new. Newton's idea was that the equability of absolute time was a direct correlate of equable motions, the paradigm for which was an inertially moving body marking off equal lengths.[13] Thus an inertially moving body is an ideal clock in Newtonian physics.[14] But as Newton himself conceded, it is quite conceivable that no actual body in the world moves perfectly equably: "It is possible that there is no uniform motion by which time may have an exact measure." (1999, 410) But this would not preclude the calculation of forces on the presupposition of this exact relation between uniform motion and time. Now, of course, if a clock is subject to too violent accelerations, it

[12] There is an additional problem with Brown and Pooley's formulation, in that it assumes some kind of gap between proper time and the integral of the line-element *ds* along the segment; but proper time was *defined* as that by Minkowski. So the authors are using "proper time" in a different sense, as the time read by a real clock in its own rest frame. But the issue of whether an actual clock will read proper time is not the same issue as whether an ideal one will.

[13] After writing this sentence, I discovered almost exactly the same sentiment expressed by Torretti: "The First Law of Motion provides the paradigm of a physical process that keeps Newtonian time, and this is enough to ensure that the latter concept is physically meaningful, even if no such process can ever be exactly carried out in the world. ... The flow of Newtonian time can therefore be read directly from the distance marks the body passes by as it moves along the ruler." (1983, 12).

[14] For a discussion of this aspect of Newton's absolute time, see Arthur (1995, 2007) and Barbour (1989).

will cease to function as an accurate clock. Brown quotes Eddington to this effect: "We may force it into the track by continually hitting it, but that may not be good for its time-keeping qualities!" (Eddington, quoted in Brown 2005, 94). But as Misner et al. remark (2003, 393), whether it is pushed beyond the point where it can still keep good time "depends entirely on the construction of the clock – and not at all on any 'universal influence of acceleration on the march of time.' Velocity produces universal time dilation; acceleration does not."

Now let me turn to the claim that the clock hypothesis has a role in the transition from kinematical to dynamical considerations. Wolfgang Rindler, after defining the clock hypothesis in his (1960), writes: "SR has no machinery to *prove* any but purely kinematic laws about acceleration. All other such laws it can merely subject to the test of invariance, which requires that a physical law shall have the same form in all inertial frames." (1960, 28–29) This is a puzzling claim. As we saw above, Special Relativity can certainly be *applied* to accelerating bodies, whatever force is the source of the acceleration. Rindler appears to mean that SR, in the form of the requirement of Lorentz covariance, is only a constraint on the formulation of dynamical laws. But in what sense does that make it "purely kinematic"? By way of justification he asks us to consider a standard clock being moved along an arbitrary n-sided polygonal path with uniform velocities along the sides and instantaneous velocity changes at the vertices. Then according to formula (6) above, "the total time increment indicated by the clock will be

$$T = \sum_{i=1}^{n} \left(1 - v_i^2/c^2\right)^{1/2} \Delta t_i \tag{7}$$

where v_i and Δt_i refer to the ith side and are measured in the reference frame." (29) He claims that this merely *suggests* taking the limit as $n \to \infty$ to get the law for a completely arbitrary motion between times t_i and t_2:

$$T = \int_{t1}^{t2} \sqrt{(1 - v^2/c^2)\, dt} \tag{8}$$

Rindler notes that this is "consistent with the relativistic demand for invariance", since (8) is equivalent to formula (2) above. But, he claims, (8) is not a unique generalization of (7), in that "more complicated laws could easily be devised which incorporate an acceleration effect and which are also invariant and reduce to [(7)] in the polygonal case" (29). Thus "it is idle to pretend that the polygonal and continuous paths ultimately become equivalent in all physical respects":

> Consider, for example, a short thin tube placed perpendicularly across the path and containing a ball at the intersection. If this tube is moved transversely over the continuous path the ball will be displaced by centrifugal forces, but this will not happen on the corresponding polygonal path no matter how large its number of sides. (29)

Thus, according to Rindler, at no point in the polygonal path is there a force on the ball, but in the curved path it experiences a centrifugal force.

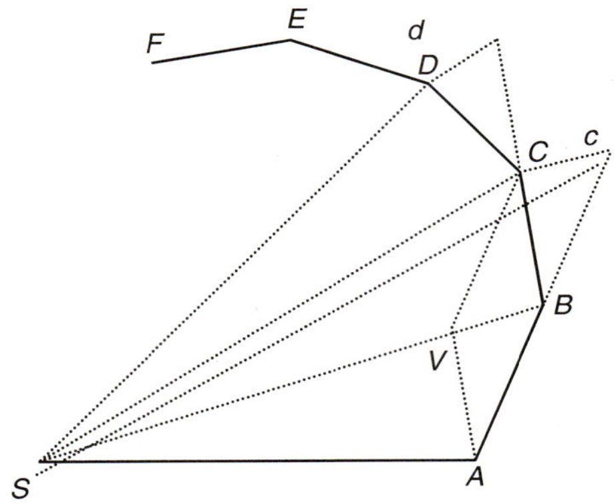

Fig. 1 Newton's proof of the area law

But this is a most unfortunate example. For the above polygonal construction is precisely that used by Isaac Newton in giving one of the very first derivations of the formula for centrifugal force![15] The success of such derivations, I contend, wholly undermines the distinction Rindler is trying to make here between a "kinematical" approach to acceleration, and one that is properly dynamical. Of course, the point is that in the polygonal model one assumes that the body (in Rindler's example, the tube containing the ball) is deflected instantaneously and discontinuously at each vertex by a discrete impulse $m\Delta v$ acting toward the centre in such a way as to produce a new inertial motion $m(v+\Delta v)$ between that vertex and the next (by application of the parallelogram law). The ball in Rindler's tube will experience each successive impulse. When one takes the limit as $n \to \infty$ and $\Delta t_i \to 0$, the successive increments of velocity effectively become elements of velocity dv directed towards the centre during an interval dt, with the result that one has effectively integrated from first principles to obtain an expression for the acceleration towards the centre, dv/dt. In the limit, the impulses experienced by the ball in ever shorter and more numerous moments, will smear out into a continuous force. Newton used precisely this procedure in his masterwork, the *Principia*, to derive Kepler's Area Law:[16]

> Let the time be divided into equal parts, and in the first part of the time let a body by its inherent force describe the straight line *AB*. In the second part of the time, if nothing hindered it, this body would (by law 1) go straight on to *c*, describing line *Bc* equal to *AB*, so that – when radii *AS*, *BS* and *cS* are drawn to the centre – the equal areas *ASB* and *BSc* would be described. But when the body comes to *B*, let a centripetal force act with a single

[15] Newton's derivation is given in the Waste Book of 1666 (ULC MS Add. 4004), transcribed in Herivel (1965, 128–131). For an explication, see Brackenridge (1985, 45–51).

[16] As I. B. Cohen mentions in his introduction to the *Principia* (Newton 1999, 71), "the issue of the mathematical rigour of Newton's polygonal analysis has been, and still remains, a subject of debate among scholars." Recent analyses include Nauenberg (1998), Pourciau (2003), and Arthur (2009).

> but great impulse and make the body deviate from the straight line *Bc* and proceed in the straight line *BC*...
>
> Now let the number of triangles be increased and their width decreased indefinitely, and their ultimate perimeter *ADF* will (by lem. 3, corol. 4) be a curved line; and thus the centripetal force by which the body is continually drawn back from the tangent of this curve will act uninterruptedly, while any areas described, *SADS* and *SAFS*, which are always proportional to the times of description, will be proportional to the times in this case. *Q.E.D.* (Newton 1999, 445) (Fig. 1).

The case is the same in Special Relativity. One can begin, as Rindler chooses to do, with the time dilation formula (6) for a time interval in one inertial frame relative to another, and consider a polygonal "orbit" created by successive impulses acting on a body. In the limit, one recovers Minkowski's formula for the proper time, (2). Or one can begin with the differential form for proper time (1), and simply integrate along the path to obtain (2), as did Minkowski. As Torretti expresses it, "The clock hypothesis implies that the time measured by our clock between any two events P and Q is none other than the proper time along the clock's worldline from P to Q." (1983, 96). Thus, I conclude, it is no more necessary to postulate the clock hypothesis as a separate assumption pertaining to dynamics than it is in classical (Galilean invariant) mechanics: in SR, it is simply the condition that defines an ideal clock. I should add: this does not mean that the CH is true by definition: it means that *if a real clock does not perform as an ideal clock*, even though the theory predicts that it should (its restorative force greatly exceeding the accelerative force etc.), then – provided there isn't some unsuspected force in operation, or similar exculpatory explanation – *there must be something wrong with the theory.*

In passing I note that, in their insistence that the clock hypothesis is a separate assumption in SR, Rindler and Brown could appeal to the authority of Einstein. For in the notes he made on Minkowski's original "*Raum und Zeit*" paper, Arnold Sommerfeld attributes a remark to Einstein that may indicate the origin of the idea that the clock hypothesis is needed – at any rate it is the first statement of it that I have been able to find. Right after mentioning Minkowski's remark that $d\tau$ is not a complete differential, and noting that this shows that the proper times of two motions connecting two world points will generally differ (as discussed in Sect. 2 above), Sommerfeld adds:

> This assertion is based, as Einstein has stressed, on the (unprovable) assumption that the moving clock actually indicates the proper time, i.e. that at each instant it gives the time that corresponds to the instantaneous state of velocity, regarded as constant. The moving clock must naturally have been moved with acceleration (with changes of velocity or direction) in order to be compared with the stationary clock at the world-point P.[17]

[17] Lorentz et al. (1952, 94). Sommerfeld's notes were included in the first German edition published in 1913 (Lorentz et al. 1913, 71). I have translated from the original German: "Dieser Aussage liegt, wie Einstein hervorgehoben hat, die (unbeweisbare) Annahme zu Grunde, daß die bewegte Uhr tatsächlich die Eigenzeit anzeigt, d. h. jeweils diejenige Zeit gibt, die dem stationär gedachten, augenblicklichen Geschwindigskeitzustand entspricht. Die bewegte Uhr muß natürlich, damit sie mit der ruhenden im Weltpunkte P verglichen werden kann, beschleunigt (mit Geschwindigskeits-

This remark appears to me to indicate that Einstein is here equating "proper time" with "time in an inertial frame", very much as Rindler did in his discussion of the clock hypothesis in SR discussed above. The fact that there is an instantaneous velocity at each instant, which is the velocity with which the body would continue its motion if (counterfactually) no force were acting on it, does not entail that there is no acceleration, any more than the fact that Zeno's moving arrow does not move in each instant of its motion entails that the arrow is not really moving. To reiterate, in the context of the global Minkowski spacetime of SR, the fact that an ideal clock indicates proper time follows straightforwardly, and is not a separate "unprovable assumption".

But what of Rindler's claim that "more complicated laws could easily be devised which incorporate an acceleration effect and which are also invariant and reduce to [(7)] in the polygonal case" (29)"? I cannot see how this can be done for classical mechanics or SR. In the polygonal model, the times are proportional to the lengths between the vertices for inertial motions, and the impulses are assumed instantaneous, so that the time increments will still add linearly; and this will apply even in the limit. Newton's construction, indeed, is perfectly general, and his proof of Kepler's Area Law is valid for a curved segment of the orbit *whatever the force law*, provided the force acts towards the centre and the orbit remains outside the centre. But it seems that Rindler has GR in mind, for in this connection he immediately refers to a proof by Møller "on the basis of the field equations of general relativity that certain idealized mechanical and atomic oscillators do, in fact, satisfy the clock hypothesis" (29–30).

In fact, it is in the context of GR that we see the motivation for thinking that the clock hypothesis might be a separate element in the theory. For it is with the publication in 1918 of Hermann Weyl's celebrated attempt at a unified theory of gravitational and electromagnetic forces (Weyl 1918b) that the status of the "clock hypothesis" is first called into question. This is the paper that introduces the idea of *gauge symmetry* into modern physics, and its significance for the clock hypothesis has been discussed in an illuminating way by Harvey Brown and Oliver Pooley in their (2001), as well as by Roger Penrose in his recent tour-de-force (2005).[18] The connection with the clock hypothesis is that in Weyl's theory (as Einstein pointed out in criticizing the theory) the proper time elapsed for a clock carried on a round-trip would not only vary with the path through spacetime, as in SR, but in such a way that if the travelling clock had encountered regions of space containing a varying electromagnetic potential in a static gravitational field, it would return to its starting point ticking *at a different rate* than one that had remained at the starting point (Fig. 2).

The leading idea of Weyl's paper is that the Riemannian geometry assumed by Einstein in his GR is insufficiently local, since it "enables us to compare, with

---

oder Richtungsänsderungen) bewegt worden sein." Unfortunately, Sommerfeld does not give a reference for the attribution of this remark to Einstein.

[18] See also the discussion of Torretti in his (1983, 189–190).

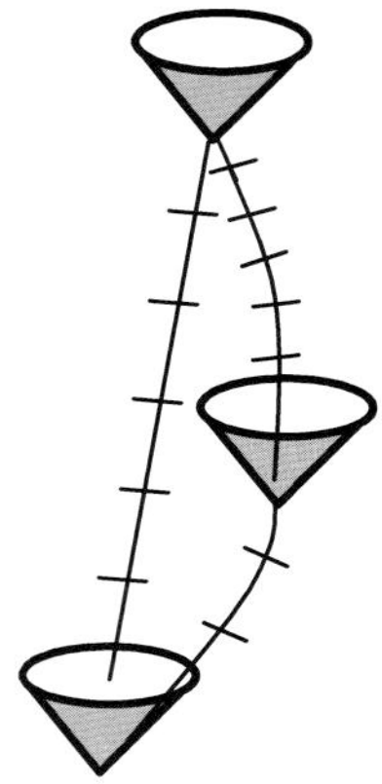

Fig. 2 The speeding up of the rate of the moving clock (after Penrose 2005, 452)

respect to their length, not only two vectors at the same point, but also the vectors at any two points" (Lorentz et al. 1923, 203). As Brown and Pooley explain, Weyl insisted instead that "only the ratios of the lengths of vectors at the same point and the angles between them can be physically meaningful" (265). The result is a *conformal geometry* in which there is no absolute scaling for spatial and temporal distances, so that the metric is only given up to a proportionality. In this scheme a Lorentz metric **g** is still required as a constraint on all the local physics, providing us with the local Lorentz group that is to act in the neighbourhood of each point. As a result, the null cones of Minkowski geometry still perform the same role that they do in Einstein's theory; this is the so-called conformal structure. Thus transformations of the form $\mathbf{g} \rightarrow \boldsymbol{\lambda}\mathbf{g}$ are permissible, where $\boldsymbol{\lambda}$ is a scalar function on the spacetime (Penrose 2005, 451). These are called *conformal rescalings*. Weyl posited some structure additional to the conformal structure (the Minkowskian null cones), namely a *gauge connection*, a bundle connection that would have the Maxwell field tensor $\boldsymbol{F}$ as its *curvature* (452). This curvature represents the (conformal) time scale change as the difference between two infinitesimal paths from a point p to a neighbouring point p'. Consequently, as Penrose explains, "in Weyl's geometry there are no 'ideal clocks'. The rate at which any clock measures time would depend on its history" (2005, 451). Of course, this turns out to be in conflict with the empirical evidence, which is why Weyl's theory was set aside.

Nevertheless, one might say that for Brown and Pooley the significance of this rival theory to Einstein's lies not in its truth or falsity, but in its very possibility. For the possibility of such a second time dilation effect of dynamical origin (Brown and Pooley 2001, 266) shows that Einstein has made two independent assumptions relevant to proper time: (a) that locally the metric is Minkowskian, and (b) that the consecutive elements $d\tau$ in consecutive LIFs (local inertial frames that are locally Lorentz) are all that contribute to total proper time elapsed, i.e. that there is no contribution over and above that of the instantaneous velocities in timelike paths that are not geodesics. Moreover, they see Weyl's argument in his 1918b as being of a piece with a pre-existing concern he had concerning the treatment of accelerated motion in SR:

> Weyl's opinion in *Raum-Zeit-Materie* (1918a) seems to have been that if a clock, say, is undergoing non-inertial motion, then it is unclear in SR whether the proper time read off by the clock is directly related to the length of its world-line determined by the Minkowski metric. For Weyl, clarification of this issue can only emerge when we have built up a **dynamics** based on physical and mechanical laws (1952, 177). (Brown and Pooley 2001, 264; Brown 2005, 115).

Now, as we saw above, this concern seems misplaced in the context of Special Relativity. Weyl himself seems to have realized this, for in his discussion of SR in his (1949) he writes that "it can be shown that the metrical structure of the world is already completely determined by its inertial and causal structure, that therefore mensuration need not depend on clocks and rigid bodies, but that light signals and mass points moving under the influence of inertia alone will suffice" (103). But, as Brown is at pains to point out, the status of SR changes between its first appearance as a global theory in 1905 (or its 1908 generalization by Minkowski), and its later subsumption into GR as holding only locally: what are accelerations resulting from gravity in the first theory are reconfigured as *inertial motions*, motions along geodesics in GR.[19] As a consequence, it is by no means a foregone conclusion that a natural clock π satisfying the condition for an ideal clock in SR, namely that it measures along its worldline the proper time $\int_P d\tau$ defined on the Minkowski metric U of each of the successive tangent spaces to its trajectory (the local Lorentz charts), will also do so in the global metric of GR. As Torretti explains, if the clock is a freely falling particle in a real inhomogeneous gravitational field,

> the several flat metrics which thus approximately hold good in the domain of each local Lorentz chart cannot be regarded as the restriction to their respective domains of a global Minkowski metric, any more than the Euclidean metrics that show up, say, in the street plans of Mannheim and Manhattan are the restrictions to these boroughs of a Euclidean metric defined on the entire Earth. On each domain U the worldline of π satisfies, within the said margin of error, the variational law $[\delta \int d\tau = 0]$; yet the law does not make any sense beyond the boundary of U. (1983, 151).

Not only must the global metric **g** be approximated on a small neighbourhood of each point by the local flat metric η, we must also stipulate that "if $\int d\tau$ is now made to stand for the length of π's worldline as determined by **g** – i.e. for the proper time measured along it by a natural clock at π – i.e. the variational principle $\delta \int d\tau = 0$ is obeyed by the freely falling particle between any two events in its history." (151) This is the *Geodesic Principle*. As Torretti observes, this was regarded by Einstein as "the core of the Hypothesis of Equivalence", or strong equivalence principle. The geodesic equation of motion for test bodies is now more commonly regarded as

[19] "The special theory of 1905, together with its refinements over the following years, is, in one important respect, *not* the same theory that is said to be a restriction of the general theory in the limit of zero gravitation (i.e. zero tidal forces, or space-time curvature). [I]n this picture, local inertial co-ordinate systems are freely falling systems. They are not in Einstein's 1905 theory." (Brown 2005, 15; cf. also p. 88).

a theorem, derivable from the vanishing of the covariant derivative of the energy-momentum tensor.[20]

So let's reconsider the logic of the argument from the failure in Weyl's theory of the clock hypothesis to the conclusion that it is an extra assumption in Einstein's. To be sure, the contrast with Weyl's theory is informative, and highlights the fact that the conforming of clocks to the clock hypothesis is a contingent matter, and also that it is implicit in Einstein's GR. But this is not sufficient to show that it is an *independent* hypothesis, rather than something that is already built into the theory. So let us dig deeper into the contrast between Weyl's theory and Einstein's. The fact that the clock hypothesis fails in Weyl's theory, despite the fact that the latter shares its conformal structure with GR, is due to the fact that Weyl's gauge connection is not a metric connection. As Brown and Pooley remark, "It is a function not only of the metric and its first derivatives, but also depends on the electromagnetic gauge field: in particular, for a fixed choice of gauge, the covariant derivative of the metric does not vanish everywhere." (267). The vanishing of the latter is the condition of *metric compatibility*, a condition which Brown and Pooley claim that "Schrödinger was right to call 'momentous'" (267). This consistency condition requires that every local Lorentz frame is an inertial frame, that is, a geodesic of the overall curved spacetime geometry. In their words,

> It means that the local Lorentz frames associated with a space-time point p (those for which, at p, the metric tensor takes the form diag $(1, -1, -1, -1)$ and the first derivatives of all its components vanish) are also local inertial frames (relative to which the components of the connection vanish at p.)" (267).[21]

Thus it is because GR obeys this condition that it becomes possible to stipulate further that the laws for the non-gravitational interactions take their familiar Lorentz covariant form relative to the local Lorentz frames. But this in turn is the content of the Strong Equivalence Principle (SEP), usually taken (e.g. by Misner et al. 1973) to be one of the essential postulates of GR:

> *in any and every local Lorentz frame, anywhere and anytime in the universe, all the (non-gravitational) laws of physics must take on their familiar special-relativistic forms.* (Misner et al. 1973, 386).

[20] See Misner et al. (1973, 471–480) and Brown (2005, 161–162). A very clear discussion and derivation of the geodesic principle is given by Stephen Lyle in his (2008, 36–46). He shows that the "principle" follows from Einstein's equations, if we assume an almost point-like particle in zero-pressure matter dust that is not jostled by other particles, has no torsion, and no electric charge, and is moving in a region of spacetime where the torsion is zero (this being a sufficient condition for the covariant derivative of the Einstein tensor to be zero) (43–44).

[21] See, for instance, Misner et al., 313. Actually, this is not quite correct, as Stephen Lyle has pointed out to me: in addition to the metric compatibility, a torsionless connection is also required in order for this to hold (private communication).

In his (2005), Brown seems to concur with this estimation of the SEP. He argues that some authors[22] have held that the "chronometric significance accorded to the Minkowski metric in SR is automatically recovered locally in GR". Against this, perhaps with the example of Weyl's theory in mind, he argues that "it is only through the SEP that such chronometric significance can be given to the tangent space geometry in the first place" (2005, 170).

This concurs with the point of view of Stephen Lyle, who shows how this chronometric significance is delivered as follows: he assumes in order, de Sitter spacetime, Schwarzschild spacetime, and the static homogeneous gravitational field, and shows in each case that by an application of the SEP one can "carry over from theories in SR that govern our clocks and rulers", to demonstrate that "the metric delivers the lengths and times that would be measured by our clocks and rulers" (2009, 5). The CH is not then an independent unprovable assumption, but a provable theorem. (In the case of de Sitter spacetime it is approximate, due to the fact that the inertial transformations, if they exist, are only approximately linear in de Sitter spacetime) (Lyle 2009).

To sum this up somewhat figuratively, it is the metric connection in Einstein's theory that threads together the various tangent spaces to a timelike spacetime path in such a way that the path has the same chronometric significance as in SR, and this is what the SEP achieves. Again, the contrast with Weyl's theory is instructive. For there the path dependence of the gauge connection is what spoils that chronometric role, and indeed makes it impossible to define an ideal clock, since "the rate at which any clock measures time would depend upon its history" (Penrose 451). It is the SEP in Einstein's GR that allows for proper time to have the same chronometric role it does in SR.

This relates in a suggestive way to another concern of Brown's, namely that GR explains something that previously should have been considered "a miracle", namely that bodies not under the action of external forces should "conspire to move in straight lines at uniform speeds while being unable, by *fiat*, to communicate with each other" (2005, 15). Famously, of course, Einstein's GR takes some of the mystery out of the numerical equivalence of inertial mass and gravitational mass by simply identifying them, the root idea behind the equivalence principle. But now we can see this as having the effect of enshrining Newton's idea that absolute time is the measure of time beaten out by a body undergoing inertial motion – which Einstein and Minkowski had simply imported into Special Relativity – in a perfectly consistent way in General Relativity. This is why the SEP manages to preserve the chronometric significance of the metric in SR: it preserves the relation of inertia to time assumed in modern physics. This relates in an interesting way to Einstein's criticism of Weyl's unified theory. For as Penrose points out, it is not just that spectral frequencies will depend on an atom's history: so will particle masses! Given the Einstein relation $E = mc^2$ together with the De Broglie relation $E = h\nu$, it follows

[22] Brown (2005, 170, n. 51) and Brown and Pooley (2001) specifically singles out Roberto Torretti, a claim which the above discussion and quotation might appear to throw into doubt. But I shall not discuss that here.

that every particle of rest mass *m* will have an associated natural frequency mc^2/h. "Thus, in Weyl's geometry, not just clock rates but also a particle's *mass* will depend upon its history." (Penrose 2005, 453). The ideal clocks of special and general relativity, by contrast, in preserving the relation of inertia to time, also preserve the constancy of rest mass in time.

4 Conclusion

I have argued here that *proper time* must be regarded as one of Minkowski's enduring contributions to physics. I have examined some confusions that still interfere with an appreciation of this, including a conflation of proper time with the coordinate time of the inertial frame of a system at rest, and the related mistaken notion that SR cannot be applied to accelerating systems. This sets the stage for a treatment of the so-called "clock hypothesis" (CH). If my arguments above are sound, then the CH is not needed as an independent postulate in SR. Insofar as it can be regarded as stating the criterion for an ideal clock in SR, it is already implicit in that theory in the invariance of proper time, as defined by Minkowski: in a spacetime whose global metric is Minkowskian, an ideal clock cannot fail to keep proper time, however it is accelerated. The argument that many real clocks will fail to satisfy the hypothesis is just the claim that many processes fail to qualify as ideal clocks; but provided we can account for that discrepancy by means of an account of the ratio of the acceleration undergone by the clock and its restorative forces, no appeal to any hypothesis extrinsic to the theory is needed. The argument that there is a contrast between a "kinematic account of acceleration" and "the dynamics of the real forces acting on the clock" seems to misconstrue the question of whether a given process can approximate an ideal clock with the question of whether an ideal clock can be defined using the theory alone; and also to come close to implying that SR is inadequate to treat real accelerations, the mistaken notion we treated in Sect. 2.

Secondly, the failure of the clock hypothesis in Weyl's unified theory of gravity and electromagnetism does not imply that it is an independent assumption in Einstein's theory. There *is* an assumption in Einstein's GR separate from the assumption of LIFs: this is the (strong) Equivalence Principle, that in every such LIF, at any spacetime point all the non-gravitational laws of physics must take on the forms they have in SR, i.e. must be Lorentz covariant. In a theory such as Weyl's, ideal clocks do not in general keep proper time. This is a result of the non-metric compatibility of Weyl's theory: the covariant derivative of the metric does not vanish everywhere. But by the same token, Weyl's theory does not conform to the Strong Equivalence Principle. Einstein's GR, on the other hand, does conform to the Equivalence Principle, and it follows from this that the clock hypothesis is no more an additional assumption in GR than it is in SR.

References

Arthur R. T. W. (1995). "Newton's Fluxions and Equably Flowing Time," *Studies in History and Philosophy of Science*, **28**, no. 2, pp. 323–351.

Arthur R. T. W. (2007). "Time, Inertia and the Relativity Principle." (preprint; *philsci-archive.pitt.edu*, deposited November 19, 2007. http://philsci-archive.pitt.edu/archive/000 03660/).

Arthur R. T. W. (2008). "Time Lapse and the Degeneracy of Time: Gödel, Proper Time and Becoming in Relativity Theory," pp. 207–227 in *The Ontology of Spacetime II*, ed. Dennis Dieks. Amsterdam: Elsevier.

Arthur R. T. W. (2009). "On Newton's Fluxional Proof of the Vector Addition of Motive Forces," forthcoming in *Infinitesimals*, ed. William Harper and Craig Fraser, University of Western Ontario Series in Philosophy of Science. Heildelberg: Springer.

Barbour J. (1989). *Absolute or Relative Motion? Vol. 1: The Discovery of Dynamics*. Cambridge: Cambridge University Press.

Brown H. (2005). *Physical Relativity: Space-Time Structure from a Dynamical Perspective*. Oxford: Clarendon Press.

Brown H. R. and Pooley O. (2001). "The Origin of the Spacetime Metric: Bell's 'Lorentzian Pedagogy' and Its Significance in General Relativity," pp. 256–272 in Callender and Huggett (2001).

Bruce J. B. (1985). *The Key to Newton's Dynamics*, Berkeley and Los Angeles: University of California Press.

Einstein A. (1905, 1923). "Zur Elektrodynamik bewegter Körper," *Annalen der Physik*, 17, 1905. Translated by W. Perret and G.B. Jeffery as "On the Electrodynamics of Moving Bodies," pp. 37–65 in Lorentz et al. (1923).

Einstein A. (1918). "Dialog über Einwände gegen die Relativitätstheorie," *Die Naturwissenschaften*, **48**, pp. 697–702, 29 November 1918 (a very rough English translation exists at http://en.wikisource.org/wiki/Dialog_about_objections_against_the_theory_of_relativity).

Einstein A. (1954, 1982). *Ideas and Opinions*. New York: Random House.

Hartle J. B. (2003). *Gravity: An Introduction to Einstein's General Relativity*. San Francisco: Addison Wesley.

Herivel J. (1965). *The Background to Newton's 'Principia'*. Oxford: Clarendon Press.

Jones P. and Wanex L. F. (2006). "The Clock Paradox in a Static Homogeneous Gravitational Field," *Foundations of Physics Letters*, **19**, no. 1, pp. 75–85, February 2006. (arXiv:physics/0604025v3).

Lorentz H. A., Einstein A. and Minkowski H. (1913). *Das relativitätsprinzip, eine sammlung von abhandlungen, mit anmerkungen von A. Sommerfeld und vorwort von O. Blumenthal*. Leipzig, Berlin: B. G. Teubner, 1913 published electronically by Ann Arbor, Michigan: University of Michigan Library (2005) at http://name.umdl.umich.edu/ATE6002.0002.001.

Lorentz H. A., Einstein A., Minkowski H. and Weyl H. (1923, 1952). *The Principle of Relativity: A Collection of Original Memoirs on the Special and General Theory of Relativity* (originally Methuen 1923) New York: Dover.

Lyle S. (2008). *Uniformly Accelerating Charged Particle A Threat to the Equivalence Principle*. Berlin/Heidelberg: Springer.

Lyle S. (2009). "Extending Bell's Approach to General Relativity," unpublished.

Minkowski H. (1908). "*Raum und Zeit*", reprinted in Lorentz et al. 1913, and in translation as "Space and Time" in Lorentz et al. 1952.

Misner C. W., Thorne K. S. and Wheeler J. A. (1973). *Gravitation*. San Francisco: W. H. Freeman and Co.

Møller C. (1955). *The Theory of Relativity*. Oxford: Clarendon Press.

Nauenberg M. (1998). "The Mathematical Principles Underlying the *Principia* Revisited," *Journal for the History of Astronomy* **29**, pp. 286–300.

Newton I. (1999). *The Principia: Mathematical Principles of Natural Philosophy*. Trans. I. Bernard Cohen and Anne Whitman. Berkeley: University of California Press.

Penrose R. (2005). *The Road to Reality: A Complete Guide to the Laws of the Universe*. London: Vintage Books.

Petkov V. (2009). *Relativity and the Nature of Spacetime*. Dordrecht: Springer.

Pourciau B. (2003). "Newton's Argument for Proposition 1 of the *Principia*," *Archives for the History of the Exact Sciences*, **57**, pp. 267–311.

Rindler W. (1966). *Special Relativity* (2nd edn.). Edinburgh: Oliver and Boyd.

Rindler W. (1977). *Essential Relativity* (2nd edn.). New York: Springer.

Sobel D. (1995). *Longitude: The true Story of a Lone Genius Who Solved the Greatest Scientific Problem of his Time*. London: Fourth Estate.

Torretti R. (1996). *Relativity and Geometry*. (First published 1983, Oxford: Pergamon Press). New York: Dover.

Weyl H. (1918a). *Raum-Zeit-Materie*. Berlin: Springer.

Weyl H. (1918b). "Gravitation und Elektrizität," *Sitzungsberichte der Königlich Preussischen Akademie der Wissenschaften zu Berlin*: 465–480. Translated by W. Perret and G.B. Jeffery as "Gravitation and Electricity," pp. 201–216 in Lorentz et al. (1923).

Weyl H. ([1921] 1952). *Space-Time-Matter* (4th edn). Translation of Weyl 1918a by H. L. Brose. New York: Dover. QC 6 W55

Weyl H. ([1949] 1963). *Philosophy of Mathematics and Natural Science*. (Princeton University Press, 1949). Republished as New York: Atheneum.

Why Spacetime Is Not a Hidden Cause: A Realist Story

Graham Nerlich

"Spacetime acts on matter, telling it how to move" (Misner et al. 1970, p. 5; Taylor et al. 1991 p. 275).

Abstract Spacetime realism requires that it is not hidden and not a cause. Its style of explanation is geometrical. It is argued that causal explanation is unworkable for cases of pure gravitation. Non-causal explanation is geometrical and exploits several identities where one might expect causal explanation. Thus a realist understanding of General Relativity is to be preferred.

1 Introduction

How does – how *could* – spacetime act on matter or tell it how to move?

The best short argument against realism runs like this: if spacetime is a real entity for General Relativity (GR) then surely the acting and the telling must be a causing – a *hidden* causing. But, equally surely, spacetime is the wrong kind of thing to *make* matter move. That's bad physics and bad metaphysics. But if spacetime causes nothing, it explains nothing either. So weed it out of the ontology of GR and settle for a codification – whatever *that* is.[1] [DiSalle 1994, pp. 321–328, 1995, pp. 275–277. Brown 2005, pp. 24–25; Brown and Pooley 2004; Torretti 2006 p. 3n. For doubts about codification (in another context) see Nerlich 2005, Sects. 2.1, 3.1]

The argument goes astray from the start. Realism doesn't need and can't admit spacetime as causing matter to move. Spacetime is not a hidden cause because not a cause.[2] Yet spacetime explains what matter does under pure gravitation. It does so rather straightforwardly. It exploits various direct identities. That is misunderstood, widely I think, perhaps because the search for causes clouds the issue.

[1] For doubts see Nerlich 2005, Sects. 2.1, 3.1.

[2] And not hidden either; see Nerlich 1994, 38–43.

V. Petkov (ed.), *Space, Time, and Spacetime*, Fundamental Theories of Physics 167, DOI 10.1007/978-3-642-13538-5_8, © Springer-Verlag Berlin Heidelberg 2010

Familiar thoughts motivate this paper. Gravity makes no sense as action across a distance by some massive things on others. It is not a force, not a cause. GR makes sense only as a local theory: it demands proximal explanation. In lots of pure gravitation situations, the *only* proximal feature available to explain anything is local spacetime structure. But surely it can't explain matter's motion by causing it. So a style of *geometrical* explanation both local and acausal surely looks at least worth consideration. Of course the idea is frightening. Ontologists abhor spacetime just as nature, it was once supposed, abhors a vacuum.

Apart from its last step, the premises of this paper's argument rest on common ground; indeed, they make up the simplest, basic ideas of GR. The step to the conclusion is no less simple and direct. Further, there are simple examples already to hand of non-causal geometrical explanation. The handedness of hands depends on whether their containing space is orientable or not. Roughly, they are handed if *there exist no paths* in the space that will smoothly map an asymmetrical object onto its mirror image, and not handed if *such paths exist.* This isn't causal explanation – space *does* nothing to hands. That is some sort of *existential* explanation. The shape of spherical space explains why there are no similar shapes of different sizes in that space – why it has no similarity geometry. A triangle with greater perimeter must contain more space than contained by a Euclidean triangle of the same perimeter, since they have greater area. That, too, is somehow existential.

What follows in this paper is familiar and obvious too. So much so, that it continues to puzzle me why it needs to be said. But that is a dangerous state of mind in which to approach the problem. I suspect that the main difficulty lies in the horror of spacetime realism. Dispelling the horror is the hard part of this work, but it is not closely examined here.

I start with some prehistory of inertial motion.

2 Cause and Classical Inertial Motion

Confusion once reigned as to what keeps an arrow flying. Galileo's giant stride towards clarity turned on the relativity of motion and the composition of velocities. He saw that "What keeps the arrow flying?" is the wrong question. Instead, ask what causes it ever to stop. Then there are genuine causal answers: e.g. gravity pulls it down to earth, or it hits something. A more precise message was fogged by the great Italian's preoccupation with circular (including horizontal) motion. This obscured the role of linearity in free motion. (Chalmers 1993).

Newton's first law of motion is clear on linearity:

> *Every body persists in its state of rest or of moving uniformly straight ahead, except insofar as it is compelled to change its state by forces impressed.* (Newton 1999, p. 416)

Thus Newton straightened out Galileo's story, but only to the extent of Corollary V

> *When bodies are enclosed in a given space, their motions in relation to one another are the same whether the space is at rest or whether it is moving uniformly straight forward without circular motion.* (Newton 1999, p. 423)

Notoriously, the rest or motion of the 'given space' is absolute. Perhaps this is why Newton hinted at the older thought[3] that a "force of inertia" ("vis insita" or "vis inertiae") causes the arrow's flying on.[4] This still leaves something to be desired but it is not the thought that *space* could cause anything.

The released bowstring pushes the arrow and causes it to fly. There, cause is force. If we look for a cause why the arrow *keeps on* flying, we look for a cause of inertial, free-fall motion: we look not for an initiating cause but for a proximal one. That a thing is moving inertially at some velocity now might be because it *was* just moving at that velocity. However, the earlier state doesn't force the later one, despite being a distinct, preceding state. A force is needed to *change* it. The structure of space is plainly no such cause even though its straights are the specified paths.

But doesn't the preceding inertial motion, the conserved momentum, *cause* the present inertial motion? Not if we accept both the first law and the relativity of inertial motion. An adroit frame-swap can transform any state of free motion to a state of rest. The effect vanishes and the cause with it. That is because the 4-acceleration vector is 0 so no force recognised in GR can be at work. Any further search for a cause of inertial states must look for an account of why things endure. I grant, more for the sake of getting along than from conviction, that it will be a causal story (e.g. Tooley 2001, p. 398). Even if there is one, space (spacetime) has no part in it. There is no need to ask why an object at rest in an inertial frame stays put; for any object in uniform motion there is a frame in which it is at rest. The thing merely endures. This is *satisfactory:* questions can rest at this point.[5]

It's remarkable that the first law says nothing at all about the causal powers of any body to which it applies. It says nothing about what causes anything to endure. The second law requires that all bodies have mass; the first mentions no property whatever. Remarkably, too, we have good *classical* reason to think that no body ever actually does escape the (gravitational) causal net or persists in its state of rest or uniform motion (although there might be some bodies on which the resultant of forces is zero, briefly or not). This suggests that the law is about *trajectories*, spatio-temporal entities, not about what might occupy them. It tells us nothing of how any such trajectory ever comes to be occupied. It need say nothing about why an occupant remains on the trajectory, but it does explain why causes are needed to drive it off. It is about the importance for dynamics of the default case: the non-causal trajectory in which there is zero acceleration. The default, in pre-spacetime talk, is rest-or-uniform-motion. I will call these *Galileo trajectories*. Their importance

---

[3] Compare Buridan and Benedetti. See *Wikipedia: the free encyclopedia* article 'The Principle of Inertia' 2.1.1.

[4] See 'A Guide to Newton's *Principia*' by Bernard Cohen, Chap. 4, 4.7 esp. p. 98 *in* Newton 1999).

[5] If all that is sound, then there is a classical non-causal process, a changing of spatial distance between two suitably inertially moving things. The motion of neither is an effect, since it vanishes under frame swaps. The changing distance between them is a covariant quantity of the Galilean (Lorentz) group: it is a real change. The change is uncaused. If so, it is odd that this was never cited (at any rate it never caught on) as an obvious exception to the rule that all changes are caused.

emerges in the relativity of motion and the composition of velocities, which, in turn, depend on the spatial and temporal symmetries of classical mechanics.

The first law really is first. It is conceptually simpler and theoretically deeper than the 2nd. Once we can decide simply when forces are on or off, we can identify the required frames of reference (candidate rest states). Inertial motion is not defined by laws of motion: rather it is a rather direct observational truth as to what trajectory is found as forces approach zero. To a large extent, Newton decided this by seeing free motion as free from *impressed* forces (impacts, pushes and pulls) and gravity. This laid a groundwork: candidate forces should have (1) observable sources, and (2) regularities governing (a) *when* and (b) *how* they are at work. This rules out arbitrary, conventional postulations of force. Only when we have the right frames of reference and, by implication, the right transformation group, may we explore accelerations relative to them in a comprehensive way; only then can forces be quantified and oriented. Then you can formulate the 2nd law and verify that 2nd derivatives are at the core of dynamics. That the 2nd law entails the first does not rob the first of first place.

3 GR Space, Time and Spacetime

Here's my strategy in a nutshell. In pure gravitation examples, GR explains what matter does by extending the idea of Galileo trajectories to 4-geodesics (straights[6]) in spacetime even though these often have no rest-or-uniform-motion image relative to frames of reference (space and time representations). Roughly, that a worldline is a Galileo 4-trajectory explains why its occupant is innocent of causal dependence, guidance etc. beyond its mere endurance (the mere extension of its worldline). It merely falls (floats) freely – free of causes and forces. Of course, the object has causal power to interact with other bodies and with force fields! However, the images of these Galileo 4-trajectories in space and time often *do* call up causal and dynamical stories about their occupants since, in that setting, they breach the first law.

This is well illustrated by Einstein's familiar example of a rotating disk in Minkowski spacetime. (Einstein 1920, Chap. XXIII; Einstein et al. 1923, pp. 115–117) Suppose an inertial frame F in which the disk's centre is at rest. Relative to F, a particle in uniform motion crosses the centre of the disk. Its 4-trajectory is Galilean. Yet, relative to a frame co-moving with the disk, the particle follows an outward spiral at varying speed. It accelerates. The first law demands the postulation of a force field throughout the frame. It will vanish at the centre of the disk and vary as a function of its radius. Relative to that frame, the particle's motion is said to be forced and caused. Plainly the path and speed of the particle are not uniform. Yet, in the

[6] I write 'straight' where you might expect 'geodesic'. Geodesics just are straights of whatever space they are in The shorter term reminds us of what matters about them for this paper.

spacetime representation, the 4-trajectory remains straight and Galilean. There is no force on the particle and no cause of its continuing motion-or-rest. The structure of spacetime explains how the trajectory is Galilean; it does not cause anything.

I place two conditions on cause: (1) if x causes y, then not (x = y); (2) for mechanics, causes are forces.

The explanatory role of spacetime in the behaviour of freely falling matter is twofold. It explains (as illustrated) how the apparent gravitational dynamics of free-fall particles in general frames of reference vanishes into the mere kinematics of geodesics in flat or curved spacetimes. It explains also by citing several *identities*. Suppose the trajectory of a cloud of test particles through flat spacetime projects it into a region of curved spacetime. There may be immanent causes for the persistence of the particles: they explain how the cloud *gets into* the curved region. Nothing in this implicates spacetime causally. The flatness of the region of spacetime does not cause the curvature of the neighbouring region that the cloud traverses. The change in shape of the cloud, the deviation of its point-parts, *is* the deviation of geodesic worldlines and not caused by it.

3.1 Free Fall in a Purely Gravitational Field

I'll enlarge that simple GR example in which the geometric structure of spacetime fully explains an observable behaviour of matter. Let's begin with an idealised cloud of matter-points (pressure free dust): it is spherical at t_0, falling freely ("under gravity alone") towards a massive object. To delete any influence from local matter, assume dust points with negligible mass, ignore gravitational forces between them, and assume there is no other interaction among the points. "Gravity" from the *distant* source is not erased; it is the curvature of spacetime. The cloud will change shape.

The origin of a space-and-time frame of reference (not inertial) floats freely at the centre of the cloud. A point at rest there will remain at rest with zero gravitational force on it. The cloud changes shape round that central point which is at rest in the frame. In the direction of the distant source, the cloud gradually stretches out fore and aft, but it contracts across the orthogonal section – it gets longer and thinner. This closely approximates classical gravity, where it has a causal, dynamical explanation. Clearly, the non-central points move, indeed accelerate in the frame. The more distant points acquire larger 3-velocities in it: some move towards the centre, others away. What accelerates them is a force demanded by the 1st law, a tidal gravitational "force".

A similar tale may be told selecting any point in the cloud as at rest.

That language, that array of theoretical concepts, is appropriate if we conceive of the frame (as we conceive of ourselves) as a spatial thing enduring in time. Spacetime is nowhere in this image. 'Spacetime' is not among the concepts in which the space and time explanation may be requested or provided.

Here the changing shape of the cloud needs an explanation. It's explained by the curvature of the spacetime 4-region. The explaining facts must be distinct from what they explain if the explanation is causal. But they are the same facts.

Each enduring point is an extended worldline. Spacetime doesn't explain the particle's extension along the straight nor what causes the straight to be occupied (and thus a worldline). Yet Galileo's insight remains – don't seek a cause for simplest dynamically-default states. That's *satisfactory* because the straight has a zero acceleration vector at every point. That's what a straight *is*. No acceleration, no force, no cause. That spacetime straights are the worldlines of simple endurance/extension is *satisfactory* for the same old reason – nothing to explain.[7] Here's where question and explanation may halt.

What spacetime explains is why a 4-straight should be the dynamical default state. It can't explain why anything is in that state.

The *identity* of the state of affairs differently presented in these descriptions explains what happens to the cloud, so long as there is a cloud. It tells us why the trajectories of the points change the shape of the cloud: the worldlines of different points lie on different straights, and these straights deviate in curved spacetime. The deviating straights project down into accelerating space and time trajectories, among them those that happen to be trajectories of particles. The deviation doesn't cause the acceleration. It's what the acceleration *is*; it *is* the change in shape made up by the trajectories of the points. The identities forbid a causal tie.

In turn, the deviation of the worldlines is not *caused* by the curvature of spacetime, since it *is* the curvature; curvature is the deviation of *all* geodesics. Flat spaces are those admitting parallels, so 'curved space' simply *means* 'space in which straights deviate'.

Spacetime doesn't cause material worldlines to lie on straights. If you like, spacetime doesn't fully explain all of this because it doesn't explain the endurance of the test particles. But the endurance doesn't cause the change in shape. Spacetime explains it through identities, not causes.

In Minkowski (1908) and Einstein (1916) this style of explanation through various spacetime identities, was without precedent and remains unique in science, both physically and metaphysically. Thus it shows the ontic type and role of spacetime as without parallel. That's its metaphysical importance.

Finally, to parody Quine – no identities without entities. Only a realist can tell this story.

---

[7] It's not so satisfactory that I assume that we will never find a deeper explanation for it or that the deeper explanation will be consistent with the one made out here. There is no explanation within General Relativity.

3.2 Light Bending

Eddington confirmed the bending of light rays near the sun as predicted in GR. The immediate observation was of dots on (several) photographic plates of the sun at eclipse. The grouping of the dots was caused by a grouping of photons. Our question is about their *separation* and how spacetime structure explains it.

To say that light rays bend round the sun is to say that in the 3-space of some frame[8], light rays do not move along straights of *that 3-space*. A tidal force, gravity, bends them, in this story.

That's causal. The story is a kind of fiction.

Once more, spacetime is nowhere in this picture. 'Spacetime' is not among the concepts in which the space and time explanation may be requested or provided.

In this case, too, the motion of photons translates up into lightlike straight world-lines. The mapping between space-and-time, and spacetime, representations is an *identity*.

The structure of spacetime *explains* why the dots on the photographic plates are separated as they are. The explanatory structure is the curvature. Spacetime curvature consists in the deviation of its straights, including lightlike ones. That, in turn, explains the separation of dots on the plate. That's how the photographic surface intersects the deviating luminal 4 straights, independently of whether the plate is there or not. Curvature does not *cause* the deviation because it *is* the deviation. The curvature tensor simply *analyses* and *measures* the deviation – an identity not a cause.

Again, flat space is, unique in having parallels. The failure of parallels *is* the curvature: it *is* the deviation of straights.

That completes the explanation. It is not causal; it is realistic – no identities without entities.

4 About Matter

I've told my story with some idealised bit-players – test particles. My cloud of dust was misrepresented as made of massless particles each of which tracks a straight in a structure unaffected by these contents. But real dust is made of small but extended specks, not particles. Even specks have some mass that will constrain spacetime structure; clothed with specks, spacetime doesn't have the same straights as it has naked.

Any spacelike cross section of a speck will be intersected by more than one time-like straight. Since these deviate in curved spacetime, the causal story within any speck is not trivial. Elastic forces inside resist the deviation of the speck's smaller

[8] Not an inertial frame, since spacetime is curved and lacks parallels. Only in the limit is spacetime flat and inertial frames locally available.

parts: internal stresses, distortions, will arise in it. As elastic wholes, specks won't lie on G-trajectories.

The causal story about specks is exhausted in the play of electromagnetic forces engaged in resisting the distortions and in any immanent causes of speck endurance. As before, spacetime explains the deviation of geodesics that change the electromagnetic forces, but it does not cause the forces. It continues to explain as before the causal-default part of the story – why this trajectory needs no cause for any geometrically simple extended thing to lie along it. At each point, its space-like acceleration vector is zero. The spacetime story is about the cause-free status *of the trajectory*. That explanation does not encroach on any theory of matter. An occupying point is irrelevant *save as an illustrative fiction.* [9]

Yet we do accept *exactly that explanation* in real if approximate cases. The orbit of Mercury is calculated treating the planet as a point (among other approximations). The observed advance of the planet's perihelion, famously, is very close to the GR-predicted Galileo-trajectory along which the idealised planet would extend. The orbit is a spacetime straight. Unknown stresses within the planet, and unobserved imperfections in its straight 4-trajectory are ignored. We fully understand why the orbit is the one we see: it's virtually a geodesic. As an illustration, it *traces* the structure of spacetime. The structure is not a hidden something (not concealed, obscured, not too small, not too fine). It is observed with highly non-trivial precision, even though we know that we see an approximation and that the unoccupied straight itself is not a visual object.

For similar reasons, in illustrative explanations, we may ignore the epicycle of feeding the small masses of the specks back into the T tensor. That will simply generate a new set of straights and these will be causal-default trajectories as before; the geometric explanation exploits just the same feature of the revised spacetime structure and its straights. It wasn't really ever about the properties of matter.[10]

5 A Parody of 'Hidden Cause'

I turn to a lively and very explicit satire on the hidden cause part of the argument in the first three sentences I began with; it is given in Brown and Pooley 2004, Sect. 1 and repeated in Brown 2005, p.24. It will be clear that it contrasts sharply with the story just told. It amusingly parodies geometric explanation as causal. In a geometrical explanation, they suggest, matter must follow something like "grooves" or "gutters" in spacetime along which spacetime "nudges" them. (Brown 2005, p.24,

---

[9] Nerlich 1979, Sect. 4; Nerlich 1991, Sects. 3 and 4.

[10] Compare (Brown 2005, p.24) that "... world-lines [of test 'particles'] follow geodesics *approximately* and then f*or quite different reasons*" from anything to do with the nature of test particles (his italics). Apart, of course from their natural tendency to persist. That leaves the story told here untouched.

p.161 for "nudge"). The thought is that the grooves *force* things to follow them. Clearly, this geometric story is causal.

I have no quarrel with the parody – but does anyone hold the view it attacks?[11]

Despite their calling this view popular, I can think of no *published* versions of anything like it, although it sometimes – too often – comes up in discussion. It is quite unworkable; how could it yield the crucial result that the geodesic followed in free fall is independent of the mass of the falling body? But something makes this mistake easy; it is exactly what makes the argument I mentioned at the beginning of the paper so plausible. Doesn't geometric explanation just *have* to be some kind of causal explanation?

Two interesting points arise: (1) the parody rests on the presupposition that test bodies would be doing *something else* if the nudge along spacetime's grooves did not turn them from it. Without that presupposition, the gutters, the nudges and the parody itself have no intelligible point. (2) This never-mentioned something else would either be a state without external cause or have such a cause. If it is uncaused, some causal default state is tacitly recognised as necessary and intelligible: why not the groove-free state - the 1st law - we began with? If it is an externally (e.g. electromagnetically) caused state, then GR tells us that the trajectories won't be geodesical and the geodesical grooves would play a totally obscure part in the parody. I conclude that the parody depends on the tacit admission that causal default states are both essential and intelligible. I warmly welcome that, of course. It does presuppose the 1st law as causal default explanation, however.

But, really, how can Brown or Pooley admit this? The rejection of causal defaults, mere kinematics, is just what their constructivism weds itself to. Everything is dynamics. Deploring mere kinematics, Brown, for instance, writes of the 1st law as a conspiracy "among force-free bodies to move in straight lines while being unable ... to communicate ... It is probably fair to say that anyone who is not amazed by this conspiracy has not understood it." He asks "by what mechanism is the rod or clock informed ... as to what this [spacetime] structure is?" (Brown 2005; p. 8, pp. 12–13) Again (24) "it cannot simply be in the nature of free test particles to 'read' the projective geometry, or affine connection..."

---

[11] I do quarrel with their ascribing the view to me on the basis of a three-sentence quotation from my 1976 book in which I said (in terms of a familiar metaphor about antennae) that action at distance plays no role in GR. There is no hint of nudges, gutters, grooves or causes, There are seven index items in the book under 'Geometric explanation'. Not one of them is mentioned by Brown or Pooley; all of them argue for, state or imply a *rejection* of the story pinned on me. No item refers readers to the passage they cite; it is about GR's being a local theory. I question whether any theory like the one parodied "has become very popular", and their citing a mere three sentences about something entirely different suggests some desperation in the search to find any that does; it also suggests that Brown's "it is one of the aims of this [his] book to rebut this and related views" is not an aim supported by significant research. Having trod what seems to me a solitary missionary path for 32 years, it is disappointing to find oneself cited as a leading spokesman for a supposedly widespread view that one has always opposed. Brown 2005 p. 23 includes the relevant claims.

After an amicable discussion, I can report that the authors have withdrawn the attribution to me.

Of course not. If my earlier arguments are good, free rods and clocks know nothing, feel nothing, 'read' nothing, *do* nothing. Their natures are completely irrelevant. Were there unextended free fall (float) strict particles, they would do the very same thing in one spacetime as they do in any other: they stay put, do nothing. They simply endure. Their worldlines extend along zero acceleration, causal default, trajectories without benefit of nudge or communication. Spacetime structure relates Galileo trajectories to each other: curvature is their deviation. That the field equation entails the law of motion in GR is a significant *formal* result. But it can't tell us what guides the point particles, since nothing can guide them. These are zero acceleration trajectories and can't be steered, guttered or grooved.

Nothing in my discussion suggests that we should know the causal default state *a priori*. My colleague, Greg O'Hair suggested that it might have been a random spatial walk. It is an *empirical, theoretical* fact that the causal default is a spacetime straight. The identities cited before are also empirical and theoretic. That makes perfect sense within a contingent geometry and mechanics. I claim that it is *satisfactory*. It is not causal.

6 Conclusion

Finally, there are two bits of unfinished business. Identity arguments are powerless to settle two remaining problems. (1) They can't relate geometric structure in one spacetime region to that in another nearby region. But I do not think that is a causal relation either; (2) Plausibly, if spacetime can't act causally on matter then matter can't act causally on spacetime. Identity arguments look impotent to tell us how matter is related to spacetime in that direction. The field equation is not an identity. Nevertheless, it is not causal either, but an equation of mutual constraint. (See Geroch 1978, p. 174, 176). One aspect of the identity of gravitational with inertial mass is that GR need only consider inertial mass. The matter side of the field equation need not be taken as a source term. This thought needs long and careful reflection on the relation between curvature of spacetime, gravitational potential energy and the mass of curved, empty spacetime and more. That is the topic of another paper – or two or three. Further, we can't set up matter and then see what happens to spacetime; nor vice versa. Indeed, it is perhaps more common to specify a metric, then look for a suitable matter tensor. So these things don't smell causal. But I hope there is something better to say about the problem than that.

References

Brown, H.R. (2005) *Physical Relativity: Spacetime Structure from a Dynamical Perspective*, Oxford, Oxford University Press

Brown, H.R. and Pooley, O. (2004) 'Minkowski spacetime: a glorious non-entity' http://philsci-archiv.pitt,edu/archive/00001661/

Chalmers, A. (1993) Galilean Relativity and Galileo's Relativity, In S. French & H. Kamminga (eds) *Correspondence, Invariance and Heuristics: Essays in Honour of Heinz Post*. Dordrecht, Kluwer

DiSalle, R. (1994) On Dynamics, Indiscernibility, and Spacetime Ontology, *Br J Philos Sci* 45: 265–287

DiSalle, R. (1995) Spacetime Theory as Physical Geometry, *Erkenntnis,* 42: 317–37

Einstein, A. (1916) On the Foundation of the General Theory of Relativity, in Einstein et al. 1923

Einstein, A. (1920) *Relativity: The Special and General Theories: A Popular Exposition*, London, Methuen, Tr. R. Lawson

Einstein, A, et al. (1923) *The Principle of Relativity*, New York, Dover

Geroch, R. (1978) *General Relativity from A to B*, Chicago, University of Chicago Press

Minkowski, H. (1908) Space and Time, in Einstein et al. 1923

Misner, C. et al. (1870) *Gravitation*, San Francisco, Freeman

Nerlich, G. (1979) What Can Geometry Explain?, *Br J Philos Sci* 30:69–83

Nerlich, G. (1991) How Euclidean Geometry Has Misled Metaphysics, *J Philos* 88:169–189

Nerlich, G. (1994) *The Shape of Space*, 2nd edn. Cambridge, Cambridge University Press

Nerlich, G. (2005) Can Parts of Space Move? On Paragraph Six of Newton's Scholium, Erkenntniss 62:119–135

Newton, I. (1999) *The Principia: Mathematical Principles of Natural Philosophy*, Berkeley, University of California Press, Trans. I. B. Cohen et al.

Taylor, E.F. et al. (1991) *Spacetime Physics* NY, Freeman

Tooley, M. (2001) Causation and Supervenience, In M. Loux and D Zimmerman (eds) *Oxford Handbook of Metaphysics*, Oxford, Oxford University Press, pp 386–434

Torretti, R. (2006) Can Science Advance Effectively Through Philosophical Criticism and Reflection? http://Philsci-archive.pitt.edu/archive/00002875

Wikipedia: the free encyclopedia article 'The Principle of Inertia' 2.1.1

Structural Explanations in Minkowski Spacetime: Which Account of Models?

Mauro Dorato and Laura Felline

Abstract In this paper we argue that structural explanations are an effective way of explaining well-known relativistic phenomena like length contraction and time dilation, and then try to understand how this can be possible by looking at the literature on scientific models. In particular, we ask whether and how a model like that provided by Minkowski spacetime can be said to represent the physical world, in such a way that it can successfully explain *physical* phenomena *structurally*. We conclude by claiming that a partial isomorphic approach to scientific representation can supply an answer only if supplemented by a robust injection of pragmatic factors.

1 Introduction: Contractions, Dilation and Structural Explanations

In this paper we defend the thesis that structural explanations are an effective way of explaining well-known relativistic phenomena like length contraction and time dilation, and then try to understand how this can be possible by looking at the literature on scientific models. In particular, we ask whether and how Minkowski spacetime's model can be said to represent the physical world, in such a way that it can successfully explain physical phenomena *structurally*. In the present, introductory section, we try to briefly justify the above thesis by providing a brief sketch of structural explanations as they are used in Minkowski spacetime, in contrast to attempts at explaining the relativistic phenomena *dynamically* (Brown, 2005). In the second section we offer a brief survey of the state of the art in the debate between the semantic and the pragmatic conception of models, with particular attention to the inferentialist conception proposed by Suárez. In the third section we argue that, in order both to solve some problems within Suárez's inferentialist approach and to account in a consistent way for the use that cognitive agents make of models, it is necessary to assume some kind of partial isomorphism between the mathematical model and the physical target. Our conclusion – the validity of which is here tested only in the specific case of structural explanations in Minkowski spacetime – makes the opposition between the pragmatic and the semantic view look much more

V. Petkov (ed.), *Space, Time, and Spacetime*, Fundamental Theories of Physics 167,
DOI 10.1007/978-3-642-13538-5_9,

apparent than real, and in fact proposes a reconciliation between the two points of view already defended with a different emphasis by Debs and Redhead (2007).

Since the publication of Einstein's original paper on special relativity (SR), phenomena like rod contractions and clocks retardations have attracted the attention of philosophers. One of the key questions that has been raised by these phenomena from the very beginning was: are they *real?*

Of course the answer to a question like this depends on what one means by the metaphysically appealing but philosophically treacherous adjective "real" in our context. If "real" means "measurable", then the answer ought to be an uncontroversial "YES" written in capital letters, as every experimental physicist working at Fermi Lab or at the LHC in Geneva could guarantee. If "real" means "invariantly true", then the answer should also be yes, written in small letters though, considering the (italicized) relativization involved in the following claim: for all possible inertial observers, it seems true to say that, "*relative to observer O*, the rod contracted a certain amount x in the direction of motion". However, if "real" means "dynamical", the vast majority of physicists and philosophers would answer the above question with a "NO", again written in capital letters. We don't need forces to account for the relativistic phenomena of contractions and dilations: after all, can't we explain such effects as, respectively, *cross sections of four-dimensionally conceived rods and projections of four-dimensionally conceived clocks* onto different, arbitrarily chosen inertial frames of Minkowski spacetime? (see Fig. 1 below)[1]

That is, we would add, we can explain such phenomena *via structural explanations*, based upon the geometrical features of Minkowski spacetime.

What are, however, structural explanations? A minimal definition of structural explanations was briefly provided by Rob Clifton:

> We explain some feature B of the physical world by displaying a mathematical model of part of the world and demonstrating that there is a feature A of the model that *corresponds* to B. (Clifton 1998, p. 7, our emphasis)

The key problem raised by this brief quotation is of course what we should mean by "*correspond*"; a verb that calls into play the general problem of how mathematical

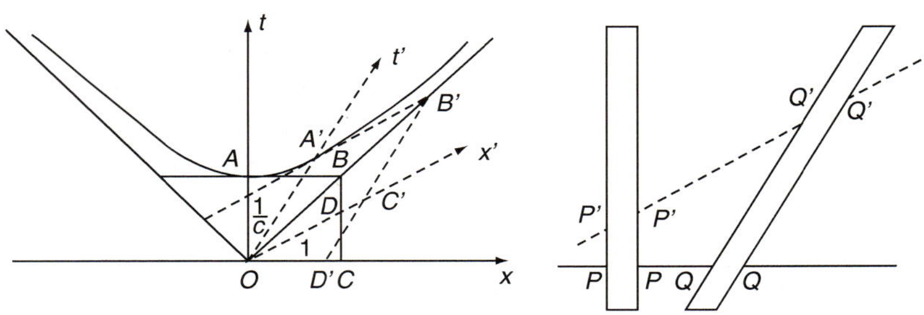

Fig. 1 Length contraction

[1] Figure 1 is taken from (Minkowski 1908). Figure 2 is taken from (Petkov 2009), p. 86.

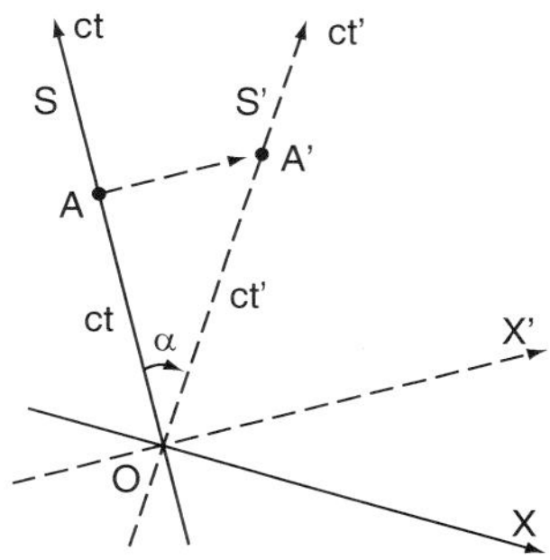

Fig. 2 Time dilation

models refer to the physical world, one of the main questions that this paper tries to tackle. Let us say at the outset that, in our understanding of structural explanations, their essential feature lies in the fact that their validity is independent of the question of what categorial framework[2] underlies the theory in question, a thesis that typically allows one to neglect attempts at explaining phenomena by invoking *causal* or *mechanistic* models.

While the idea of structural explanation has mostly been developed by having quantum mechanics in mind (Dorato and Felline, forthcoming), also SR and the structure of Minkowski spacetime have already been regarded as a template of a theory providing structural explanations:

> Suppose we were asked to explain why one particular velocity (in fact the speed of light) is invariant across the set of inertial frames. [...] [The Lorentzian] causal explanation is now seen as seriously misleading; a much better answer would involve sketching the models of space-time which special relativity provides and showing that in these models, for a certain family of pairs of events, not only is their spatial separation x proportional to their temporal separation t, but the quantity x/t is invariant across admissible (that is, inertial) coordinate systems; further, for all such pairs, x/t always has the same value. This answer makes no appeal to causality; rather it points out structural features of the models that special relativity provides. It is, in fact, an example of a structural explanation (Hughes 1989, pp. 256–257)

The following example will show in what sense structural explanations of physical phenomena in Clifton's sense can avoid any appeal to causality or forces. Suppose that we want to understand why it is the case that clocks in relative motion measure a time that is dilated with respect to the time measured by clocks at rest in the chosen inertial frame. The typical explanation that is provided in most textbooks is repeated from Feynman's lectures (Feynman et al. 1963, I vol., 15–6). Take a light ray that goes up and down between two mirrors (see Fig. 3, left). Each round trip of the light ray (supposing it is originally emitted from the bottom) gives us *one* beat of our clock. But in the moving frame, which is endowed with the same kind of

[2] A categorial framework is the set of fundamental metaphysical assumptions about what sorts of entities and what sorts of processes lie within a theory's domain (Hughes 1989).

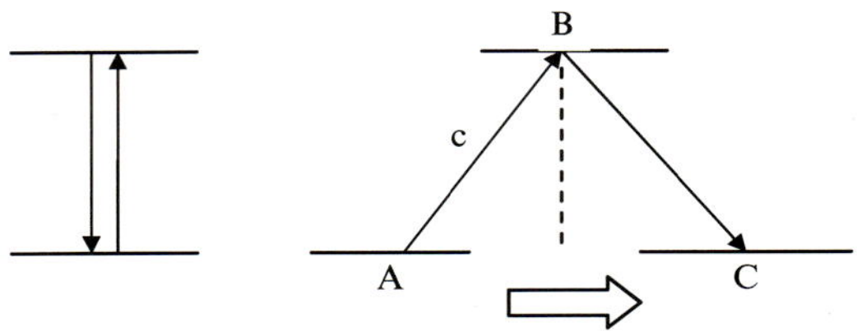

Fig. 3 Clocks in relative motion

light clock, the observer at rest will observe that the light ray originating from A is reflected at B and then comes back at the bottom mirror at point C. (see Fig. 3) *Since the light ray must zigzag along a longer path, the measured time will be longer*. The explanation in question is obviously geometrical/algebraic, and therefore structural in the sense specified by Clifton. It is geometrical because in order to show the dependence of the dilation on the relative velocity, it relies on Pythagoras' theorem: if the hypotenuse in Fig. 3 has a length c, and the moving clock travels a distance 2u from emission to reception of the pulse, the height of the triangle is $(c^2 - u^2)^{1/2}$ which is obviously smaller than c, so that $2(c^2 - u^2)^{1/2} < 2c$.

Despite examples of this kind, Harvey Brown (2005) and some co-workers (especially Dr. Oliver Pooley, a former, brilliant student of his (see Brown and Pooley 2006)) have instead been arguing for some years that length contractions and clock dilations, so far often regarded as purely *kinematical* effects, need a *dynamical*, presumably *quantum* explanation, in terms of Lorentz covariant laws, an explanation that does *not* require a privileged inertial frame.

The following brief (and admittedly incomplete) criticism of Brown's dynamical proposal (but see Dorato 2007; Norton 2008) will however serve to illustrate in what sense structural explanations of length contraction rely on the *structure* of Minkowski spacetime, and as such do not presuppose dynamical effects. SR tells us that the amount of contraction of a body depends on the arbitrary choice of the measuring frame, and therefore on the relative velocity between the two inertial frames. If this is agreed upon, it is not clear why we should grant the deformation a *dynamic* significance, rather than a simple geometrical/structural significance. If relative to frame f the contraction F of object O is $F(f)$, relative to frame f′ is $F(f')$, relative to f'' is $F(f'')$ and so on, the implication *that there is no intrinsic shape* of the body O is quite natural, since length or shape in the special theory of relativity are *non-invariant* notions.

One could even be tempted to conclude that there is no fact to be explained. However, to the extent that relativistic phenomena do need an explanation, as we believe it is actually the case,[3] structural explanations do suffice, in virtue of the following *geometric* and *topological* aspect of Minkowski spacetime. If we conceive

[3] Recall that there is a relational and objective matter of fact for all observers about the contraction of a ruler *relative to an inertial wordline O*(see above).

spacetime and the physical world four-dimensionally, as recommended by Minkowski in his original paper (1908), we should conceive four-dimensionally also physical objects. Such four-dimensionality ought to be regarded as one of their key objective features, derived by the fact that we model them as "inhabitants" of Minkowski spacetime.[4]

The main fact to rely upon at this point is that four-dimensional entities can be "sliced" in different ways: according to the frame of reference that we happen to choose, we obtain a different spatial section of a four-dimensional entity, in the same sense in which, by slicing a four-dimensionally conceived electromagnetic field, we obtain different but separate electrical and magnetic fields. The geometrical aspect provided by the "slicing" (a cross-section) is what makes the explanation of length-contraction *mathematical*, and therefore, in Clifton's sense, *structural*: we are simply locating length contraction (the phenomena to be explained) in the mathematical model of Minkowski spacetime (the explanans). Notice that causation, mechanical models or dynamical forces are never called into play and they seem to be wholly superfluous.

Nevertheless, in defence of his claim of the necessity of a dynamical understanding of special relativity, Brown (2005), Brown and Pooley (2006), and Brown and Timpson (2006) have often appealed to the distinction between principle and constructive theories (Einstein 1919), and to the fact that SR, in Einstein's own opinion, is to be conceived as a principle theory. For the sake of brevity, Brown's argument can be summarized by the following two premises:

1. Geometrical explanations provided by SR with the help of the structure of Minkowski spacetime cannot be regarded as explanations typical of constructive theories
2. Principle theories lack the explanatory power of constructive theories

It then follows that, according to Brown, and contrary to what we argued so far, *we still lack a genuine understanding of the phenomenon of length contraction or time dilation.* A thorough understanding of the latter phenomena could only be provided by a 'constructive theory' in Brown's (controversial, from our point of view) sense. According to Einstein, a constructive theory is a theory that, like statistical mechanics, is capable of constructing, or giving a deeper account of, physical phenomena – phenomena that the principle theory instead constrains only via very general empirical principles which do not depend on hidden levels of description.[5]

---

[4] For the importance of questions of dimensionality in Minkowski spacetime, see Petkov (2007). For a defense of a fourdimensional metaphysics, which here we take for granted, see Sider (2003).

[5] "One can distinguish various kinds of theories in physics. Most of them are constructive. These seek to construct a model of the more complex phenomena out of a relatively simple formalism taken as a basis. Thus the kinetic theory of gases seeks to reduce mechanical, thermal, and diffusional processes to the movements of moleculesñi.e., to construct them out of the hypothesis of molecular motion. When one says that we have succeeded in understanding a group of natural processes, one always means that a constructive theory has been found that comprehends the relevant processes" (Einstein 1919, transl. by Don Howard). It is possibly not irrelevant to remark that none so far has been able to provide any such constructive theory, neither for SR nor for GR.

In arguing for premise (2), Brown often relies on Balashov and Janssen (2003) characterization of the different way in which explanations are provided in principle theories and constructive theories respectively. This reliance, however, creates two sorts of difficulties, which here can only be sketched (but see Felline, (Forthcoming)). The first difficulty originates from the fact that Brown uses Balashov and Janssen' characterization in order to claim that principle theories typically provide Deductive-Nomological (DN) explanations, while constructive theories typically rely on model-based explanations. However, this claim, as some others based on this characterization, is *unwarranted*: as we hope to have shown above by illustrating the role of structural explanations in Minkowski spacetime, also principle theories are capable of providing perfectly acceptable model-based explanations, which (explicitly at least) do not mention any physical laws in their premises. To the extent that principle theories rely on structural explanations, it is not true that the latter are only based on DN explanations. Clearly, *the sense in which structural explanations rely on "models" is certainly different from the sense in which "model" is used, say, for referring to the billiard-ball model typical of the kinetic theory of gases*, the standard example of a constructive theory.

The second problem is that Brown misunderstands the way structural explanations function in the context of SR, by saddling them with an implausible *causal* type of substantivalism associated to Minkowski spacetime, a (dirty-water) substantivalism that he himself correctly rejects, unfortunately together with the baby (structural explanations) (Brown and Pooley (2006)). In order to understand the origins of this unfair characterization of structural explanations, it is important to keep in mind his premise (2). Since according to him only constructive theories provide genuine explanations, and since only such explanations rely on "models", Brown is led to think that also geometrical explanations, if genuine, must function like the billiard-balls model of the kinetic theory. In order words, he is led to presuppose that since the representation provided by the *model* used in constructive explanations typically include a strong form of *ontological commitment* toward the target entities and processes, the same must hold for structural explanations within Minkowski spacetime.

A *constructive* model of spacetime conceived in this ontic sense would then represent spacetime as a substance which exists independently of, and acts on, things, events and processes immersed in it. A *principle* theory explanation would instead not be ontically committed at all. In support of the fact that our reading of Brown's approach to "model" is plausible, consider how well it fits with his (misleading) understanding of structural explanations, within what he calls the 'orthodox' view of SR – i.e., *the 'constructive version' of the geometrical explanations provided by SR and illustrated above in a non-causal, non-metaphysical way.* According to Brown, in what he regards as the 'orthodox', geometrical explanation of the kinematic behaviour of bodies, Minkowski spacetime ought to play a causal role! But since, as Brown correctly notes, Minkowski spacetime cannot have the function of 'shaping' rods by causing in this way the Lorentz contractions, the unwarranted conclusion is that structural explanations cannot be effective. However, why saddling structural explanations with this causal-metaphysical baggage, when their main purpose is to do away with unwanted metaphysical assumptions?[6] Thinking that in a physical theory like SR a geometrical/structural explanation can work *only* if it does so causally is reducing *ad absurdum* this view.

Here we will not further discuss Brown's arguments against the success of structural explanations in SR. For the aim of this paper, it is important to have shown how a poor understanding of the relation between mathematical models and the physical world may prevent philosophical progress and cause misunderstandings. In the next section we will therefore briefly present the debate on the nature of models, in order to see how one should properly understand the effectiveness of the geometrical explanations provided by the structure of Minkowski spacetime. How should we explain such explanations, given that the universally agreed-upon *background dependence* (or causal inertness) of Minkowski spacetime makes Brown's causal reading rather implausible?[7]

2 The Debate on Models: The State of the Art

The Semantic View (SV) has until very recently dominated the discussions in the theory of scientific models[8]. According to this view, a scientific model is a set-theoretic structure

$$S = \langle U, O, R \rangle$$

i.e. an abstract triple consisting of (a) a non-empty set U of individuals called the domain of the structure S, (b) an indexed set O of operations on U (which may be empty), and (c) a non-empty indexed set R of relations on U. Within the SV, the relation between a model and its target is traditionally defined as a dyadic relation of isomorphism or "embedding a physical theory in a mathematical structure" (French 1999, p. 188), or as a weaker relation of similarity (Giere 1988). However, both these relations have been found problematic and other morphisms have been put forward, like a relation of partial isomorphism (French and Ladyman 1999).

The plausibility of such accounts as an explication of the concept of scientific representation has recently been challenged (see Suárez 2003; Frigg, 2006). According to Frigg, for instance, while it can be considered a requirement for the accuracy of *some kind* (or, with Frigg's term, *style*) of scientific representations, isomorphisms (or other dyadic relations of morphism) are in general *not sufficient* to account for the way cognitive agents (scientists) utilize models in order to perform "surrogative reasoning" about the target (physical) system. In other words,

[6] For the discussion of another improper, metaphysical use of Minkowski spacetime, see Dorato (2006).

[7] Given that also the metric field is causally inert even though it is surely correlated to matter (a causal reading of the metric field is very controversial to say the least), how could one regard a causal reading of Minkowski spacetime as plausible?

[8] See Suppes (1967), Suppe (1977), van Fraassen (1980), Giere (1988).

to argue that a model *M* represents a target *S* iff *M* is isomorphic to *S* is not sufficient to explain the complex relations holding between representation and the different uses that are made of models. For instance, the relation of isomorphism is *symmetric*, while the relation of representation clearly isn't. Isomorphisms can hold only between two abstract, already mathematized structures, not between an abstract structure and a physical target, so that we must focus on *data which have already been mathematised* (Suppes' models of data (1962)); furthermore, the same abstract model can be multiply realized, so that there is a problem of under-determination of reference, accompanied by the fact that the same target can exemplify many different structures. Finally isomorphisms, *unless partial*, seems incapable of explaining cases of *misrepresentation* (think of gas molecules represented as billiard balls: of course, they are not literally "balls").

In order to overcome all of these problems, a more pragmatic approach has been proposed, one that focuses more on the *use* that cognitive agents make of models. In particular, Mauricio Suárez proposes a deflationary approach to scientific representation, according to which: "[r]epresentation is not the kind of notion that requires, or admits such [universal necessary and sufficient] conditions. [...] [F]inding necessary conditions will certainly be good enough." (Suárez 2004, p. 771, see also van Fraassen 2008).

Suárez correctly claims that if a theory is meant to account for our deep-grounded intuitions about scientific representations, it must also account for the fact that a scientific representation is not just the product of an arbitrary convention between agents. Consequently, there must be something in the model *M* that makes it the case that a cognitive agent can legitimately use *M* to perform surrogative reasoning about the target. But here, let us note, some potential room is made for a more conciliatory view between the model-theoretic and the pragmatic camp. Scientific models, we are told, do more than merely denote an object, as they allow us to draw relevant conclusions about their target: in other words, they are *informative* about it. Suárez claims that it is in virtue of their informativity that scientific models are *objective*. This word is clearly ambiguous between "intersubjectively shared" and "representing properties of the target existing independently of the model". Given his anti-representationalist approach, one expects Suárez to opt decisively for the former, merely *epistemic* alternative, while rejecting the latter, more ontological version, which would take us back toward some sort of isomorphic view of representation.

Suárez's proposal is that it is exactly the capacity to allow for "surrogative reasoning" that determines the objectivity of scientific models. Scientific representation is therefore characterized by the following criterion:

[inf]: A represents B only if (a) the representational force of A points towards B, and (b) A allows competent and informed agents to draw specific inferences regarding B. (ibid., p. 771)

The *representational force* of a source (or, more simply, the *force of a model*) is "the capacity of a source to lead a competent and informed user to a consideration of the target", in virtue of "a *relational and contextual property* of the source, fixed and maintained in part by the intended representational uses of the source on the part of agents" (ibid., p. 768, our emphasis).

Which relational and contextual property of the source is Suárez talking about? The answer is that he cannot be more specific about it, since more specificity would push him toward admitting the possibility of some sort of well-defined relation *always* existing between model and target, and he denies exactly this point. And yet, it is "[inf]'s part (b) that has the important function of contributing to the *objectivity* that characterises scientific representation. In contrast to part (a), the above mentioned relational property in no way depends on an agent's existence or cognitive activity. It requires the model A to have the *internal structure* that allows informed agents to correctly draw inferences about the physical target B, but it does not require that there be any agent who actually does so." (ibid., p.774). This is still very vague, as the reader will recognize, but the idea is that the "internal structure of the model" is *not* fixed by the context of inquiry, and it is this non-contextuality which, according to our author, guarantees the objectivity or non-arbitrariness of the model.

Suárez explains how the concepts of informativity and objectivity are related with an example. Consider a piece of paper and two pens writing on it, and stipulate that they represent respectively the sea and two ships sailing on it. Compare then this representation with the opposite one, in which the paper represents the ships and the pens represent the sea. Suárez claims that "the ships-on-sea system is more "objectively" characterised by the first denotational arrangement than by the second" and with this he means that the second representation "is certainly less informative, since the relative movements of pens and paper can not allow us, for instance, to infer the possibility that the two ships may crash" (p. 8).

To see why Suárez's proposal cannot be satisfactory, consider the following situation. If it is true that, with the referential conventions: ships = paper, pens = sea, one cannot draw the conclusion that the two ships may crash, it must be admitted that in this non-standard representation one can draw other interesting conclusions. One can for instance conclude (under the assumption that the paper is rigid) that whatever movements the ships can perform, they will always remain at the same distance one from the other, or that if you burn one of them also the other will burn. These conclusions are obviously *false*, since, after all, the piece of paper is unique, and the example has it that there are *two separate ships*; but if degree of faithfulness and accuracy of the representation to the physical reality are irrelevant for objectivity, then one cannot argue that one representation is more objective than the other: the two are on a par. One can at most claim that one representation allows to draw *more* inferences than the other.

It then seems that we cannot discriminate between objective and non-objective representations without providing some more specific characterization of the kind of relation that must hold between target and model. The above example of the pens, which might simply be badly chosen, suggests, however, something relevant to our main purpose: it is the good degree of "faithfulness" to the world of one of two representations that allows a more accurate and reliable surrogative reasoning; it is this accuracy that, in its turn, measures the objectivity of a model. After all, a very good reason for more or less informativity or inferential power is given by the fact that one representation captures key features of the world (target) much better than the other.

Suárez rejects this account from the very beginning. He does so because, he argues, if objectivity were defined in terms of "truthfulness" or accuracy, it would be very difficult (if not impossible) to account for cases of misinterpretation, inaccuracy, idealizations, etc.[9] However, the notion of surrogative reasoning must be obviously further characterized: the requirement that a user can draw inferences about the target is too weak to grasp the objectivity of representation, since one can draw irrelevant, or even wrong inferences; and if one has to restrict one's attention to *successful* inferences, one might be in need of an explanation of this success that is not circular, i.e., that is not given in terms of informativity.

In a word, the problem with Suárez's account is that, remaining as it does wholly on the epistemic terrain, there is no possibility of knowing, for instance, how the mathematical models hook up with the physical world, and how to claim that structural explanations are genuine accounts of *physical* phenomena. And for us this is a serious drawback, since we take structural explanations to be genuine explanations of physical phenomena.

In the following, final section, we will argue that the step from Suárez's inferentialism to a more substantive account of representation, however, is not too long. In particular, Suárez allows for additional, contextual necessary conditions that a scientific representation must satisfy, but denies that these conditions can contribute to the objectivity of representation, which for him is a non-contextual notion. On the contrary, we will now see how insisting on the pragmatic aims of the users of the model can solve a series of difficulties of the isomorphic account of representation, so as to achieve that genuine informativity of the model about the world that Suarez was striving to capture. This will also entail a *rapprochement* between the two allegedly opposed camps that we have presented in this section.

3 Minkowski's Model and Structural Explanations

In a nutshell, our argument so far can be summarized as follows: (a) the relativistic contractions and dilations need to be explained; (b) structural explanations provide a genuine explanation of these physical phenomena; (c) structural explanations are not to be cashed out in terms of causal or mechanical or DN type of explanations; (d) it then becomes highly plausible that such explanations require *some* form of morphism between the model and the target, so that the semantic view, supplemented with a robust doses of pragmatism, seems the only explanation for their genuine explanatory character. Note that the first three premises have been argued

[9] Batterman (2010) argues that asymptotic behaviour cannot be captured by any kind of isomorphic relation.

for in the first section. In order to defend the conclusion (d), we will begin from the same starting point that is considered central by the proponents of the pragmatic view, namely that of giving a more precise account of the way cognitive agents use scientific models.

Going back to our case study, we can introduce the problem of explaining structural explanations in these terms: which contextual, necessary conditions should a mathematical model meet in order to serve as a provider of a structural explanation? While we will keep on considering structural explanations in Minkowski spacetime as our case study, we will leave open the question to which extent the remarks offered in this section can be suitably generalized.

First, we have seen that a structural explanation typically does not require any specific ontological commitment about what is represented by the structure, i.e., in our case it is neutral with respect to the traditional division between substantivalism and relationism about Minkowski spacetime.[10] It follows that a cognitive agent can use models of Minkowski spacetime to provide a consistent structural explanation, without being committed to a specific ontology of the controversial type illustrated by Brown (i.e., a causal form of substantivalism).

Second, in a structural explanation the physical *explanandum* B is understood in terms of the relational properties of its formal counterpart A. Remember that structural explanations make essential use of the mathematical laws and principles defining the model. Of course, such laws and principles ought to *codify* physical postulates of the theory: in our case, of the main characteristics of the mathematical model is the invariant quantity ΔS, which represents the speed of light in all inertial frames:

$$\Delta S = \sqrt{(t_p - t_q)^2 - (x_p - x_q)^2 - (y_p - y_q)^2 - (z_p - z_q)^2}.$$

The main point is, however, that in order to transfer *knowledge* about the relational properties of the model A into knowledge about the relational properties of the physical explanandum B, we must assume that the relational properties and laws exemplified by the model A are also (at least in part) relational properties and laws exemplified by B. Without this exemplification of the relations and laws of A by B, no transfer of knowledge from A to B could ever occur, and any inference drawn in the model would not be about the physical world. As a consequence, objectivity would be lost. In other words, the performance on the part of an agent of a structural explanation of a physical phenomenon presupposes the assumption of a *partial isomorphism* (French and Ladyman 1999) between the relations exemplified by the model A and those exemplified by the target B.

However, how can an isomorphism exist between an abstract and a concrete structure? In order to solve this difficulty, imagine a concrete sphere designed on a board with chalk: this has certain structural features and properties that we can

[10] Whether it is also neutral toward a form of spacetime structural realism is of course much less clear. But see below.

determine by studying spheres in an abstract and idealized manner (for instance, we imagine that any point of the surface is really equidistant from the center). What matters for us is that *it is only if the Earth and the sphere on the board exemplify approximately and partially the same structure of the abstract sphere that we can suitably transfer knowledge from the latter to the two former, concrete, physical objects*. It is in virtue of such an assumption of partial isomorphic correspondence that structural explanations are considered effective. The correspondence is partial because the Earth is not exactly spherical, of course, but depending on our cognitive aims, we can decide to go ahead with the given approximation and treat the Earth as if it were a perfect sphere.

Going back to questions debated in Sect. 2, it is in virtue of an assumption of partial isomorphism of this kind that the Minkowskian model can be considered *informative* about the physical world, and therefore, whenever the right contextual situations obtain, enable informed users to give a genuine structural explanation of a physical phenomenon. Furthermore, notice that it is the use that the speaker does of a certain model that solves many of the problems of the isomorphic account of scientific representation. First of all, it is the fact that *we intentionally use a model* to represent something physical that makes the relation of isomorphism, which is per se symmetric, *asymmetric*. By using x to represent y, we thereby automatically select one of the two directions of correspondence of model and world. Secondly, insisting on the particular interests of the user, one can obviously avoid all the underdetermination claims, since one uses particular aspects of the model for particular purposes, selecting particular aspects of reality as it is more convenient for the aim at hand. Thirdly, as we have seen, also the problem of correspondence between an *abstract* and a *concrete* structure is overcome: the *ante rem abstract* structure (the abstract sphere) is exemplified by the *concrete* physical system, in such a way that the former partially exists *in re* in the latter (the concrete sphere on the board).

This remark is linked to another important point concerning the reality of Lorentz contractions, with which we opened the paper. If conceived as structural explanations, the geometrical explanations provided by SR do not conceive of relativistic phenomena like Lorentz contraction as merely perspectival, in the sense of unreal. As already argued in Sect. 1, Lorentz contraction should be conceived as real in the same sense in which the structure of spacetime is real. Minkowski spacetime is real to the extent that it is genuinely exemplified by physical fields and events. Likewise, Lorentz's contractions are real to the extent that they are exemplified by physical systems in reciprocal motion, as illustrated by a geometrical explanation of the relevant phenomena. In order for this conception of relativistic phenomena to be justified, however, one necessarily has to acknowledge that Minkowski spacetime models, and in particular the invariance of the spacetime interval, actually captures a feature of reality, in the sense that the latter is a concrete instance of the former. The difference between a mere mathematical explanation and a mathematical explanation *of a physical phenomenon* is therefore given by the existence of an exemplification relation: without the assumption of this exemplification relation between the target and the model, the geometrical accounts provided by SR would end up being a mere logico/mathematical derivation of the *explanandum* – something more similar to a Deductive Nomological explanation, realized thanks to a mere mathematical law belonging to a mere mathematical structure.[11]

Of course, by insisting on the importance of structural explanations in certain contexts, we are not denying that, *in other contexts,* causal explanations may not be more appropriate. It is not necessary for our main claim to argue that the isomorphic approach works in all possible contexts, but only that it works in our particular case study, i.e., as an explanation of structural explanations in SR.[12] On the contrary, as already anticipated above, we accept a contextual/pragmatic dimension of explanation and therefore of the use of models. And we believe that there is no general set of necessary and sufficient conditions that in all possible contexts can tell us where a causal rather than a structural explanation is appropriate. On the contrary, it is not impossible that the two types of explanation can coexist even in the same theory. Obviously, depending on the kind of allowed surrogative reasoning (causal explanation vs. structural explanation), one imposes different conditions for rationally believing in the informativity of the representation.

4 Conclusion

We would like to end this paper with a general consideration about the theory of scientific representation. The point of departure of Suárez's defence of the inferentialist view *vis à vis* the semantic view was that only the former accounts for the way cognitive agents use models in science. But it is not clear at all why defenders of the isomorphic approach could not make room for a decisive pragmatic component of scientific representation (see Debs and Redhead 2007). Consider the question: what justifies a rational agent to interpret the product of her surrogative reasoning as a piece of *knowledge* about the target? Since *knowing* p entails p, we must assume that there is some sort of truthlikeness in the representation allowed by a model. Clearly, if the fact that the model enables us to draw correct inferences about the target is objective in more than an epistemic sense, this fact must receive an explanation in terms of the existence of some objective relationship between the model and the world. We have argued that *what kind* of relationship holds is contextual (in the context of the geometrical/structural explanations it is the assumption of a partial isomorphism that needs to be assumed), but this contextuality does not go against a genuine 'objectivity' (in the stronger sense) of the resulting scientific explanation, because some kind of grasp into physical reality is still always necessary in order to account for the rationality of any kind of surrogative reasoning. This conclusion leaves space to the idea that the two accounts of representation end up being much closer than their defenders tend to admit. Possibly, they are the two inseparable sides of the same coin.

---

[11] One can recognize this charge against the effectiveness of geometrical explanations also in Brown's works (see Felline Forthcoming).

[12] For problems with the isomorphic approach, see Batterman (2010).

Acknowledgements The authors thank Vesselin Petkov for his careful reading of a previous version of this paper, which eliminated a mistake. Laura Felline's research is founded by a Master and Back scholarship from Regione Sardegna. She would also like to thank all the members of Alophis – Applied Logic and Philosophy of Science – group for daily support and help.

References

Balashov, Y., Janssen, M. (2003), "Presentism and Relativity" *The British Journal for the Philosophy of Science* 54(2): 327–346.

Batterman, R. (2010), "On the Explanatory Role of Mathematics in Empirical Science" *The British Journal for the Philosophy of Science* 61: 1–25.

Brown, H. (2005), *Physical Relativity. Spacetime Structure from a Dynamical Perspective*. Oxford University Press, Oxford.

Brown, H., Pooley, O. (2006), "Minkowski Spacetime: a Glorious Non-Entity", in Dieks D. (ed.) *The Ontology of Spacetime*, Elsevier, Amsterdam, pp. 67–89.

Brown, H., Timpson, C. (2006) "Why special relativity should not be a template for a fundamental reformulation of quantum mechanics", in Demopoulos W. and Pitowsky I. (eds.) *Physical Theory and its Interpretation: Essays in Honor of Jeffrey Bub*, pp. 29–42. Springer, Dordrecht.

Clifton, R. (1998), "Structural Explanation in Quantum Theory" Eprint: http://philsci-archive.pitt.edu/archive/00000091/00/explanation-in-QT.pdf, unpublished.

Debs, T., Redhead, M. (2007), *Objectivity, Invariance and Conventions: Symmetries in Physical Science*, Harvard University Press.

Dorato, M. (2006), "The irrelevance of the presentism/eternalism debate for the ontology of Minkowski spacetime", in Dieks D. (ed.) *The Ontology of Spacetime*, Elsevier, Amsterdam, pp. 93–109.

Dorato, M. (2007), "Relativity Theory between Structural and Dynamical Explanations", *International Studies in the Philosophy of Science*, 21(1): 95–102.

Dorato, M., Felline, L. (forthcoming), Scientific Explanation and Scientific Structuralism, in Bokulich A., Bokulich P. (eds.), *Scientific Structuralism*, Boston Studies in Philosophy of Science.

Einstein, A. (1919), "What is the Theory of Relativity?", *Times* (London). 28 November 1919. Reprinted in Einstein (1982, 227–32) *Ideas and Opinions*, Crown Publishers, Inc., New York

Felline, L. (Forthcoming), "Scientific Explanation between Principle and Constructive Theories".

Feynman, R., Leighton, R. Sands, M. (1963), *The Feynman Lectures on Physics*, Addison Wesley, Redwood City, California.

French, S. (1999), 'Models and Mathematics in Physics: The Role of Group Theory', in Butterfield J., Pagonis C. (eds.) *From Physics to Philosophy* Cambridge University Press, Cambridge, pp.187–207.

French, S., Ladyman, J. (1999), 'Reinflating the Semantic Approach', *International Studies in the Philosophy of Science* 13: 103–121.

Frigg, R. (2006), "Scientific Representation and the Semantic View of Theories", *Theoria* 55: 37–53. http://philsci-archive.pitt.edu/archive/00002926/01/Scientific_Representation.pdf.

Giere, R. (1988), *Explaining Science. A Cognitive Approach*. University of Chicago Press, Chicago.

Hughes, R.I.G. (1989), *The Structure and Interpretation of Quantum Mechanics*, Harvard University Press, Cambridge.

Hughes, R.I.G. (1993), *Theoretical Explanation*, Midwest Studies in Philosophy XVIII: 132–153.

Minkowski, H. (1908), "Space and Time," as reprinted and translated in Petkov V. (ed.) *Minkowski Spacetime: A Hundred Years Later.* Springer, Berlin, pp. xiv–xlii.

Norton, J.D. (2008), "Why constructive relativity fails", *British Journal for the Philosophy of Science*, 59(4): 821–834.

Petkov, V. (2007), "Relativity, Dimensionality, and Existence", in Petkov V. (ed.) *Relativity and the Dimensionality of the World*, Springer, Berlin, pp. 115–135.
Petkov, V. (2009), *Relativity and the Nature of Spacetime*, 2nd ed. Springer, Berlin.
Sider, T. (2003), *Four-dimensionalism. An ontology of persistence and time*. Clarendon Press, Oxford.
Suárez, M. (1999), "Theories, Models, and Representations", in Magnani L., Nersessian N. J., Thagard P. (eds.) *Model-Based Reasoning in Scientific Discovery*. Kluwer, New York, pp. 75–83.
Suárez, M. (2003), "Scientific Representation: Against Similarity and Isomorphism", *International Studies in the Philosophy of Science* 17: 225–244.
Suárez, M. (2004), "An Inferential Conception of Scientific Representation", Philosophy of Science (Symposia), 71: 767–779.
Suppe, F. (1977), "The search for Philosophic Understanding of Scientific Theories", in Suppe F. (ed.) *The Structure of Scientific Theories*. University of Illinois Press, Urbana, pp. 3–232.
Suppes, P. (1962), Models of data. in Nagel E., Suppes P., Tarski A. (eds.) *Logic, Methodology and Philosophy of Science: Proceedings of the 1960 International Congress*. Stanford University Press, Stanford, pp. 252–261.
Suppes, P. (1967), "What is a Scientific Theory?", in Morgenbesser S (ed.) *Philosophy of Science Today*. Basic Books, New York, pp. 55–67.
van Fraassen, B. C. (1980), *The Scientific Image*. Oxford University Press, Oxford.
van Fraassen, B. C. (2008), *Scientific Representation: Paradoxes of Perspective*. Oxford University Press, Oxford.

Relativity of Simultaneity and Eternalism: In Defense of the Block Universe

Daniel Peterson and Michael Silberstein

Abstract Ever since Hermann Minkowski's now infamous comments in 1908 concerning the proper way to view space-time, the debate has raged as to whether or not the universe should be viewed as a four-dimensional, unified whole wherein the past, present, and future are regarded as equally real or whether the views espoused by the possibilists, historicists, and presentists regarding the unreality of the future (and, for presentists, the past) are more accurate. Now, a century after Minkowski's proposed block universe first sparked debate, we present a new, more conclusive argument in favor of the eternalism. Utilizing an argument based on the relativity of simultaneity in the tradition of Putnam and Rietdijk and explicit novel but reasonable assumptions as to the nature of reality, we argue that the past, present, and future should be treated as equally real, thus ruling that presentism and other theories of time that bestow special ontological status to the past, present, or future are untenable. Finally, we respond to our critics who suggest that: (1) there is no metaphysical difference between the positions of eternalism and presentism, (2) the present must be defined as the "here" as well as the "now", or (3) presentism is correct and physicists' current understanding of relativity is incomplete because it does not incorporate a preferred frame. We call response 1 deflationary since it purports to dissolve or deconstruct the age-old debate between the two views and response 2 compatibilist because it does nothing to alter special relativity (SR), arguing instead that SR unadorned has the resources to save presentism. Response 3 we will call incompatibilist because it adorns SR in some way in order to save presentism a la some sort of preferred frame. We show that neither 1 nor 2 can save presentism and 3 is not well motivated at this juncture except as an ad hoc device to refute eternalism.

1 Introduction

As Ladyman et al. [12] wisely note, the following are distinct but frequently conflated, deeply related questions in the metaphysics of time:

1. Are all events, past, present and future, real?
2. Is there temporal passage or objective becoming?

V. Petkov (ed.), *Space, Time, and Spacetime*, Fundamental Theories of Physics 167,
DOI 10.1007/978-3-642-13538-5_10,

3. Does tensed language have tenseless truth conditions?
4. Does time have a privileged direction?

This paper will focus almost exclusively on question (1). In the philosophy of time, this major question has captivated philosophers for decades now. This problem stems from two competing notions of time. The first, originally suggested by Heraclitus, is called presentism.[1] Though we will later present the presentist position more clearly so that it can be made relevant to a more thorough and modern treatment of presentist/eternalist debate, a good starting definition for presentism is the view that only the present is real; both the past and the future are unreal.[2] This view is close to, but not the same as, possibilism, which states that the future is unreal while both the past and the present are real. Both of these stances claim to adequately capture the manifest human perception of time. We tend to view ourselves as occupying a unique temporal frame that we call the present that always moves away from the past towards an uncertain future.

However, with the advent of relativity, a different stance, whose primary ancient proponent was Parmenides of Elea, provided a viable alternative to Heraclitean presentism. This new stance, eternalism, was translated into the language of relativity by Hermann Minkowski in 1908 to suggest that time and space should be united in a single, four-dimensional manifold. Thus arose the notion of a 4D "block universe" (BU) in which the past, present, and future are all equally real. This view is called eternalism, and two arguments by Putnam [16] and Rietdijk [17] allegedly show that special relativity (SR) with its relativity of simultaneity (RoS) implies that only the BU perspective is correct.

This paper proceeds as follows. First, we examine the basic structure of the RoS eternalism argument suggested by Putnam, Rietdijk, and more recently Stuckey, Silberstein, and Cifone [23, 24, 27, 28] (hereafter SSC) and present our own novel interpretation or version of the argument for eternalism. Following our proposal, we suggest various points of contention that presentists and possibilists might exploit or have exploited in seeking to either refute eternalism or collapse the presentism/eternalism dichotomy. We have compiled a reasonably exhaustive taxonomy of possible outs that the presentist or possibilist could take to avoid the argument from RoS for BU.[3] After elaborating our own version of the argument, we respond to each counter-argument and show that these objections do not dismiss RoS's problems for presentism.

[1] Recent defenders of presentism include Bourne [1], Craig [5], and Smith [6], whom we take to be our primary presentist opponents for the purposes of this discussion.

[2] What is meant here by "real" is the topic of great debate (see Dorato [10] and Savitt [20] for more on this issue), and we will later clarify our criteria for reality in such a way that much of the vagueness that arises from an imprecise definition of 'real' is dismissed.

[3] One possible refutation of the RoS argument, derived from the work of Harvey Brown (Brown [2]; Brown and Pooley[3]), suggests a kind of re-interpretation of Minkowski space-time as a codification of the behavior of matter as opposed to representing the geometrical structure of space-time. Our response is to be found in Appendix A but has not been integrated into the paper at large because the objection does not fit smoothly into our primary taxonomy.

2 The Argument from the Relativity of Simultaneity

2.1 General Outline and Definition of Terms

Before presenting our RoS argument against presentism, we first provide a general outline of RoS arguments for eternalism and give preliminary definitions of some relevant terms. The general form of the arguments against presentism utilized by Putnam, Rietdijk, and SSC goes as follows:

1. Define presentism.
2. Define the term "co-real".[4]
3. Show that the consequences of the definition of "co-real" and RoS contradict presentism.
4. Conclude that presentism is false from the conjunction of 1 and 3.
5. Conclude that eternalism is true from the rejection of presentism.

To begin with, we must provide our own definitions for the terms that form the foundation of our revamped version of the RoS argument. The first term to be defined is "presentism". Presentism is a kind of realism that takes as real only those events[5] which occur in the present. For instance, since we are sitting next to our friend Joe who is currently reading a paper, the event of his reading a paper and the event of our writing this paper are both real while the event of Joe's leaving to eat dinner is not real because it has not happened yet and the event of our leaving to eat lunch is not real because it has already happened. In terms of simultaneity, then, one can define presentism as the view that the only real things are those which are simultaneous with a given present event. Eternalism, by contrast, is the view that all events past, present, and future are equally real. Thus, Joe's reading, our typing, Joe's leaving for dinner, and our leaving for lunch are all equally real despite the fact that one of these events has already occurred while another has yet to occur. Eternalists hold that all events are equally real, regardless of whether or not said events are simultaneous.

There are two elements, then, that are important for establishing both presentism and eternalism: reality and simultaneity. The debate presupposes that there is a

---

[4] The actual term "co-real" appears only in the SSC papers, but since these present the most recent incarnation of the RoS argument against presentism, we follow their terminology here. It should be noted that Rietdijk does not provide an analysis of the term 'reality' in his paper, and while Putnam does discuss some basic assumptions about reality that are necessary for his argument to go through, they are not argued for or supported in any great detail.

[5] We use the term "events" here to bypass any concerns that may arise due to the identity of individuals like those raised by French and Krause [11] or issues of endurance and perdurance. Such issues as identity and endurance/perdurance, while interesting, need not directly bear on this debate, and so we invoke events that are assumed to be of infinitely small extension and duration (as such they should be fully understood only in terms of their identifying coordinates) to bypass such debates. We are not committed to the claim that such events are in some way the atomistic components of what exists in space-time; rather, we simply invoke them to avoid begging the question on issues like identity and endurance/perdurance.

unique (non-equivocal) sense of the term reality that both sides share. The dispute therefore is over whether or not present events have some ontologically privileged status qua their property of "existing at time some time t where t is in the present". To this end we will first minimally characterize the terms "reality" and "simultaneity" for use in the context of our revamped argument. Before beginning, we should emphasize that we are being purposefully vague with our first characterization of reality here so as to determine reality's most general non-equivocal properties which we will build upon later in this paper. Two events which "share reality" as we characterize it share a single, unique feature (i.e., the same ontological status with respect to realness); this uniqueness, we believe, is the absolute minimal criterion an event would have to satisfy for it to be considered "real" in any meaningful sense of the word.

To better understand the minimal sense of reality at work here, we define two separate notions: the "reality value" and "reality relation." "Reality values" or "R-values" can be thought of as representing the ontological status of any given event. Within space-time, every event can be assigned an R-value that denotes its ontological status, and there is a one-to-one and onto mapping of possible R-values onto ontological statuses. In the interest of defining reality generally, we will not attempt to enumerate how many R-values exist, but one could easily take reality to be binary and thus assert that, for any event, if its R-value is 1, that event "is real", and if its R-value is 0, that event "is not real." One could use higher values to denote other states, such as "possibly real", "real in the future", etc., but, as previously stated, we will not attempt to enumerate all such possible R-values here.[6] It should be pointed out that our uniqueness criterion on reality translates into this system simply as the claim that every event has a single unique R-value. This seems intuitive since an event with an R-value of both 1 and 0, on our scheme, would be both real and unreal, which would be a contradiction.

Our other sense of reality as expressed in the "reality relation" will be essential to our discussion of co-reality. The reality relation can be recast as the idea of "equal reality" and exists between any two or more events that can be considered "equally real." Translated in terms of R-values, a reality relation exists between any two events that have the same R-value. For instance, if events A and B are equally real, then the R-value of event A is the same as the R-value of event B. One should notice here that our definition of "equally real" does not assume that two equally real events are both "real"; equally real events A and B may have whatever R-value you please as long as the R-values are the same for both A and B. This relation explains what a presentist means when she says, "The present is the only thing that is real" since the presentist will hold that events in the future and the past will have different R-values from events in the present.[7] Thus, our purposefully limited characterization

[6] See Appendix B for a more nuanced view of R-values and possible objections to the RoS argument that one might raise based on our naive characterization of R-values described here.

[7] To reiterate, what we have characterized here is the minimal position a presentist must take with regard to a characterization of reality. It might be objected that, at this point, we have not actually defined "what reality is." We will cash out a richer notion of reality later in the paper so that we are

of the "equally real" relation has been defined so as to be useful in a definition of co-reality.

As for simultaneity, if it is possible for one to construct a hyperplane of simultaneity (i.e. a four-dimensional manifold in space-time constructed in such a way that all of the events connected by this manifold are space-like separated from one another) between any two or more events, then these events are said to be simultaneous. Such simultaneous events are required to be space-like separated events that appear to be simultaneous in some subluminal inertial reference frame. Light-like and time-like separated events cannot have a hyperplane of simultaneity constructed between them. Also, a hyperplane of simultaneity may be drawn between any two space-like separated events, meaning that the space-like separation of events A and B is necessary and sufficient for their simultaneity.

Combining the criteria of equal reality ("equally real" means that two events have the same R-value) and simultaneity ("simultaneous" means that two events are space-like separated such that a hyperplane of simultaneity can be constructed between the two events) gives us the relation of "co-reality", which refers to, as the name suggests, two events that are equally real and "simultaneous." The presentist perspective can be restated in terms of this "co-reality" as the stance that "simultaneity between events is a necessary and sufficient condition for the reality (that is, for both events sharing the R-value 1 corresponding to "real") of these events if at least one of these events occurs in the 'present'". For the presentist, any two space-like separated points are thus co-real as we have defined "co-reality". Our restatement of presentism in terms of co-reality here is the assumption that we alluded to in step 1 above.

Our previous examples should make our notion of co-reality more explicit. For instance, the presentist takes Joe's paper reading and our paper typing to be co-real events because they are space-like separated, meaning that there exists some frame in which these two events are simultaneous. However, our paper typing and our leaving for lunch are time-like separated, so there is no sub-luminal frame in which

---

careful not to beg the question against critics like Savitt and Dorato; for now, we are characterizing reality only to a minimal degree in an attempt to determine the properties of the "co-reality" relation, and as such we need only endorse the minimal sense of reality that bears upon our discussion of co-reality.

The presentist might object to our characterization of her conception of reality, but to refuse the characterization of reality we have provided here would be to take an anti-realist stance since a non-unique or equivocal conception of reality would make the idea of "reality" a useless concept for the purposes of this debate. Thus, the presentist cannot argue against our minimal characterization of reality and remain a committed presentist, and the same goes for the eternalist. In the words of Dolev, if one denies this minimal ontological assumption then "neither the tensed nor the tenseless view has the final word in the metaphysics of time."

The presentist could argue against us on the grounds that it is relations, perhaps, that are fundamentally real and not events; this, however, would simply lead us to re-atomize our space-time such that these relations become the fundamental ontic units which assume R-values and the relation of "equally real" connects two such lesser relations. Therefore, even if one makes an argument that forces us to change the fundamental ontic units of our setup, our basic characterization of R-values and "equal reality" can stand unadulterated.

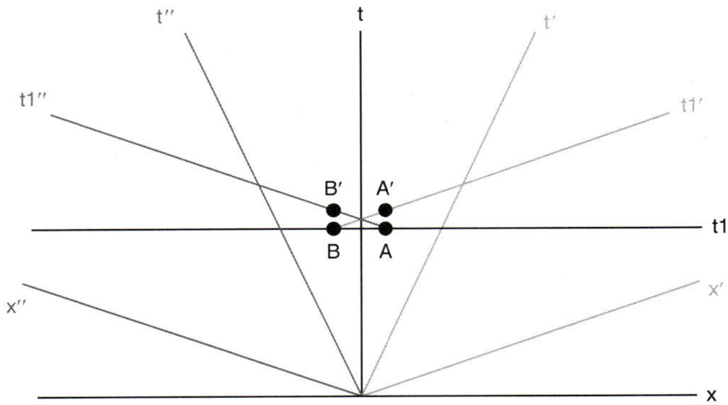

Fig. 1 RoS proof space-time diagram

these two events are simultaneous and they are therefore not co-real. These two criteria of reality and simultaneity as we have defined them are necessary and sufficient for our use of "co-real", and so we turn next to our RoS argument that utilizes this notion of "co-reality" to reveal the tension between presentism and relativity.

2.2 *RoS Argument*[8]

Consider the following situation: our friends John and Josephine stub their toes at the same time in my stationary reference frame.[9] The event of John stubbing his toe is labeled A in Fig. 1 and the event of Josephine stubbing her toe is labeled B in Fig. 1. At a later time (but again, simultaneously in our rest frame), both Josephine and John shout in pain from stubbing their respective toes. John's shout of pain is labeled A′ while Josephine's shout of pain is labeled B′ in Fig. 1. I note that in my frame, both toe-stubs occur at time t1 in Fig. 1. Thus, events A and B are simultaneous and co-real as per the previously-established criteria.

[8] One could argue that, having already defined "co-reality" as we have, the RoS argument has already been made for us: any two space-like separated points are equally real, and space-likeness is not transitive (i.e. A and B could be space-like separated and B and C space-like separated but A and C time-like separated), so we must conclude that any two events (time-like, light-like, or space-like separated from each other) are equally real. The RoS argument in 2.2, however, is a bit more nuanced than the argument just proposed, and it makes it easier for one to determine which definitions and assumptions about reality play what role in the argument. As such, we hope the reader will bear with the exposition for this longer argument.

[9] We are assuming that these "toe-stubs" in this example are the kind of events described in footnote 5 for the reasons stated in that footnote.

Now, some time before this the alien battle cruisers P and D pass each other directly over our heads. The primed axes refer to the frame for battle cruiser P and the double-primed axes refer to the frame for battle cruiser D. Both of these battle cruisers tell a different story from ours. For battle cruiser P events B and A' occur at the same time, and thus B and A' are equally real per co-reality. For battle cruiser D, however, events B' and A occur at the same time, and thus B' and A are equally real per co-reality. We now introduce the symbol "r" to stand for "shares an R-value with" or "is equally real with". The following three statements are true (at least from someone's perspective):

$$ArB$$
$$BrA'$$
$$B'rA$$

From the previously established criteria for equal reality, we can establish two important facts about co-real events α, β, and γ. First, if $\alpha r\beta$ is true, then $\beta r\alpha$ is true since R-values are unique. Thus, the operator "r" is symmetric. This fact must be true since equal reality is an equivalence relation.[10] The second important fact about equal reality is that the co-real operator is transitive, even across frames. That means that if $\alpha r\beta$ is the case and $\beta r\gamma$ is the case, then $\alpha r\gamma$ must also be the case. This follows directly as consequence of our definition for equal reality.[11] Thus, applying the properties of transitivity and symmetry to the above relations, we arrive at the result that:

$$ArA'$$
$$BrB'$$

[10] One might object that, for historicists and possibilists in particular, the "co-real" relation is not an equivalence relation. For instance, right now the Norman Invasion is "real" to us because it is in our past, and so the historicist/possibilist would want to say that such an event is as real as our writing this paper; however, at the time of the Norman Invasion, we were not yet born, so we were "not real" at that time. The equal reality relation only holds one way.

However, one can respond to this claim by citing the fact that the equal reality of simultaneous events is an equivalence relation in historicism and possibilism even if the "equal reality" relation in general is not. Two events that happen at the same time must be equally real if it is temporality alone that bestows metaphysical status on events. The above argument only necessitates the treatment of "equal reality" as an equivalence relation for cases where the two "equally real" relata are space-like separated and thus simultaneous. In such a case, equal reality is an equivalence relation even for historicists and possibilists. Thus, the fact that equal reality is not an equivalence relation in general does not mean that equal reality is not an equivalence relation in the case of simultaneity; in fact, the opposite is true.

[11] This feature of co-reality is perhaps not intuitive, but a simple conceptual argument can show why equal reality, as we have defined it, must be a transitive property. If two events A and B are co-real in a given frame, this means that they share an R-value. Likewise, co-real events B and C must also share a unique R-value. Since the uniqueness criterion on reality implies that the R-value shared by A and B must be the same R-value shared by B and C, it then follows that A and C must have the same R-value as well, and thus they must be equally real.

Generalizing from this result, then, one can conclude that a prior event (the stubbing of a toe) is as real as a later event (a shout of pain). If the first event (A, for instance) occurs in the "present", then A′ occurs in the future and the RoS argument suggests that the future is as real as the present. Likewise, if A′ occurs in the present, then A occurs in the past and the RoS argument suggests that the past is as real as the present. Both of these conclusions contradict the presentist assertion that the present is real while the past and future are not since past, present, and future must share the same ontological status by the above argument. Since presentism in conjunction with relativity and our other basic assumptions leads to a contradiction, presentism must be false given our assumptions. Finally, since variations of this argument would answer equally well anyone who would argue that only the past is real or only the future is real, the only conclusion left for a realist is that eternalism must be correct since both presentism and possibilism must be discarded. We have thus achieved our goal of constructing a rigorous argument for eternalism from RoS in the tradition of Rietdijk, Putnam, and SSC though our argument provides a more detailed analysis of the assumptions about the nature of "is real" that go into the RoS argument.

3 Presentist Points of Contention

There are several points in the above argument for eternalism that presentists (or anti-realists, for that matter) could attack or have attacked. The goal of this section is to provide a basic taxonomy of points of contention presentists utilize or could utilize to respond to both the argument presented above and eternalism in general.

3.1 Deflationary Objections: No Presentist/Eternalist Distinction

The first attack on the RoS argument which works equally well against any argument trying to prove or disprove eternalism is that there is, in fact, no metaphysical or empirical distinction between the views supported by presentists and those supported by eternalists. This collapse of the dichotomy between presentism and eternalism is most ardently argued for by Savitt [20] and Dorato [10] in recent papers. Both of these papers utilize semantic arguments to suggest that the distinction between presentism and eternalism boils down to a difference in definitions for "real" which translates, in various contexts, to differences in tensed versus tenseless existence claims. These two authors claim that presentism and eternalism are both essentially either vacuously true when viewed with the proper definition of existence in mind (for instance, to say that the present is the only thing that "exists now" is tautological since "now" is defined in terms of the present) or analytically false when viewed with the improper sense of existence in mind (for instance, to say that the present is the only thing that "exists tenselessly" is to ignore the past and future that are assumed in the phrase "exists tenselessly"). These two authors go on to attack defenses of eternalism that rely on modality and other semantic considerations, leading them to the conclusion that the problem posed by the presentist/eternalist debate is truly a non-starter by way of a "Wittgenstein-like" or "Austin-like" deflation.

In an earlier paper , Dorato [9] discusses various other semantic arguments against eternalism specifically in an attempt to show that eternalism is as problematic as presentism. The first contention Dorato raises is against the eternalist perspective that "the past, present, and future are all real at the same time", which he views as meaningless since one cannot say anything about the relationship between the past, present, and future at a given time since all three temporal regions cannot be simultaneous. There must be a temporal separation between the past, present, and future for them to be well defined, so any statement about how the past, present, and future interact at a given time collapses this distinction and thus becomes meaningless. The second argument against eternalism on semantic grounds is that an eternal truth like "event A takes place at time t" may be timeless, but the object of this statement, event A, is not necessarily as timeless as the statement about it. Dorato thus believes that eternalism confuses the following two statements:

1. "X is the case at t" is an eternal truth
2. X exists eternally

And thus, since eternalism makes this error, it is a deeply flawed and confused view. These two linguistic objections to eternalism, as well as the much larger objection that there is no metaphysical presentist/eternalist dichotomy, will be addressed later in this paper.

3.2 Compatibilist and Incompatibilist Objections

Two other groups of people who reject the RoS argument for BU are the compatibilists and incompatibilists. Compatibilist philosophers of time attempt to hang presentism on a given relativistic invariant (like the fact that all inertial frames agree on the ordering of time-like events, or "proper time").[12] Incompatibilists, on the other hand, invoke some preferred frame or other entity with which to adorn Minkowski space-time in hopes that this new frame will provide a suitable place to hang presentism and becoming. These positions constitute a shift in the definition of "co-reality" as it we presented previously. Both compatibilists and incompatibilists would reject our definition and propose another, though various compatibilists and incompatibilists will propose differing versions of "co-reality". There are essentially two ways philosophers can and do object to the RoS argument:

[12] It should be noted that we do not disagree with the compatibilist assertion that to be real something must be "real in all frames"; in fact, we embrace this idea, and it is a central aspect of our definition for reality that frame-invariant properties like time-like separation are necessarily "real" features of space-time.

1. Reject our characterization of simultaneity in our definition of co-reality (redefine simultaneity, compatibilist and incompatibilist objection)
2. Reject our characterization of reality in our definition of co-reality (reject transitivity of co-reality, compatibilist objection)

Option 1 can and has been argued for on several different grounds. It has most famously been argued that either (a) our notion of simultaneity is not a suitable criterion for reality because the present refers to only the "here and now", not simply the now, or (b) simultaneity is relative to some preferred foliation of space-time.[13] Objection (a) is raised most famously by Stein [26, 27] in his response to Putnam, and objection (b) has been raised by various philosophers and physicists who have rather disparate views as to what the preferred foliation of space-time is and from whence it issues.[14] We will address both of these objections to the RoS argument individually in the following sections.

Compatibilist option 2 is typically raised either by those like Savitt [21] and Dolev [8] who believe that an argument for a transitive notion of reality has not and cannot be convincingly made especially within the framework of SR or by anti-realists (including solipsists) who believe that the phrase "reality" should only pertain to one's own frame (or, worse yet, only to oneself). The first of these objections is then the only one particularly relevant to the presentist/eternalist debate because an anti-realist would no sooner be a presentist than an eternalist. The transitivity of "is co-real with" is objected to on this view precisely because it leads to the view that presentism is wrong. Thus, it seems like any presentist interested in saving her stance would object to the transitivity of co-reality implied by our definition of reality as many before her have chosen to do.

4 Response to Objections

4.1 Defining Terms: Establishing a Presentist/Eternalist Distinction

Dorato and Savitt claim that there is no metaphysical or empirical distinction between the eternalist and presentist perspectives by critically examining the terms "is", "exists", and "real" used in several definitions of reality and in doing so point out the shoddy conclusions that linguistic sloppiness engenders in the presentist/eternalist debate. Our goal in this section is to provide an original account of reality which supports a metaphysical/empirical distinction between the presentist

[13] The first of these objections, (a), is a compatibilist objection while the second objection, (b), is an incompatibilist objection.

[14] See Cifone (2004, PhilSci Archive) for specific examples of proposed preferred foliations to space-time.

and eternalist positions. Such a reasonable definition is sufficient to counter Dorato's and Savitt's deflationary claims.[15]

Our definition of reality relies upon two concepts: "definiteness" and "distinctness". For an event to be real, we posit, the event must be both definite and distinct. We take a definite event to be one which is meaningfully determined. A useful example of the distinction between definite and indefinite can be found in quantum mechanics.[16] With respect to a particular variable like spin in the x-direction, a pure-state quantum system may be in an eigenstate or a superposition of eigenstates. If there exist a multitude of systems in the same eigenstate, an x-spin measurement on any of these systems will always yield the same value. Thus, we say that an eigenstate of x-spin is property-definite with respect to spin in the x-direction. However, if the system is in a superposition with respect to x-spin, different systems prepared in the same x-spin superposition may give different x-spin values when measured. There is no way to predict the value of the x-spin of such a superposition after measurement given any information about the system prior to measurement, and as such, the superposition of x-spin is said to be property-indefinite with respect to x-spin. Generalizing from our characterization of property definiteness, we define event-definiteness as definiteness with respect to at least one property. Thus, if an event is property-definite with respect to at least one property, we say it is event definite and thus "real".

We should note here that our event-definiteness criterion is an objective criterion of a system, and as such, unlike property-definiteness, a system must be indefinite with respect to all of its properties to be considered indefinite qua system. Therefore, quantum superpositions are not objectively indefinite, for there exists some property with respect to which this superposition is definite by the very nature of superpositions; it is only the x-spin value of such a superposition that is indefinite. If a given event is definite with regard to any property, it is taken to be objectively definite and thus may be real (as long as it meets our distinctness criterion as well, that is).

It should also be pointed out that event-definiteness is a frame-independent property of events in the universe; though different observers may disagree about the state of a given system (as Rovelli [18] points out in his paper on relational quantum mechanics), they will all agree about whether or not it is definite simpliciter. One might take issue with our assertion of the frame-independence of definiteness; for instance, some postulate that quantum collapse is hyperplane-dependent, and thus

[15] For an alternative response to such a deflation by way of logical and linguistic analysis, see pages 14–17 of Sider [22].

[16] We are not claiming that quantum superpositions are unreal or non-existent simplicter; rather, we are providing an example in an instrumental spirit of how a property might be indefinite and thus suggesting how one might generalize from this example to form an idea of general indefiniteness. This indefiniteness, if made general and applicable to all properties, would make an event effectively unreal. However, superpositions themselves are by their very nature in a determinate state with regard to some property, so they are obviously not wholly unreal in this sense.

an observer in one frame will see a quantum system as having some definite property "y" while an observer in another frame might observe y to be indefinite. However, even if collapse is so dependent, the fact remains that each of these observers will observe there to be some definite property "z", and thus, by our definition, one must take the quantum superposition to be definite qua system. That is, there is no frame of reference from which one can observe the quantum system in question to be without any definite properties. Therefore, our definition of definiteness directly implies the kind of transitivity we exploit in our RoS argument.[17]

The other criterion for an event to be real is that it must be distinct. A distinct event must be in some way different from other distinct events (a la Leibniz, call it the discernability of non-identicals). Such a criterion for the distinctness of events is different from a criterion that requires the distinctness of particles. While it may be that two completely indistinguishable particles can both be distinct, the issue of concern here is the reality of events, and it is the case that two completely indistinguishable events cannot be distinct per the identity of indiscernables; or if you prefer, two completely indistinguishable events cannot be numerically distinct. This criterion of distinctness may be viewed as a more pragmatic concern (we have no reason to take event B to be numerically distinct from event A if all of B's properties are identical to A's). Such a criterion of reality keeps one from treating as real two (allegedly distinct) "events" that might seem to be different but are truly one and the same event – the differences are purely perspectival as in the Lorentz transformations of SR. For example, as per Newton's third law of motion, there is no need for us to count as distinct both the event of a car hitting a wall and the event of the wall hitting the car; they are simply two different ways of viewing the same singular event.[18]

Having established these two criteria for reality, does there appear to be a difference between the presentist and eternalist positions? The answer is "yes" because the distinctness and definiteness of the past and future are not analytic. The presentist claims that past and future events lack both/either definiteness and distinctness simpliciter while the eternalist says all events past, present and future

[17] We should point out here that presentists who claim that there simply are no past or future events can be treated as taking such event as indefinite on our picture here, since a non-existent event cannot have any definite properties. Thus, our account of definiteness provides a criterion for reality that explains this possible presentist stance.

[18] One might well wonder what purpose introducing "distinctness" as a criterion here serves above and beyond the work already done by definiteness. Distinctness is important in this discussion because it allows for nuances within possible presentist positions. We believe there may be presentists who concede that some future events are determined in that they have some definite property, yet who may still reject that the future and present are "equally real". They could do so by way of distinctness, claiming that there are an infinite number of events (one of which will be actual, the rest of which will not be) which are all "definite" in some sense but indistinguishable. The future would thus be definite but not distinct, and so the presentist could write it off as unreal. For the purposes of this discussion, it is in our interests to give as many reasonable possibilities to the presentist as we can, and so we have included distinctness in our discussion for the sake of completeness.

possess both definiteness and distinctness. The first fact to note about the future is that it is unknown to us. One might even be tempted to say that it appears indefinite since it seems (at least on some stochastic accounts of quantum outcomes) that there is no way for us to know the future (in principle) no matter how much we know about the present. Such stochastic accounts of objective quantum indefiniteness (as opposed to subjective quantum indefiniteness for deterministic interpretations) should not be confused with what we will call O- (objective) indefiniteness and S- (subjective) indefiniteness more generally. O- and S-indefiniteness are best understood as a different kind of indefiniteness entirely which will be made clearer by an appeal to the idea of "Newton's god" (NG), an entity in the 5th or higher dimension "looking down" at her space-time "sensorium".

Depending on whether the future is O-indefinite or S-indefinite, NG would observe different things as she looked down on her "sensorium". If the future and past are S-indefinite only, NG would physically *see*[19] the past, present, and future – all of space-time, a 4D BU. NG would see events in the past, present, and future – a static multi-colored marble of world-lines/tubes, if you will. If the future and past are truly O-indefinite, however, NG would not be able to see the future or past from her 5th-dimensional perch, but only a continually temporally evolving present. If the future is truly O-indefinite, it does not matter whether NG is observing us flipping a coin or measuring the spin of an electron with stochastic outcomes; either way, she will not observe the future outcome, and likewise if the future is merely S-indefinite then in both the classical and quantum case NG will observe the future outcome. In the O-indefinite case, NG may be able to predict the outcome just as any one of us may be able to predict the outcome of a coin flip, but NG will not be able to observe this future outcome.

The eternalist, presentist, and possibilist positions become clear and distinct given this characterization of O- and S-indefiniteness. Eternalists believe that the future and past are only S-indefinite; though beings within space-time may not be able to observe the past or the future, a being outside of space-time would be able to easily observe them. Thus, NG sees a 4D BU when she looks "down on" the universe. The presentist, on the other hand, holds both the past and future as truly O-indefinite and thus believes that NG would see an evolving 3D time-slice of the universe when she looks "down on" her "sensorium".[20] Finally, the possibilist takes the future to be O-indefinite but the past S-indefinite only, thus leading to the belief that NG would see a growing BU when she looks "down" on the universe. Diagrams of these various NG perspectives may be found in Appendix C.

Another way of viewing our "Newton's god" argument is in terms of "where" time is in the presentist picture compared to the eternalist picture. In the presentist picture, NG is still constrained by time. The fact that NG is removed from spa-

---

[19] When discussing what NG "sees" we are only invoking the traditional physical sensory modalities of this entity. We make no claims about other ways of knowing or omniscience that one in NG's position might be able to employ by means other than perception.

[20] On some presentist views, she might even see a point. See Stein [26] for more on this view.

tial strictures does not entail her separation from some notion of time in which she must still continue to exist. It is possible, then, for NG to remove herself from space without removing herself from time on the presentist picture. On the eternalist picture, however, NG is free from the strictures of temporality. It is unclear what the character of the 5D universe NG inhabits is (the 5th dimension could be conceived as some sort of second-order time, a 4th-order space, or some phenomenology of dimensions we do not experience); however, the point is that NG is free from time as well as space as it exists in the BU since the two are inextricably linked, and thus time has the same ontological status as space. The eternalist does not have to argue that time behaves the same way as space does, simply that time and space are inextricably linked, which is a stance that the presentist rejects since the presentist views universe as 3D.

There may be some who believe that NG is not a suitable tool for dealing with the presentist/eternalist distinction; in particular, one might find our NG question-begging since a god's eye point of view might seem to violate basic tenets of SR; however, one must note that by hypothesis NG is removed from the 4D-manifold (space-time) that she observes. Such a being would be constrained to see a space-time that conforms to special relativity even though this "god-frame" itself would not so conform. SR can only make claims about perceptions of space-time from within space-time, and since this "god-frame" is outside of space-time, this relativistic objection does not obtain. Even without positing the existence of NG or even a position from which NG could look, we have already shown that the presentist/eternalist distinction can be stated in terms of the separability of space and time, and so if this objection to NG as question-begging is simply that one cannot remove oneself from space without removing oneself from time as well, then the objection has already conceded our point to us. Using our novel argument for the eternalist position, Dorato's two previous objections to eternalism can be ignored as well. Nowhere in our argument do we claim that the past, present, and future are all "simultaneous", nor is there any confusion between eternal truths about existence and the eternal persistence of events. First, an appeal to some sort of "second order" time is completely unnecessary for our formulation of the eternalist position, and as such the accompanying language of the "past present, and future existing simultaneously" has been discarded. As noted above, Newton's god's frame need not necessarily be conceived as some sort of second order time; further, it is merely a thought experiment to show that Dorato/Savitt type arguments are dependent on verificationism of a sort special relativity need not entail. In the following passage Dainton [7] paints a suggestive picture of what it means to take Newton's god's perspective of the BU seriously:

> Imagine that I am a God-like being who has decided to design and then create a logically consistent universe with laws of nature similar to those that obtain in our universe. Since the universe will be of the block-variety I will have to create it as a whole: the beginning, middle and end will come into being together. Well, assume that our universe is a static block, even if it never 'came into being', it nonetheless exists (timelessly) as a coherent whole, containing a globally consistent spread of events. At the weakest level, "consistency" here simply means that the laws of logic are obeyed, but in the case of universes like our own, where there are universe-wide laws of nature, the consistency constraint is stronger:

> everything that happens is in accord with the laws of nature. In saying that the consistency is "global" I mean that the different parts of the universe all have to fit smoothly together, rather like the pieces of a well-made mosaic or jigsaw puzzle (119).

It would be absurd to argue, therefore, that two perspectives as different as these are, are in fact, metaphysically and empirically equivalent in principle; such a claim could only be sensible if one assumes a spatiotemporal-anthropocentric verificationism, and there is no non-question begging reason to do so. For this reason, Dorato's and Savitt's grander claims must be dismissed. The most these two authors can suggest is that a better definition of reality is necessary before the presentist/eternalist debate can be undertaken, and so, with such a definition provided, Dorato's and Savitt's deflationary claims can be rejected. Dorato and Savitt are right to point out concerns with definitions of terms (such as "real") in arguments such as ours, but generally speaking this is the most that linguistic analysis can contribute to the presentism/eternalism debate. The most such appeals can do is determine that certain positions in the debate are "unspeakables" or that the language used must be clarified for the debate to proceed.

4.2 *The Transitivity of Reality*

Our new definition of an event's reality as a combination of definiteness and distinctness also has implications for the second compatibilist objection to the RoS argument, namely that there is no good reason why reality or the "is co-real with" relation ought to be transitive. The first response to this claim is that any relativistically invariant relational property must be transitive across all reference frames. For example, consider the property of "light-likeness along direction x ".[21] Any two events that are light-like separated in some direction share this property, and all observers in all frames will agree that two events are light-like separated if they are so due to the fact that the speed of light in a vacuum is a universal constant. Thus, if event A is light-like separated from event B and event B is light-like separated from event C in the same direction, then event A must be light-like separated from event C (in this same direction). This deduction is true even if one adds different relativistic frames into the equation. For instance, if event A is light-like separated from event B in direction x in a frame moving with velocity v and event B is light-like separated from event C in direction x in a frame moving with velocity u where

[21] The "x" in "along direction x" in this property should be a four-dimensional vector pointing from one event to the other. We include this condition to rule out the following, non-transitive case: consider a light beam shot out from a spaceship at A, reflected off of a mirror at B, and returned to the ship at C. A and B are light-like, B and C are light-like, but A and C are time-like. However, this non-transitivity arises from the fact that the direction of the light is changed at B, and so the vector x shifts at this point. We thank Gordon Belot for bringing this objection to our attention.

u is not equal to v, it is still the case that event A and event C are light-like separated in a frame moving with velocity w no matter what the value of w.[22] Thus, from this simple example, one can see that a relativistic invariant quantity is transitive across inertial frames.

There are two other relativistic invariant properties aside from "light-likeness" that we would like to discuss now. The first of these is number. All observers, no matter their frame, will agree on the number of events that occur. Thus, no matter what frame an observer is in, it will never be the case that she will see an event take place that another observer does or could not see. Though observers may disagree about some of the properties of an event, no observer will see a "novel" event; that is, there is no event simpliciter that one can only see if one is in a certain reference frame. This means that the very existence, the very definiteness of an event-as-such must be a relativistic invariant, and thus as per our pre-established criterion, definiteness must be transitive across frames.

Another relativistic invariant is the space-time interval between two events. This separation is defined by the Minkowski space-time metric as: $s^2 = t^2 - x^2 - y^2 - z^2$ where "s" is the space-time interval, "t" represents time, and "x", "y", and "z" are spatial coordinates in 3-space. Because the interval between events is an invariant, it is always possible for observers in different frames to distinguish between different space-time events in a consistent manner. Because of this, no observer will confuse two events that are seen as distinct in another frame. Thus, the invariance of the space-time interval implies that distinctness is a relativistic invariant. Thus, as per our pre-established criterion, distinctness also must be transitive across frames.

Now, since reality in our formulation has definiteness and distinctness as necessary and sufficient conditions and since both definiteness and distinctness are relativistic invariants, it follows that reality, the conjunction of definiteness and distinctness, should also be a relativistic invariant. Finally, as has already been established, any relativistic invariant must be transitive across frames, and therefore our "equal reality" relation must be transitive across frames. This argument suggests that, as a logical consequence of special relativity combined with our definition for reality, reality must be frame-independent. This logic provides more than sufficient reasoning to support objectivity in our co-reality definition, and so the weight now falls on the shoulders of Savitt and the presentists to explain why "is real for" should not be transitive if they want to continue pushing this point.

4.3 Against the Point Presentist

There have been several arguments against the "here, now" presentist as Stein[23] presents him. This variety of presentist holds the present to consist of a single point

[22] Within relativistic limits, of course.

[23] Bourne [1] points out that Stein was not assuming a "common sense" notion of simultaneity when he attempted to redefine simultaneity within relativity as the "here" and the "now". It seems

in space-time and defines the "now" as both temporal and spatial. There have already been several excellent responses to Stein's view, most notably those provided by Cifone (2004) and Petkov [14]. We will here reiterate and rephrase Cifone's and Petkov's points to show that the "point" presentists, as they are traditionally called, do not hold a viable position.

The first argument against point presentism comes from Cifone. As previously discussed, it is easy enough to see how anti-realism can be reduced to a form of point presentism, but the opposite seems true as well. Point presentists can be taken to be essentially solipsists since what exists at only one point (presumably, the point where the point presentist currently exists) is all that exists. This is not an argument in itself, and there are ways around point presentist solipsism, but these views are almost equally bad. If there is more than one "point present" in the world (that is, if he rejects solipsism), what is required for a point to be "the present"? Is there some "present-maker" that defines the present, that selects it out from all possible "presents"? And if there is, what would such a "present-maker" be? What is more, if there are a large number of "presents" that all compose reality, why do none of them agree with each other? For if the present is only a single point, it follows that multiple "nows" will not count other "nows" as real. There will be no agreement among different observers in different frames, let alone different observers in the same frame, as to what constitutes reality. Thus, it seems that the point presentist loses all semblance of self-consistency when he explains his position and runs the risk of having his position collapsed into absurdity.

Perhaps most damning to the point presentist, however, is Petkov's response. Petkov points out that a point presentist reduces reality to a single, 0-dimensional point. If point presentism is correct, he asks, why does the universe appear to be four-dimensional, as evidenced by the aforementioned 4D space-time invariants? The universe defended by presentism which lacks the 4D-manifold in favor of a 3D universe seems unable to support or explain phenomena like length contraction and time dilation, but it appears nearly impossible to reconcile a 0-dimensional view of space-time with such phenomena. Such a view, Petkov argues, reduces to solipsism. After all, consider two observers A and B. If A and B are distinct observers, any observation event by observer A will not be real to observer B since only observer B's "here and now" are real to him. This solipsism leads to the loss of realism that Cifone (2004) points out. Petkov also claims that only a 4D view is supported by special relativity by refuting the 3D picture of the world as well. His argument is that the phenomena of length contraction and time dilation, both of which allow different observers to hold ontologically distinct and correct beliefs about the 3D properties of an object, cannot be as completely described by a 3D worldview as by a 4D block universe view. He compares the situation to looking at a 2D plane; one can certainly describe the plane as a series of lines in the x-direction for different, constant values in the y-direction, but this "complete" description of the phenomenon does not

---

that Stein's original point was not so much that simultaneity had a different nature than previously thought but rather that the conception of simultaneity that comes to play in everyday discourse has no currency in special relativity.

change the fact that it is a 2D plane and not a 1D line that is being described. If a 3D world is inadequate, then, it stands to reason that lower dimensional representations of space-time would likewise be inadequate. Thus, the 0D description of the world presented by the point presentist must be incorrect. If one is to believe in the point presentist as a viable alternative to the eternalist and the traditional presentist, the point presentist must provide physical support for a 0D universe or else abandon his view.

Before leaving point presentism, however, there is one perspective similar to Stein's that advocates changing the definition of simultaneity in order to save the presentist from the RoS argument. This more recent shift is presented by Bourne [1] and ought to be addressed here since it is a challenge to the notion of simultaneity we employ, a challenge that adheres to the logic that Stein originally used when proposing point presentism (see previous footnote). Bourne argues that simultaneity is absolute within space-time. According to Bourne, the notion of absolute reality does not translate into the language of relativity because no one can determine whether or not two events are simultaneous by observations within a frame. He turns simultaneity on its head in presentism, not by defining "what is real" by "what is present" but rather "what is present" by "what is real." Bourne appeals to a linguistic analysis in terms of conjunction, instead of observables in the world as the basis for reality and thus simultaneity. In short, Bourne's reinterpretation of simultaneity insists that simultaneity is absolute by ruling out the possibility of determining simultaneous events (or, it seems, reality) by observation alone.

Bourne's reinterpretation of simultaneity shows to what extremes presentists must go to rescue their philosophy of time from the RoS argument. By the time Bourne is finished with simultaneity, there is nothing resembling the common-sense notion of simultaneity left. Not only is simultaneity dictated as absolute without empirical evidence or verification (for surely one cannot appeal to physical grounds for such an argument), but simultaneity has now also been removed from the realm of science altogether. There is no longer any observation that can determine if two things occur at the same time! Not only does this assertion fly in the face of common-sense views of simultaneity, it also poses dire consequences for science and human knowledge when combined with presentism. If Bourne's simultaneity gives us no access to a distinctively "real" character for "real" events, how can any empirical evidence help in determining which things are real and which things are not? Does linguistics then pose a better means to come to truths about the natural world than science does for Bourne? If we are planning on choosing a metaphysics of time that best accounts for the phenomena at hand without making any wild metaphysical claims, it seems clear that Bourne's reinterpretation of simultaneity does not save presentism since even the claim that the past, present, and future are all equally real is a more conservative claim than that simultaneity and reality are both phenomena to which no one has empirical access.

It is, however, possible that one can reinterpret Bourne's claims about the simultaneity in physical terms; such a reinterpretation of Bourne's simultaneity would necessitate a preferred foliation of space-time.[24] Though we will not address Bourne's revised notion of simultaneity directly any further since he does not explain his simultaneity in terms of preferred foliations of space-time in any satisfying way, we will address preferred foliation presentists generally in the next section.

4.4 Preferred Foliations in Space-Time

A slightly tougher objection to RoS is raised by those suggesting that space-time has a preferred foliation. Such a foliation would run counter to current beliefs not only about eternalism but about relativity as well, for one of the chief tenets of relativity as it is traditionally interpreted[25] is that there exists no preferred reference frame. The good news for the eternalist is that there is very little physical evidence[26] to support such a preferred foliation, but it such preferred foliations may be postulated. Assuming that such a foliation is found, then, does our RoS argument for BU still follow?

The first response to the preferred foliation objection is that no preferred foliation theory as it currently stands, even if it were proven to be true, provides the necessary physical mechanisms that would be needed to explain why such a frame would be preferred. Until physical motivation for a preferred frame is provided, one cannot abandon the RoS argument. Perhaps there is some way in which the "now" transforms as it goes into other frames. Perhaps the "now", though it is dependent on its preferred space-time foliation, is still present or still has metaphysical influence on other frames. Until physical motivation for a preferred reference frame is provided, one simply cannot know these things. After all, we do use CMBR ("cosmic time") as a pragmatic preferred frame in physics but it does not impugn BU any more than proper time does. In a purely relativistic context, the claim that the Big Bang occurred 14 billion years ago is completely frame dependent; there are other possible, equally valid choices to be made. The point is that none of these invariant features internal to SR changes the fact that M4 unadorned has no resources to

[24] Bourne explicitly endorses such preferred-foliation presentists in his book, though he does so in a different section from the one in which he advocates his radical revision of simultaneity.

[25] Other non-standard interpretations, like the Lorentz interpretation, yield the same results as the M4, no preferred frame interpretation of space-time, so it should be pointed out that it is not the physical results of special relativity that are threatened by the preferred frame but rather the currently-held understanding of special relativity which is under fire. See Appendix A for more information on the rejection of the geometrical special relativity interpretation.

[26] There are those who claim that at the end of the day, a correct theory of quantum gravity or a correct interpretation of quantum mechanics (such as Bohmian mechanics) might yield an absolute preferred frame. While technically true, recent work by Callender [4] and Monton [13] suggests that: (a) an absolute preferred frame is not a likely consequence of future theorizing in either case and (b) even if these preferred-foliation theories do pan out as expected, they will run into all the problems outlined in this section.

construct an absolute and objective preferred frame and that RoS implies the equal reality of all events. On our view, one can always conventionally define a preferred frame such as cosmic time; however, unless one can show that a preferred frame a la a physical mechanism is the cause of physical effects like Lorentz contraction and time dilation (as opposed to mere relativistic effects), a pragmatic-preferred frame of this sort does not refute BU.[27]

Callender's [4] objection to the preferred foliation view, however, is perhaps stronger. Callender proposes a problem he calls the "coordination problem". The idea is that even if there is a preferred reference frame,[28] there is no reason to believe that this reference frame would provide anyone with a suitable "now" upon which to base presentism. One must in some way prove that the physical preferred frame is precisely the same as the metaphysical preferred frame posited by the presentists. How would one be able to make such an association? And, perhaps more importantly, even if it were possible for one to argue that the physical and metaphysical preferred frames were, in fact, one in the same, how would this alter the presentist's conception of the present?

Let us try to cash out what it would mean to live in a universe in which a preferred frame forms the basis for an absolute reality. Imagine two twins who are born in such a preferred foliation of space-time. The absolute simultaneity of the preferred frame mandates that these two twins will agree on their ages at all points in time (twin 1, Alice, will turn 21 when twin 2, Bob, turns 21, etc.). However, if Bob decides to take a trip and leave the "real" foliation of space-time, the "absolutely simultaneous" events (picked out based on the preferred frame) involving Alice and Bob describe Alice and Bob as being different ages (Alice, perhaps, is 23 while Bob is only 22); however, whenever Alice and Bob interact directly with each other by shaking hands, giving each other a high five, etc., they will agree that they are both the same age. According to the preferred frame presentist, then, Bob's leaving Alice's frame changes his ontological status. His age and size physically change as he travels around the universe, yet Bob is completely unaware that he is undergoing these changes.

---

[27] Another objection to such a move comes from John Mather, winner of the 2006 Nobel prize in physics with George Smoot for their discovery of the blackbody form and anisotropy of the cosmic microwave background radiation, in a talk given at Swarthmore College in October 2007. In his talk, Mather suggested that there may be many "preferred frames" provided by the CMBR depending on how the source of the CMBR is moving. If there are, in fact, a multitude of "preferred frames", any idea of "reality" that could be grounded in CMBR would be useless for presentism because our uniqueness criterion would be violated. There would be many "real" frames that one could choose. It should also be noted that Mather himself does not believe that the CMBR frame should be treated as anything more than a useful frame for doing calculations; that is, like the proper time frame, the CMBR frame is not "real" in some special way but is rather just a helpful tool for physical calculations.

[28] Specifically, Callender is concerned with a preferred reference frame that might emerge from robust violations of the locality principle in Bohmian mechanics (and other modal interpretations) or preferred frames required for instantaneous collapse in some collapse accounts of quantum mechanics.

This situation produces several problems for the presentist since she must explain why changing one's velocity should cause one's views about oneself to be more or less in line with "reality." When I get in my car and drive to the store, for instance, I have changed my inertial frame; am I now closer to the "real" frame or farther from it? Either way, I don't experience the immediate world differently, nor do I perceive any differences in myself, yet my ontological status has changed. What, then, is the basis for calling such a velocity shift a "shift into (or out of) delusion" since I notice no difference in myself when I speed up or slow down? The other problem for the preferred frame presentist is a related concern: if the preferred frame is what's "real" but I experience the world in exactly the same way whether I'm in the preferred frame or not, why should I care about "reality"? What makes reality a meaningful concept to me if it is not linked with any physical, psychological, or epistemological change? For a preferred frame presentist, reality has no important implications other than to save presentism. Again, reality becomes distantly removed from our experiences, and though we may be able to convert all of our dimensions, temporal and spatial, into our "real" dimensions according to the preferred frame, these real dimensions will be no more important to our lives than our dimensions according to any other frame.

In the end it seems like the preferred foliation proponent is providing a view that is perhaps as inimical to the presentist as to the eternalist. One of the major reasons why presentists hold the position they do is that it seems to agree with the human manifest experience of time. If this experience were hung on some preferred frame due to microwave background radiation or preferred frames as posited by some Bohmians and collapse theorists, it would be possible for a "now" to exist that was completely alien to human experience. Does the phrase "now" even have any meaning when it has been removed from human perceptions of time? The burden falls to the presentists here to prove that a meaningful "now", a physical preferred foliation of space-time, and an identical metaphysical preferred foliation of space-time are all compatible, and since no such reconciliation of all three of these space-time features has been provided by the presentist camp, we are forced to conclude with Saunders [19] that the burden of proof in the presentism/eternalism debate lies entirely on the shoulders of presentists because M4 unadorned does not have the resources to ground the presentist's preferred frame, at least nothing not ad hoc, merely pragmatic, or perspectival.

4.5 The Spatial Presentist: Absurdity in Incompatibilist Presentism?

Having answered the presentist objections to the RoS argument in turn, we would like to propose another argument along the same lines as the RoS argument which, we believe, should serve as a preemptive criticism against incompatibilist presentist arguments to come. Suppose that there exists a new kind of realist called a spatial presentist. The spatial presentist believes not that all events occurring simultaneously are real but that all events that occur in the same place are real. Perhaps there

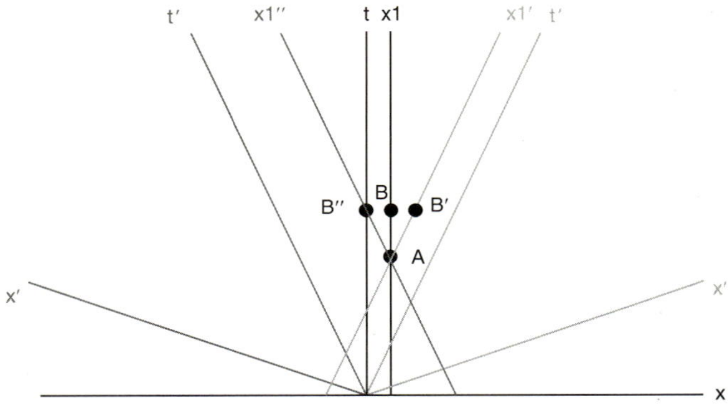

Fig. 2 Spatial presentist argument

is a sphere (infinitesimally small, for our purposes) that the spatial presentist has set aside, following which he claims that "the only things that are real are those in this sphere". One might ask, then, what would be real after the creation of the sphere at an event A in the above diagram, which shows, from relativistic considerations, what events will be observed to fall inside the sphere by observers in different inertial reference frames.

From Fig. 2, it is clear that we are left in a situation directly analogous to the temporal presentist situation previously established in our RoS argument, for the above space-time diagram shows a property we will call the relativity of same position or RoSP. One can simply rotate our Fig. 1 and make a RoSP argument to disprove spatial presentism in the same way that the RoS argument disproves temporal presentism. The arguments are completely symmetrical in the same way that RoSP is symmetrical with RoS.

But what does this show? Only that if an incompatibilist presentist of the non-spatial variety wants to assert that temporal presentism and temporal presentism alone is correct by proposing some new feature of space-time, she must be careful that her argument and mechanism establish presentism but do not allow for spatial presentism.[29] This is yet another burden that the incompatibilist presentist must carry. The symmetry between RoS and RoSP suggests that incompatibilist presentists must establish a physical basis for temporal asymmetry so that spatial presentism does not become as viable and defensible a position as presentism itself, for reconciliation between spatial and temporal presentism must lead to point presentism, which is an unappealing position for reasons previously discussed.

[29] This is, of course, assuming that the presentist in question is not a point presentist or some new form of presentist who wishes to tie the conception of the "now" together with some more evolved conception of the "here."

5 Conclusion

Though the traditional formulations of the Putnam, Rietdijk and SSC's RoS argument for the block universe may that leave the argument open to attacks by philosophers of language and presentists, we have reformulated the argument with more specific definitions that make eternalism the likely victor over presentism. Thus, the task before the presentist in defending herself has become even grander; she must (1) find a way to dispel the RoS argument, (2) show why presentism is more likely than eternalism, and (3) integrate temporal asymmetry as fundamental to her argument lest her argument run the risk of establishing an obviously false view (spatial presentism) as well as it establishes her temporal presentism.[30] It is clear from our previous discussion that the most common presentist argument that "space and time are not perceived to act in the same way" is not sufficient to shoulder the weight of a full presentist defense, and thus a more developed presentist argument addressing all of our concerns must be proposed before presentism can escape from the jaws of the RoS argument. Even the retreat into the position of Savitt and Dorato that there is no significant difference between presentism and eternalism seems a difficult one to hold in light of our definitions for definiteness and distinctness. And so, in conclusion, we echo Saunders in stating that while eternalism in itself may not have been conclusively proven correct by our arguments, the burden falls upon the presentist to show why eternalism is not much more probable.[31]

6 Appendix A: Against the Dynamical Interpretation of Special Relativity

A number of philosophers have defended a dynamical interpretation ("constructive" in Einstein's language) of SR of late (e.g. Brown [2]). In the following passage Callender [4] claims the latter interpretation is a potential problem for the RoS argument for BU:

> In my opinion, by far the best way for the tenser to respond to Putnam et al. is to adopt the Lorentz 1915 interpretation of time dilation and Fitzgerald contraction. Lorentz attributed these effects (and hence the famous null results regarding an aether) to the Lorentz invariance of the dynamical laws governing matter and radiation, not to space-time structure. On this view, Lorentz invariance is not a space-time symmetry but a dynamical symmetry,

[30] We would like to note at this point that there is an obvious reason why spatial presentism has never caught on in the philosophy of time: it does not agree with our perceptions of reality. However, if one wants to dismiss spatial presentism on these grounds but remain a presentist, one's workload is not lessened since one must now provide a link between these particular experiences and reality.

[31] We would like to thank Mark Stuckey, Michael Cifone, David Baker, Gordon Belot, and the audience at the 3rd International Ontology of Space-time Conference at Concordia University in 2008 for comments on previous versions of this paper.

> and the special relativistic effects of dilation and contraction are not purely kinematical. The background space-time is Newtonian or neo-Newtonian, not Minkowskian. Both Newtonian and neo-Newtonian space-time include a global absolute simultaneity among their invariant structures (with Newtonian space-time singling out one of neo-Newtonian space-time's many preferred inertial frames as the rest frame). On this picture, there is no relativity of simultaneity and space-time is uniquely decomposable into space and time. Nonetheless, because matter and radiation transform between different frames via the Lorentz transformations, the theory is empirically adequate. Putnam's argument has no purchase here because Lorentz invariance has no repercussions for the structure of space and time. Moreover, the theory shouldn't be viewed as a desperate attempt to save absolute simultaneity in the face of the phenomena, but it should rather be viewed as a natural extension of the well-known Lorentz invariance of the free Maxwell equations. The reason why some tensers have sought all manner of strange replacements for special relativity when this comparatively elegant theory exists is baffling (2006, 3).

See also Harvey Brown's book *Physical Relativity* [2] and his essay "Minkowski Space-time: A Glorious Non-Entity" [3] co-authored with Oliver Pooley for a more developed argument for this stance.

First, Brown [2, p. 7] himself is clear that he is not defending either the ether or a preferred-frame, unlike Lorentz himself. We grant that SR is neutral about the ontology of space-time, but we think there are good reasons for preferring the kinematical over the dynamical interpretation, though we cannot pursue them here.[32] We do want to note that we are not convinced of Callender's claim that the dynamical interpretation of SR necessarily refutes the RoS. At least in the case of Brown, who again, does not claim to be defending absolute simultaneity, while his arguments may lead to space-time relationalism, they do not obviously entail the falsity of RoS as such. So until someone provides a cogent argument from space-time relationalism to the falsity of the RoS, our argument remains intact. Second, even granting an absolute frame, Brown's dynamical interpretation does not obviously save the presentist since she must still face some of the problems raised in Sect. 4.4. For example, even if there is an absolute space-time and a universal moment of the present, there is no reason to believe, as per Callender's objection discussed in Sect. 4.4, that such a present lines up with human experience of the present. What is more, as long as Lorentz contractions and dilations exist, one observer traveling at relativistic velocities may observe his present to be different from the present of those around him. Does that mean that, since he is dealing with past or future versions of these other beings, that they are not real since they are not actually experiencing the present simultaneously with the relativistic observer? There seems to be a suggestion of some sort of frame-dependent solipsism, which would constitute an anti-realism that presentists would reject as readily as eternalists.

Finally, if we are to take seriously the implication that quantum mechanics (our best theory of matter) is to special relativity what statistical mechanics is to

[32] See Michel Janssen's "Drawing the line between kinematics and dynamics in special relativity" in the Phil. Sci. archive (reference number 3895) for good arguments favoring the kinematical interpretation. See also Petkov [15] where the kinematic interpretation of Minkowski space-time realism has consequences not easily or obviously accounted for by the dynamical interpretation.

thermodynamics, then had not quantum mechanics better be able to explain (in some robust sense of the word) the key features of SR such as Lorentz invariance? Obviously, this condition has not been met and merely interpreting Lorentz invariance to be restricted to dynamical laws only hardly does the trick.

7 Appendix B: Objection to RoS Argument by Meta-Time

One of the points we believe we have established in this paper is that the eternalist perspective does not require any meta-time or generally any 5th dimension to be coherent; however, one might object that our R-value, R-relation language itself begs the question against eternalism and refutes our repeated assertion that no eternalist meta-time is necessary. The objection might go as follows: Suppose that one is committed to simply a binary ontology of R-values such that an R-value of 1 represents "real" and an R-value of 0 represents "unreal". The eternalist perspective here seems straightforward (all R-values are 1, or, perhaps less likely, all R-values are 0), but the presentist perspective is not so straightforward. At time t_1, only events at time t_1 have an R-value of 1 while all other space-time events have an R-value of 0. At time t_0, only events at t_0 have an R-value of 1 while all other space-time events have an R-value of 0. Thus, if t_1 is not the same as t_0 (that is, as long as space-time has a temporal dimension), R-values must change with time, meaning that there must be some sort of extra dimension posited to account for this notion of change. Thus, one might object that the only way for one to meaningfully capture the presentist perspective using R-values is to assume some sort of meta-time, and thus the eternalist is only right if one assumes meta-time, which is to give the eternalist his conclusion from the start. Thus, the RoS argument seems to both beg the question and assume meta-time.

Our response to this objection is to note that the objector has taken a fairly narrowly view of R-values (though, to be fair, this naïve view is essentially the one we advocate in this paper for the sake of simplicity). There is no reason why R-values cannot be tweaked to suit the presentists' notion of reality. Let us allow a different kind of R-value, then, one more in keeping with predicates such as Goodman's infamous "grue". We now define a series of R-values that can take any value we would ascribe to events occurring at a certain time from the beginning of time to the end of time. Each R-value represents the predicate "is real at time x and is unreal elsewhere" where "x" is the R-value of the event. Such an R-value scheme is static; there is no meta-time required to account for changes in R-values because no matter what time we perceive it to be, the R-value of every event will remain the same. Thus, by re-characterizing R-values in terms of the time of various events, we can avoid this objection to the RoS argument.[33]

[33] One might wonder about how we would treat the perspectives of presentists who expect the present to have a certain duration instead of being instantaneous. The answer would be to transform the R-value into an R-vector, with the first entry representing the time at which the event

Still, there remains a further issue: time alone is not enough to provide us with proper R-values, especially not in the relativistic context we assume for the RoS argument to go through. Since time is a frame-dependent quantity, it seems that, by allowing grue-like R-values that are based on temporal coordinates, we are forced to give up on the objectivity and frame-independence of R-values. What this shows, however, is not that R-values are not objective but that time itself is not the proper quantity to base an R-value on. Instead, the R-value used for the presentist ought to be the proper time or the space-time interval from some fixed point. Such quantities retain the essential character of our R-value time-dependence while still providing the objectivity we seek from an R-value. Thus, if the presentists' R-values are defined as the proper time at which a given event is real, then the R-value has the character we expect and the RoS argument goes through without begging the question or assuming meta-time.

One final remark ought to be made, however: why should we not stop by just defining the presentists' R-values in terms of time instead of proper time? We lose objectivity in doing so but seem to better capture our intuitions about time and the present. The answer, we believe, comes back to what the R-values are to represent: reality and ontological status. It seems that, if anything ought to be frame-independent, it ought to be the ontological status of an event. If events are capable of having some frame-independent properties, such as a proper time coordinate, distance via the space-time interval, and the kind of event-definiteness and distinctness we previously discussed, then it would seem ridiculous to say that the fundamental ontological status of the event, which ought to be its most crucial, fundamental, essential property, is somehow less objective than these frame-independent properties. This disagreement may boil down to which intuitions are most important to capture: intuitions about reality, or intuitions about the behavior of time. Given the fact that human intuitions concerning time seem at best incomplete and at worst wrong in many cases, we believe that it is reasonable to prefer capturing the former intuition to capturing the latter. As such, we believe that defining presentist R-values in terms of proper time, as we suggest in this section, is the best way to nuance R-values so as to reconcile them with presentism: not only does it allow the RoS argument to follow as we've characterized it, it also reconciles our intuitions about reality with presentism in the most reasonable way possible.

---

switches from "unreal" to "real" and the second entry representing the time at which the event switches from "real" to "unreal" again. This vector scheme could account for even absurdly complex presentist/possibilist/historicist positions that have events blinking into and out of existence many, many times by simply characterizing each shift from real to unreal (and back again) in terms of the time at which the shift occurs and characterizing the R-vector of the event in terms of these times.

8 Appendix C: A "God's Eye" View of Space-Time on Different Theories of Time

Fig. 3 Eternalist perspective on space-time

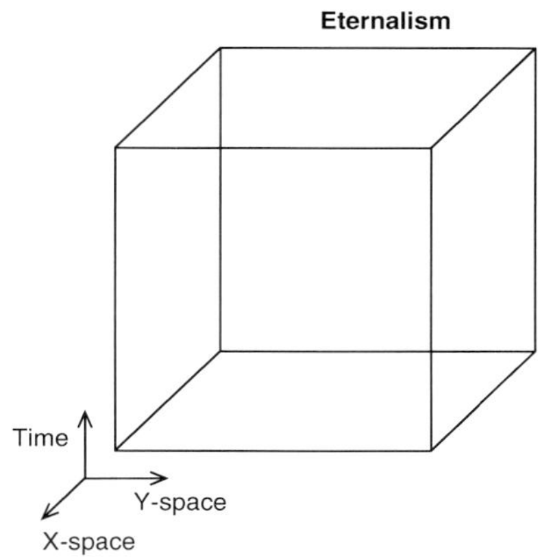

Fig. 4 Presentist perspective on space-time

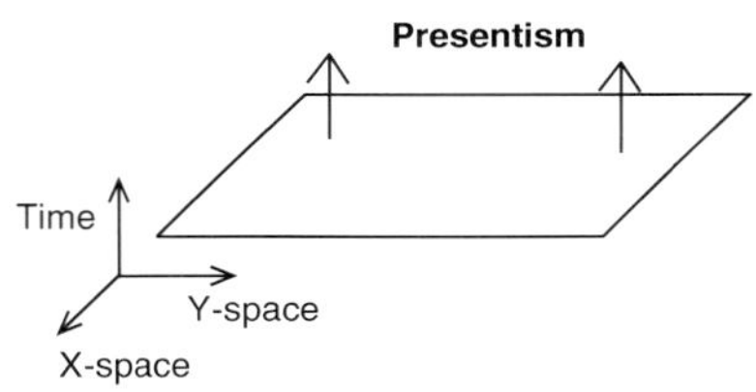

Fig. 5 Possibilist perspective on space-time

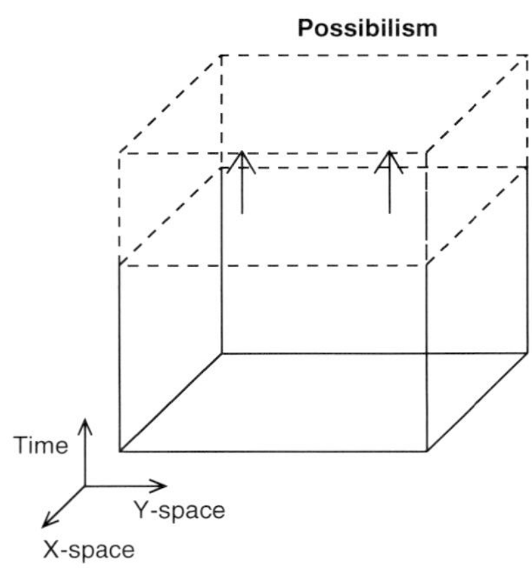

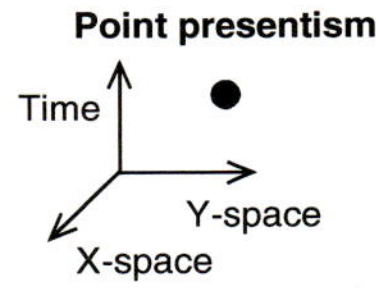

Fig. 6 Point presentist perspective on space-time. This perspective, idealized here as a single point in space-time, is the most difficult to represent visually since it should have an infinitesimal size. The single dot of the present is the only thing in space-time that exists on this view of space-time, making it a much more limited and precise view of the present than the more general form of presentism previously represented

References

1. Bourne, C. *A Future for Presentism.* New York: Oxford University Press, 2007.
2. Brown, H. *Physical Relativity: Space-time Structure from a Dynamical Perspective.* Oxford: Oxford University Press, 2005.
3. Brown, H. and O. Pooley. "Minkowski Space-time: A Glorious Non-Entity" in The Ontology of Spacetime, Ed. D. Dieks. Utrecht, the Netherlands: Elsevier, 2006. pp 67–89.
4. Callender, C. "On Finding 'Real' Time in Quantum Mechanics" (draft) Absolute Simultaneity. Eds. W.L. Craig and Q. Smith. Oxford: Oxford University Press, 2007.
5. Craig, W. L. *Time and Eternity: Exploring God's Relationship to Time.* Wheaton: Crossway Books, 2001.
6. Craig, W. L. and Q. Smith. *Einstein, Relativity, and Absolute Simultaneity.* London: Routledge, 2007.
7. Dainton, B. *Time and Space.* Montreal: McGill-Queen's University Press, 2001. p. 119.
8. Dolev, Y. *Time and Realism: Metaphysical and anti-Metaphysical Perspectives.* Cambridge: MIT, 2007.
9. Dorato, M. "Absolute becoming, relational becoming and the arrow of time: Some non-conventional remarks on the relationship between physics and metaphysics". Studies in the History and Philosophy of Modern physics, 37, 2006. pp 559–576.
10. Dorato, M. "The Irrelevance of the Presentist/Eternalist Debate for the Ontology of Minkowski Spacetime" in The Ontology of Spacetime, Ed. D. Dieks. Utrecht, the Netherlands: Elsevier, 2006. pp 93–109.
11. French, S. and D. Krause. *Identity in Physics: A Historical, Philosophical and Formal Account.* Oxford: Oxford University Press, 2006.
12. Ladyman, J. et al. *Everything Must Go: Metaphysics Naturalized.* Oxford: Oxford University Press, 2007.
13. Monton, B. "Presentism and Quantum Gravity" in The Ontology of Spacetime, Ed. D. Dieks. Utrecht, the Netherlands: Elsevier, 2006, pp 263–280.
14. Petkov, Vesselin. "Is There an Alternative to the Block Universe View?" in The Ontology of Spacetime, Ed. D. Dieks. Utrecht, the Netherlands: Elsevier, 2006. pp 207–228.
15. Petkov, V. "On the Reality of Minkowski Space". Foundations of Physics 37, 2007. pp 1499–1502.
16. Putnam, H. "Time and physical geometry". Journal of Philosophy, 64, 1967. pp 240–247.
17. Rietdijk, C. "A rigorous proof of determinism derived from the special theory of relativity". Philosophy of Science, 33, 1966. pp 341–344.
18. Rovelli, C. "Relational Quantum Mechanics". Aug 1996. quant-ph/9609002.
19. Saunders, S. "How Relativity Contradicts Presentism" in Time, Reality, and Experience Ed. Craig Callender. Cambridge: Cambridge University Press, 2002. pp 277–292.
20. Savitt, S. "Presentism and Eternalism in Perspective" in The Ontology of Spacetime, Ed. D. Dieks. Utrecht, the Netherlands: Elsevier, 2006. pp 111–127.

21. Savitt, S. "Being and Becoming in Modern Physics", The Stanford Encyclopedia of Philosophy (Fall 2007 Edition), Edward N. Zalta (ed.), forthcoming URL = http://plato.stanford.edu/archives/fall2007/entries/spacetime-bebecome/
22. Sider, T. *Four-Dimensionalism*. Oxford: Oxford University Press, 2001.
23. Silberstein, M. et al. "An Argument for 4D Blockworld from a Geometric Interpretation of Non-relativistic Quantum Mechanics." In Relativity and the Dimensionality of the World, Ed. V. Petkov. Heidelberg: Springer, 2007. quant-ph/0605039.
24. Silberstein, M. et al. "Why Quantum Mechanics Favors Adynamical and Acausal Interpretations such as Relational Blockworld over Backwardly Causal and Time-Symmetric Rivals" in a focus issue of Studies in the History and Philosophy of Modern Physics on Time-Symmetric Approaches to Quantum Mechanics. H. Price and G. Bacciagalupi, Editors. Volume 39, Issue 4, 2008. pp. 732–747.
25. Silberstein, M. and M. Cifone. "Static for Dynamism: Special Relativity and the Block Universe", International conference on the ontology of space-time, Concordia University. Montreal, Quebec, Canada, 2004. May 11–14.
26. Stein, H. "On Einstein-Minkowski Space-Time". The Journal of Philosophy, 65, 1968. pp 5–23.
27. Stein, H. "On Relativity Theory and Openness of the Future". Philosophy of Science, 58, 1991. pp 147–167.
28. Stuckey, M. M. Silberstein, and M. Cifone. "Reconciling Spacetime and the Quantum: Relational Blockworld and the Quantum Liar Paradox" in Foundations of Physics, Volume 38, Number 4, 2008. pp. 348–83.

Minkowski Space-Time and Thermodynamics

Friedel Weinert

Abstract The paper discusses both geometric and axiomatic approaches to Minkowski space-time and argues that different inferences to the nature of space-time follow from them. Whilst the geometric approach leads to the traditional block universe view, the axiomatic approaches have as a philosophical consequence a dynamic view of space-time. Both approaches lead to opposite but equally consistent views of four-dimensional space-time. It is a case of underdetermination.

1 Introduction: Geometric and Axiomatic Approaches

Ever since Minkowski published his four-dimensional representation of space-time, the dominant view in physics and philosophy has been that time is a fourth dimension such that human perception of change and the passage of time are a mere illusion, due to our particular slicing of space-time. But four-dimensional space-time is a block universe. This conclusion takes the form of an inference from the measurable and observable evidence. Traditionally the block universe was inferred from the stipulation of relative simultaneity as a consequence of the Special theory of relativity (STR) (Eddington, Einstein, Gödel). But newer defenses infer a static block universe from the well-known relativistic effects: length contraction, time dilation, the twin paradox. The argument states that such relativistic effects would be impossible in a three-dimensional world. As they occur and are observed, it is legitimate to infer (a) that the physical world is four-dimensional, and not just a mathematical model, and (b) that this four-dimensional world is static and timeless. (Lockwood 2005; Petkov 2005, Chap. 4) Yet it is by no means clear that Minkowski himself was a believer in the block universe. In his 1908 Cologne lecture on 'Space and Time' he speaks of a four-dimensional physics but concedes that a 'necessary' time order can be established at every world point. The conception of the block universe, however, focuses on Minkowski's geometric approach, which is based on his world postulate. But an alternative view has been in circulation since the 1910s according to which the nature of space-time has to be based on the behaviour of light. (Robb 1914; Cunningham 1915; Carathéodory 1924;

V. Petkov (ed.), *Space, Time, and Spacetime*, Fundamental Theories of Physics 167,
DOI 10.1007/978-3-642-13538-5_11,

Schlick 1917; Reichenbach 1924) These axiomatic approaches constitute a light geometry, according to which the behaviour of signal propagation, under thermodynamic aspects, forms histories of trajectories in space-time. It is the assertion of this paper that they give rise to a different inference regarding the nature of space-time. If we built our inferences to the nature of space-time on other aspects of the physical world, which nevertheless may fall within the domain of the Minkowski space-time conception – dissipation and energy flows – we arrive at a dynamic conception of Minkowski space-time.

Note that this alternative view does not deny the four-dimensional reality of space-time. It is true that, on the geometric approach, signals are represented by their word lines in space-time and there is no need for dynamic aspects. On the other hand, we cannot simply assume that the mathematical representation of space-time is isomorphic to the physical nature of space-time and that we can infer the nature of space-time from the geometric approach. The alternative approach envisaged here makes use of thermodynamic aspects. If we accept the four-dimensionality of the physical world, and then inquire whether it is 'static' or 'dynamic', it is clearly important to consider both kinematic aspects of the physical world, as enshrined in the equations of the STR, and dynamic aspects, related to questions of energy flow, entropy and dissipation, since signal propagation in Minkowski space-time is constrained by these aspects. This point about representation should justify a consideration of space-time from an axiomatic approach.

The paper will explore the compatibility of Minkowski's space-time representation of the Special theory of relativity with a dynamic conception of space-time by investigating axiomatic approaches to the STR, as they were developed by Robb (1914), Carathéodory (1924) and Reichenbach (1924). A central feature of these accounts is to regard the propagation of optical signals as constituting histories of space-time relations. As it turns out this propagation involves invariant sequences between events, which become central for the understanding of time. It will be argued that one root of a dynamic conception can be located in the thermodynamic and entropic features of the propagation of signals in space-time. If we accept that the geometry and nature of space-time have to be inferred from a range of measurable and observable phenomena (cf. Huggett 2006; Petkov 2005), and that the inference is legitimate on both the axiomatic and geometric approaches, the paper concludes that the question of the ontological nature of space-time is at this stage a case of underdetermination by the evidence.

2 Axiomatic Approaches to Space-Time

Let us now consider what effect a chosen representation has on our understanding of space-time. Since Minkowski's introduction of the conception of four-dimensional space-time, a minority view has scraped a meagre existence in the shadows of the majority view. The majority view is the Parmedian block universe, aptly expressed in Einstein's words: 'From a "happening" in three-dimensional space, physics becomes (...) an "existence" in the four-dimensional "world".' (Einstein 1920, 122) Although Einstein's early commitment to the block universe was inspired by Minkowski's world postulate, in his later years Einstein wavered in his support for the Parmedian view. He began to consider thermodynamic aspects of the propagation of signals in space-time. This alternative view, which is notable for its Heraclitean ancestry, had its predecessors in the axiomatic approaches adopted by Robb (1914), Carathéodory (1924) and Reichenbach (1924). It avoids the binary choice into which McTaggart's metaphysical speculations seem to lure us: either we accept a dynamic A-series or the static B-series, but in either case time is unreal. The alternative view offers the conceptual possibility of a dynamic space-time, which is nevertheless rooted in the B-series. This view is worth exploring because it allows us to fully accept the consequences of the theory of relativity, without endorsing the Parmedian view of the block universe.

But how is this schematic programme to be cashed in? What does it mean that space-time trajectories have a history? To answer this question we do well to look at some attempts to construct axiomatic accounts of space-time, which do not start from Minkowski's 'absolute world postulate'; in Einstein's words it is a 'four-dimensional continuum described by the "co-ordinates" x_1, x_2, x_3, x_4, (which) was called "world" by Minkowki, who also termed a point-event a "world-point". (Einstein 1920, 122) Reichenbach, Robb and Carathéodory developed, apparently independently of each other, such axiomatic accounts, which start from a basic 'before-after' relation between null-like related events. Although these events are represented in geometric terms, they are crucially based on optical facts, like the emission and absorption of photons. The propagation of these signals constitutes an invariant conical order under the Lorentz transformations. The null-like and time-like trajectories between space-time events form the Minkowski world lines of light signals and material particles, respectively. The propagation of these signals constitutes a history of space-time relations, which may include *both* kinematic and dynamic aspects.[1] It may be said that the axiomatic approaches lead to the view of a 'growing block universe'.

2.1 *Robb's Account*

These axiomatic attempts reverse the usual tendency to 'spatialize time'. Robb starts with the thesis that 'spacial relations' may be analyzed in terms of the time relations 'before' and 'after' or, as he concludes, 'that the theory of space is really a part of the theory of time'. (Robb 1914, Conclusion) Essential for this conception is the notion of conical order, which is analyzed in terms of the relations of 'before' and 'after' instants of time. An instant (an element of time) is the fundamental concept, rather

[1] Huggett (2006, 47) defines a 'relational state as a specification of the totality of relations, mass and charges of bodies at a time.' See also Penrose/Percival (1962, §2)

than the space-time event. Furthermore the 'before/after' relation of two instants is an asymmetrical relation. In this way Robb builds a system of geometry, in which we encounter the familiar light cones of the Minkowski representation of space-time. Robb reverses the Minkowski approach in terms of geometrical relations and starts from physical facts, an approach, which is reflected in Einstein's later reservations about the block universe.

> If a flash of light is sent out from a particle P at A1, arriving directly at particle Q at A2, then the instant A2 lies in the α-subset of instant A1, while the instant A1 lies in the β-subset of A2. Such a system of geometry will ultimately assume a four-dimensional character or any element of it is determined by four coordinates. (...) It appears that the theory of space becomes absorbed in the theory of time. (Robb 1914, 8–9)

Here the α-subset is the future light cone of instant A1 and the β-subset is the past light cone of A2. (Fig. 1) After 21 postulates and over 100 theorems defining the light cone characteristics, Robb eventually defines the familiar conditions of the space-time interval, *ds*. The most interesting aspect of Robb's axiomatic system is that it regards Minkowski's contribution as 'merely analytical' and treats the geometry as a 'formal expression' of optical facts, like the propagation of signals in space-time. Thus Robb unwittingly opens up the possibility of considering kinematic space-time relations with respect to other physical aspects of space-time, since his declaration that 'a before-after relation of two instants is an asymmetrical relation' (Robb 1914, 5) will be based on thermodynamic aspects of electromagnetic radiation. Robb's intention is to clarify notions like the conventionality of simultaneity by avoiding attempts to define 'instants of time at different places'. By declaring that events are instantaneous which occur at the same instant, Robb anticipates the notion of relative becoming and local temporality, which have recently been mooted.

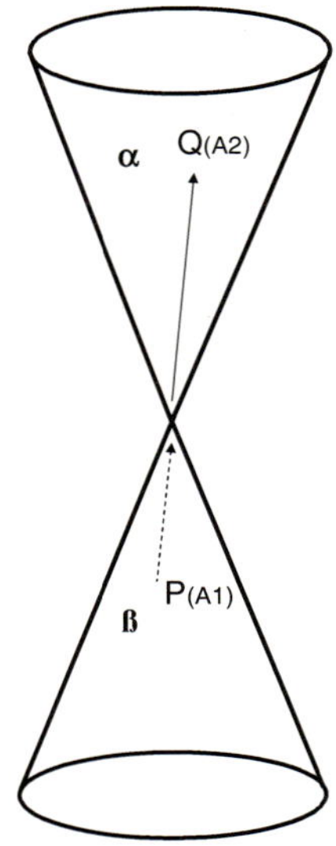

Fig. 1 'Corresponding to any point in space, there is an α-cone of the set having that point as vertex, similarly there is also a β-cone of the set having the point as vertex. If A_1 be any point and α_1 the corresponding α-cone, then any point A_2 is after A_1, provided $A_1 \neq A_2$ and A_2 lies either on or inside the cone α_1. (Robb 1914, 5–6)

'The present instant, properly speaking, does not extend beyond here.' (Nature **107**, 1921, 422) But in the end Robb is still puzzled about time:

> Though space may be analyzable in terms of time relations, yet these remain mysterious; events occur in time, yet any logical theory of time itself must imply the Unchangeable. (Robb 1914, Conclusion)

2.2 *Carathéodory*

In 1916 Einstein encouraged Constantin Carathéodory to consider the problem of closed world lines in the General theory. (Hentschel 1990, 352–354) 10 years later, and without referring to Robb, Carathéodory (1924) started with the STR and took a similar approach but with fewer axioms and postulates. Carathéodory aims at a simplification of Einstein's theory: it is to be based on temporal relations (earlier, later, and simultaneous) but these temporal relations are based on the behaviour of light signals. Carathéodory proceeds to define axioms of temporal succession and of light propagation. These axioms provide the concept of a 'light clock', which allows to measure time-like relations between events in space-time. These axioms are followed by axioms of topological space, which are reminiscent of Robb's conical order and hence allow the introduction of coordinate systems. Finally, he introduces Einstein's principle of relativity. Thus topological spaces consist of light cones, which are constituted by what Carathéodory calls 'normal light propagation'. As is to be expected Carathéodory defines equivalent topological spaces by the use of normal light propagation, satisfying relativity and symmetry requirements. Carathéodory, in fact, constructs what Reichenbach (1924) calls a 'light geometry', whose axioms are based on empirical facts.

> The propagation of light in (our topological space) $\mathfrak{R}$ is to be called 'normal' if, amongst all possible representations of the space $\mathfrak{R}$ by three parameters, there exists at least one coordinate system x, y, z, which satisfies the following condition:
>
> If we interpret x, y, z as right-angled coordinates of a Euclidean space, then of two simultaneously emitted light signals, which run through the two closed light polygons and whose end points coincide with the origin O of the coordinates x, y, z that signal is to arrive earlier, which describes the shorter (in a Euclidean sense) polygon. If the two polygons are of equal length, the signals are to arrive simultaneously.
>
> This shows that in a space of normal light propagation there exists a natural measure for both distances and angles, which depends solely on temporal measurements from the light polygons. (Carathéodory 1924, Sects. 9 and 10; translated by the author)

As noted earlier, it is one of the advantages of these axiomatic approaches, based as they are on 'optical facts', that they permit an easy transition from kinematic to dynamic considerations. This is reflected in Carathéodory's observation that Liouville's theorem also applies to the transformation of the topological space with coordinates x, y, z, t to primed coordinates. Carathéodory expresses the non-tilting of light cones in Minkowski's representation, which is a consequence of the constancy of c in Minkowski space-time, in the statement:

> If two media **A** and **B** move relative to each other with normal light propagation, then every linear light ray of one medium will be transformed into a linear light ray of the other medium. (Carathéodory 1924, Sect. 25; translated by the author)

Liouville's theorem in classical mechanics states that a volume element along a flowline conserves the classical distribution function $f(r, v)drdv$:

$$f\,(t + dt, r + dr, v + dv) = f(t, r, v) \tag{1}$$

(Kittel and Kroemer 1980, 2nd edition, 408; Albert 2000, 73f) In other words, if we consider trajectories in phase space, which include both position and momentum of particles, then the equation of motion of such systems can be expressed in terms of its Hamiltonian, H. H expresses the conservation of total energy of the system. Liouville's theorem then states that the volume of the phase space, which an ensemble of trajectories occupies, remains constant over time. Translated into the language of four-dimensional light cone structure, Liouville's theorem shows that the volume of the phase space regions is invariant over time even though the expansion of the trajectories within this volume can start from different initial states. But an immediate consequence of this theorem is that even though the *volume* is preserved the *shape* of this phase space region is not preserved (see Fig. 2) and this implies a dynamic evolution of the trajectories within this region. For two shapes cannot differ from each other without an evolution of the trajectories. It also implies that a reversed evolution of the trajectories will preserve the volume but not the shape and hence that reversed trajectories need not be invariant with respect to the shape of the phase space region.

The main purpose of these axiomatic approaches is to develop the STR as a light geometry, whose axioms are based on empirical facts. It does not start with an assumption of the existence of the four-dimensional Minkowski 'world' – which is pseudo-Euclidean and in which the linear homogeneous functions x_1, x_2, x_3, x_4 permit a rotation to primed functions x'_1, x'_2, x'_3, x'_4 by the transformation rules of the Poincaré group. The axiomatic approaches start with 'optical facts', like the propagation of light signals. But then it follows that they must be subject to

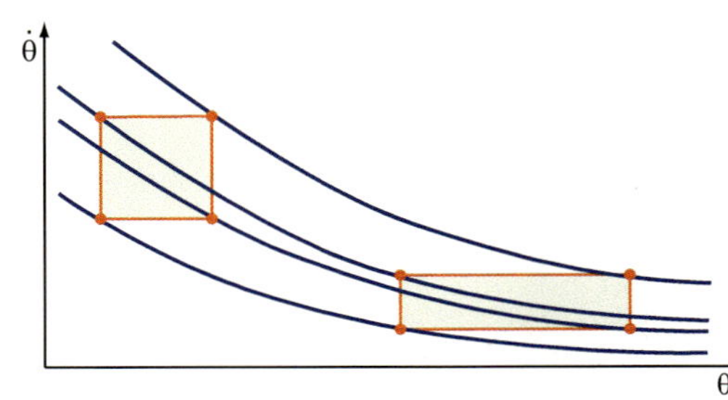

Fig. 2 Liouville's Phase volume invariance theorem. Source: Stöckler (2000, 4th edition, 206); cf. Davies (1974); Reichenbach (1956, 76); Albert (2000, 71–72, 103)

entropic constraints. According to the Robb-Carathéodory representation, the four-dimensional world does not 'exist' but it 'happens' through the propagation of time-like signals between successive events in space-time. These approaches therefore reverse Einstein's famous step from a 'happening' in the three-dimensional world to 'existence' in a four-dimensional world. (Einstein 1920, 122) As the world lines propagate through space-time, they form a history of space-time relations in a conical order. But does this really remove the puzzle about time, so forcefully expressed in Robb's concluding remarks? What did Minkowski mean when he conceded that a 'necessary' time order can be established at every world point? What does it mean that space-time trajectories have a history? In order to answer these questions we must turn from purely kinematic to dynamic considerations. We have two reasons for this transition. As Carathéodory's application of Liouville's theorem to light cone structures shows, we can introduce the thermodynamic language of phase space and speak of the flow of points in phase space. We need to investigate the implications of this shift in perspective.

3 Towards Dynamics

An essential aspect of the geometric view of STR is that it only deals with kinematic relations. The axiomatic approaches remind us that energy considerations are important in the STR and belong to a proper consideration of the four-dimensional world.

3.1 Dynamic Aspects

For a consideration of dynamic aspects it is important to introduce some physical grounding to the asymmetric kinematic relations as the axiomatic approaches of Reichenbach, Robb and Carathéodory emphasize. The axiomatic approaches seek a physical grounding to the asymmetric relations between space-time events in 'optical facts'. For the question that needs to be addressed is: Even if the 'before-after' relation, which is central in the axiomatic approaches, constitutes an asymmetric relation between space-time events, how does this linear order lead to a dynamic view of space-time? Here we want to consider some entropic aspects, because light propagation and signal propagation can be characterized in terms of energy flows and dissipation, processes which are subject to such entropic constraints.

3.2 Provisos

Note that the argument is not to be confused with the usual thermodynamic arguments for or against the arrow of time. Although Eddington held that the increase in entropy established a global, cosmological direction of time, several objections have been raised against the identification of entropic processes with the global arrow of time: (1) Popper (1956, 1957) pointed out that the arrow of time cannot have a stochastic character, which it would 'inherit' from an association with the second law of thermodynamics in its probabilistic interpretation. On Boltzmann's probabilistic interpretation of the 2nd law the increase in entropy is merely overwhelmingly likely, and therefore would in principle allow a reversal of the arrow of time. But even without invoking the law of entropy, Popper held that 'it is absurd to link entropy to the arrow of time because of the existence of thermodynamic fluctuations.' (Popper 1957) Such reversible behaviour has been observed in highly viscous liquids (*Physik Journal* June 2008, 21–22) and can be 'engineered' through the recovery of phase correlations in quantum mechanical *which-way* experiments. (2) The application of the entropy concept to the whole universe is problematic because the entropy concept is best defined for closed systems in thermodynamic equilibrium but the universe as a whole has no environment. (Uffink 2001; Drory 2008) An entropy-free method of obtaining a temporal order is to define a global intrinsic temporal orientability of space-time.

> A relativistic space-time $<M, g, \nabla>$ is said to be temporally orientable if there exists a continuous nonvanishing vector field on M which is timelike with respect to g. (Earman 1974, 17; cf. Cf. Huggett 2006, 234)

The metaphorical arrow of time is then seen as an expression of the geometrical time-asymmetry of the universe. (Aiello et al. 2008) (3) Alternative models for the 'arrow of time' on a global scale have been proposed, for instance the expansion of the universe from the big bang. (Gold 1966; Earman 1974; Earman 2006)

It seems at first that the entropy-free approach is more satisfactory for a global arrow of time but even a global arrow would have to be based on entropic considerations, like for instance Penrose's Weyl curvature hypothesis. Such concerns have no impact on the interpretation of Minkowski space-time. In fact it shows that we should clearly distinguish between the 'passage' and the 'arrow' of time. Space-time observers may perceive a 'passage' of time, in the sense of a local direction of temporal processes, even in ignorance of a global arrow of time. Concerns about the global 'arrow' of time belong to questions of the topology of time. The physical passage of time is compatible with different topologies, for instance linear or circular conceptions of time. Such questions do not address the argument of the block theorist who infers the block universe from the geometric interpretation of space-time phenomena. The definition of temporal orientability appeals to continuous *time-like* vector fields but this does not address the question of time within Minkowski space-time, which according to the axiomatic approach is based on the behaviour of clocks and light signals, and, as we shall argue, the flow of energy. Although these aspects do not involve the 'global' arrow of time, they are concerned with the passage of time in Minkowksi space-time.

It is worth noting that in these discussions often implicit presuppositions about the nature of space-time are at work, such as substantival or relational approaches. For the geometric approach to Minkowski space-time implicitly favours a substantival reading of space-time, whilst the axiomatic approaches, introduced above, implicitly favour a relational understanding of space-time. The following considerations will embrace a relational view of space-time, according to which space is the order of coexisting events in space-time and time is the order of the succession of co-existing events. The notion of order is crucial in this context. The Leibnizian view of order is of course pre-relativistic so that the 'order of coexisting events' presupposes absolute simultaneity but not Newtonian absolute space and the 'order of successive events' presupposes a unique temporal axis for all observers but not absolute time in the Newtonian sense. To speak of space-time relationism means to subject the order of coexisting events to the condition of relative simultaneity and the constancy of c and to speak of the order of successive events means to confine this order to null-like and time-like relations between events in space-time. The Leibnizian order becomes the conical order of events. This move to space-time relationism is possible because, in spite of the notion of relative simultaneity, space-time observers can agree on a number of invariant relationships between events in space-time.

3.3 Inferences to the Nature Space-Time

The Leibnizian characterization of space and time in terms of the order of events and the relations between them does not restrict us to a consideration of kinematic relations and material bodies. It is a common misunderstanding that relationism is limited to occupied space-time events. (Friedman 1983) A 'liberalized relationism' admits a system of both actual and possible relative trajectories. (Teller 1991; Weinert 2006) It is easy to see an alliance between the axiomatic accounts of four-dimensional space-time and space-time relationism. The axiomatic accounts are based on the fundamental 'before-after' relations between space-time events, whose physical manifestation is the propagation of optical signals. Although the traditional relationist speaks of the order of 'events', 'processes' or 'material objects' in the physical universe, a contemporary relationist is not restricted to purely kinematic relations to constitute physical time. The space-time relationist will consider both kinematic and dynamic 'processes', which will help observers in inertial motion with respect to each other to identify physical time. As the propagation of signals constitutes the grounding of the 'before-after' relation in the axiomatic approaches, it is appropriate to consider entropic aspects of this propagation. The exchange of signals is clearly of great importance in Minkowski space-time, as is well illustrated in the famous twin paradox. As one resolution of the twin paradox in Minkowski space-time shows – it appeals to the relativistic Doppler effect and abstracts from the short periods of acceleration and deceleration of the space-travelling twin – the propagation of signals – their emission and reception – plays an important part in a consideration of four-dimensional space-time. This feature becomes prominent in the axiomatic approaches.

The question of the nature of space-time is a matter of admissible inferences, which inertial observers in space-time would draw from their respective experiences. An influential tradition, from Einstein and Gödel to the present day, has inferred the block universe from the measurable and observational relativistic effects.

Such inertial observers, who are attached to reference frames, should also be aware of the propagation of signals, since this is their way of communicating. Such observers would not be far removed from the original concern of Einstein about the coordination of distant clocks. If Reichenbach, Robb and Carathéodory were inertial observers they would direct their attention to thermodynamic properties of signal propagation, which could serve as their basis for inferences about space-time. Whilst the geometric view infers the block universe from the relativity of simultaneity and more recently from other relativistic effects, the axiomatic view will consider dynamic properties of signal propagation, which are considered as the physical basis of the geometric relations. More importantly, as we shall argue below, it will focus on certain invariant relationships between events in space-time, which are crucial for the appreciation of physical time.

4 Irreversibility, Regularity and Invariance

In this section we shall consider which inferences about the nature of space-time follow from a shift to dynamic aspects.

4.1 Reichenbach & Grünbaum

Reichenbach distinguished the topological question of time order ('before-after') from the dynamic question of time direction. (Reichenbach 1956, 26) He claimed that entropic considerations 'will enable us to solve the problem of the direction of time, a problem that cannot be solved in the framework of Einstein's theory of relativity, because it requires a transition from strictly causal relations to probabilistic relations.' (Reichenbach 1956, 25–26) Reichenbach turns to the statistical interpretation of entropy:

> The direction of physical processes, and with it the direction of time, is thus explained as a statistical trend: the act of becoming is the transition from improbable to probable configurations of molecules. (Reichenbach 1956, 55)

Further, Reichenbach points out (1956, 60) that the statistical form of the second law defines a value of S for both equilibrium and non-equilibrium states. This entropic approach has been criticized as 'yielding the wrong result somewhere in space-time'. (Earman 1974, 22) This objection is based on a geometric view of a global temporal orientability of space-time. The entropic approach harbours some interesting results from the point of view of the axiomatic method and space-time relationism. In his later years Einstein himself grew more aware of dynamic aspects of signal propagation in space-time when he objected to Gödel's interpretation of Minkowski space-time in terms of a block universe and the denial of the objective passage of time. (Fig. 3)

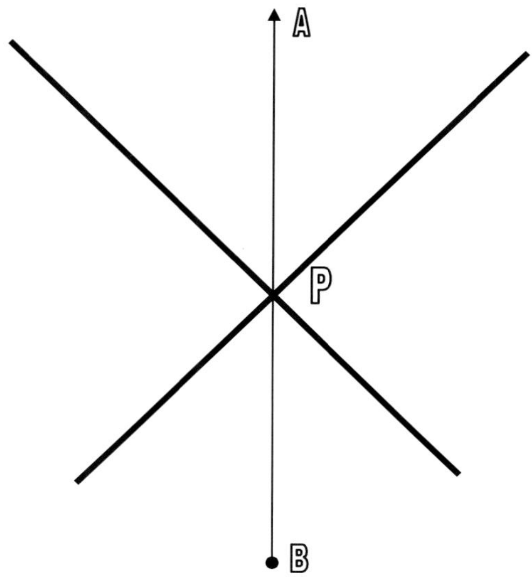

Fig. 3 Einstein's consideration of the (local) direction of time in response to Gödel's idealistic interpretation of the special theory of relativity. A *time-like* world line exists between events A and B, which lies within, not outside, the light cone. A and B are linked by an irreversible signal. Einstein (1949, 687)

Einstein wrote: 'What is essential is the fact that the sending of a signal is, in the sense of thermodynamics, an irreversible process.' (Quoted in Denbigh 1981, 40) The most interesting result, on Reichenbach's entropic approach, is that it is the majority of branch systems which show an increase in entropy. It is the sectional nature of time direction, which is appealing to the space-time relationist. 'The direction in which most thermodynamic processes in isolated systems occur is the direction of positive time.' (Reichenbach 1956, 127; cf. Denbigh 1981, Chap. 6.3) Grünbaum took up this suggestion but reduced it to de facto irreversibility.

For Grünbaum the direction of physical time is grounded in de facto irreversible processes. (Grünbaum 1967, 1955) Grünbaum makes an explicit distinction between physical time and human perception of time. The anisotropy of physical time is not to be confused with a 'transient now' or human perception of becoming ('river of time'). Grünbaum agrees with Reichenbach that the positive direction of physical time is the direction of entropy increase in the majority of branch systems. The emphasis on de facto irreversible processes means that they are contingent and compatible with the time reversal symmetry of the basic mechanical laws. He thus rejects Popper's argument that 'thermodynamic behaviour cannot constitute a basis for the anisotropy of time.' But he also distances himself from Reichenbach in 2 ways:

1. Grünbaum does not assume that entropy is defined for the whole universe. To be fair to Reichenbach, he holds that the overall entropy of the universe can only be inferred from the entropic behaviour of branch systems. 'The universal increase of entropy is reflected in the behaviour of branch systems, so to speak; and only this reflection of the general trend in many individual manifestations is visible to us and appears to us as the direction of time.' (Reichenbach 1956, 131)

2. Grünbaum does not assume parallelism of entropy increase in branch systems and the universe. Thus Grünbaum is truly committed to the sectional nature of the passage of time in local neighbourhoods. Space-time relationism does not require that entropy increase occur in *all* physical systems. Grünbaum's notion of de facto irreversibility has been characterized as weak T-invariance. This weak T-invariance must satisfy the

> requirement that its time inverse (although perhaps improbable) does not violate the laws of the most elementary processes in terms of which it is understood. (Landsberg 1982, 8)

This take on things implies that the T-invariance of physical laws is compatible with asymmetric solutions, if appropriate boundary conditions are taken into consideration. (Price 1996, 88–89, 96; Denbigh 1981, Chap. 6.2) Under these conditions T-invariance turns into weak T-invariance.

Whilst the entropic approach satisfies the space-time relationist's need for an empirical grounding of time in the behaviour of certain physical systems, in Reichenbach's version it also suffers from some weaknesses. For instance, Reichenbach's characterization of branch systems as 'systems that branch off from a comprehensive system and remain isolated from then on for some time' (Reichenbach 1956, 118) is relatively ill defined and neglects that no subsystem is ever totally isolated from the more comprehensive system. Reichenbach claims that the entropic approach can solve the problem of time. This claim has several important aspects, which should be carefully distinguished: (a) It indicates dynamic and regular features of signal propagation in Minkowski space-time. Reichenbach points out that the entropic approach confirms common sense in its intuition that 'time flows' and that 'becoming occurs'. (Reichenbach 1956, 17)

> The concept of *becoming* acquires a meaning in physics: The present, which separates the future from the past, is the moment when that which was undetermined becomes determined, and 'becoming' means the same as 'becoming determined.' (Reichenbach 1956, 269; cf. Torretti 2007)

But Reichenbach has a tendency to define the entropic behaviour of space ensembles (ensembles in branch system) as *the* direction of time. But such an identification falls foul of the famous reversibility objections to the classic conception of the 2nd law. It is therefore advisable, in line with Reichenbach's point about the sectional nature of time direction, to consider certain physical processes as indicators of an objective physical passage of time. (b) For this approach to have any chance of succeeding it must be recognized that entropic relations are frame-invariant in the STR (Einstein 1907). This aspect is particularly important because many physical parameters become frame-dependent in the STR and could not serve as a basis for the identification of physical time beyond proper time. (c) Once we appreciate the importance of invariance for the measurable passage of time, we realize, as we shall discuss, that there are other invariant relationships between space-time events, which could serve as candidates for the identification of objective physical time.

The emphasis on the sectional nature of time direction in the work of Reichenbach and Grünbaum seems to survive in latter-day attempts to save a notion of 'relational becoming' (Dorato 2006), which regards proper time – time along a world line or local temporality – as the only legitimate notion of time in the STR. (See Dieks 1988; Harrington 2008; Stein 1991) These approaches retain the welcome separation of the notion of becoming from the 'presentism/eternalism debate' (Dorato 2006, Sect. 1) but they also neglect the importance of invariant relationships. Even the idea of local time – clock time along a world line as real – prevents us from noticing the invariant features across reference frames. As the axiomatic approach implies, such invariant relationships are essential for a proper appreciation of the notion of time. For it is not sufficient to register regular pulses in one reference frame, regular pulses must be invariant across reference frames in inertial motion with respect to each other for the notion of physical time to make sense across different reference frames. It is therefore important to consider these aspects of invariance.

4.2 Time & Invariance

For a reader of the relevant literature, inspired by space-time relationism, it is surprising to find many authors affirming the reality of a static block universe in the same breath as the asymmetric propagation of electromagnetic signals in space-time. (Davies 1974; Lockwood 2005; Petkov 2005) It seems a puzzle that, on the one hand, no time passes in Minkowski representations of null-like related events but, on the other hand, space-time travellers communicate by the propagation of light signals. (Shallis 1983, 62) But this puzzle disappears when we realize that these aspects are matters of representation in the geometric and axiomatic approaches respectively. Any association of the arrow of time with entropic processes is regarded with a considerable amount of suspicion, not just for the reasons cited above, but also because it is one of the scandals of modern physics that there is still no consensus on the precise meaning of the 2nd law of thermodynamics. (See Duncan and Semura 2007; Leff 2007; Aiello et al. 2008) On the other hand, both the axiomatic approaches and relationism about time require a physical grounding, where this physical grounding is a matter of appropriate choice. As Saunders points out, a question that is even more important than objective becoming is whether change is real. (Saunders 1996, 20–21) This depends on an appropriate physical grounding and entropy seems to be a favourite candidate. (See Wald 2006; Davies 1974) But for the 'passage' of time in Minkowski space-time even regular change must have invariant aspects. In other words a symmetry transformation between inertial frames in Minkowski space-time must leave invariant features. For a dynamic view of Minkowki space-time, invariance is an important aspect of the temporal relations between events.

We can distinguish several invariant relationships in Minkowski space-time:

- Traditional replies to the block view have relied on the invariance of c and the space-time interval ds. The invariance of c means that light cones in Minkowski space-time do not tilt, a fact, which Carathéodory related to Liouville's theorem. The invariance of ds means that observers will disagree about the spatial and temporal lengths between events in space-time from their respective reference frames, but that the space-time interval, which captures the famous union of space and time, which Minkowski announced in 1908, remains invariant for all time-like related observers.

- Simulations of the molecular dynamics of relativistic gases have shown that the temperature of a moving body does not depend on its state of motion. It is possible to define a relativistic temperature from statistical data (and to construct a thermometer), which respective observers in Minkowski space-time could in principle use to determine time across their respective frames. Bodies appear neither hotter nor cooler if a relativistic temperature $T = (k_B \beta_j)^{-1}$ is adopted [where k_B is the Boltzmann constant and β_j is a numerical distribution parameter, which in these experiments took the value $\beta_j = 0.702\,(m_1 c^2)^{-1}$]. The experimenters concluded that 'the temperature of classical gaseous systems can be defined and measured in a Lorentz invariant way.' (See Cubero et al. 2007) In principle it would be possible to read time off these thermostats but in practice it is inconvenient and other methods are preferable.
- But signal propagation offers other possibilities of determining the passage of physical time in Minkowski space-time. Signal propagation is a thermodynamic and therefore anisotropic process both for inertial and accelerating observers in flat and curved space-time. (Petkov 2005) It turns out that entropy and the spreading of energy states are also relativistically invariant. (Einstein 1907; Pauli 1981, Sect. 46–9) What follows from this invariance is that the convergence and divergence of signals is frame-independent, in local neighbourhoods.

The central point in these invariance aspects is that the direction of the energy flow runs in the same direction for all observers, who could possibly communicate through such means. So even though two observers do not agree on the reading of their respective clocks they will agree on the divergence of their signals from their point of origin. They therefore have a physical grounding for their time measurements.

> (...) with the energy flow pointing to the same direction all over the spacetime, we can legitimately say that $\sigma > 0$ [σ is entropy production per unit volume] corresponds to a dissipative decaying process evolving from non-equilibrium to equilibrium as $e^{-\gamma t}$ and $\sigma < 0$ corresponds to an antidissipative growing process evolving from equilibrium to non-equilibrium as $e^{\gamma t}$. The two processes, which in principle are only conventionally different, turn out to be substantially different due to the future-directed energy flow that locally expresses the global time-asymmetry of the universe. (Aiello et al. 2008, 287)

As the quote suggests, it is helpful to introduce a 'spreading metaphor' to capture the essence of the second law. According to this metaphor the entropy symbol, S, is a shorthand for spreading of energy, which includes spatial spreading of energy and temporal spreading over energy states. This entails a picture of dynamic equilibrium in terms of continual shifts from one microstate to another. (Leff 2007, 1748) In order to quantify the spreading metaphor, a spreading function $\Im$ is introduced, which is a function of a system's energy E, its volume V and particle number N.

Connecting the spreading function to entropy S, Leff writes:

> For a constant-volume heating process that proceeds along a given $\Im$ curve, $dE = \delta Q$ is the (inexact) heat differential. Equation (22) - $(\partial\Im/\partial E)_{V,N} = 1/T$- implies that $d\Im = dE/T = \delta Q/T$, in analogy with the Clausius entropy form $dS = \delta Q/T$. Thus, with the temperature definition (22), the spreading function $\Im$ shares the important mathematical property $d\Im = \delta Q/T$ with entropy S. (Leff 2007, 1763–1764)

With these considerations in mind we can return to our earlier observation that histories in space-time must include both kinematic and dynamic considerations. We considered (a) that apart from the time reversal invariance of the dynamic laws, there are energy flows, pointing in the same direction in local neighbourhoods in space-time and (b) that these energy flows are associated with the statistical version of the 2nd law, which expresses both regularity and weak T-invariance.

Prior to Einstein, all approaches to time agreed that time was a universal parameter, irrespective of the question of whether it only existed in the mind or in the physical world and irrespective of the question whether it existed in the absence or the presence of physical events. The notions of absolute and relational time, whatever their differences, nevertheless express the requirement for regularity and invariance in physical systems. While in pre-relativistic notions of time regularity and invariance were frame-independent notions, it is important to note that the STR calls for different combinations of regularity and invariance.

Proper time is regular for respective observers, attached to inertial frames, but it is not invariant across different coordinate systems. Hence the STR allows only for certain invariant regularities. The importance of STR, under the present perspective, resides in its distinction between frame-dependent and frame-independent parameters. The invariant relationships between space-time events therefore acquire considerable importance for a dynamic view of Minkowski space-time. It is these regular and invariant relationships, which are based on specific physical processes, which give rise to an objective passage of physical time.

5 Conclusion

The early block theorists held that two observers in Minkowski space-time could not establish the 'march of time' because of the problem of the relativity of simultaneity. Later block theorists held that the well-known relativistic effects establish the reality of the four-dimension space-time, in which the passage of time reduces to a mere human illusion. But clearly if the two observers agree on certain regular and invariant temporal directions of physical events, even only locally, they may conclude that physical time passes and generally that the four-dimensional world evolves into their local future. The identification of these time directions is not based on a global definition of time-orientability of relativistic space-times or the slicing of four-dimensional space-time by conscious observers. It is based on asymmetric physical processes, like the energy flow and propagation of signals from the source into the future light cones of observers. From the observable dissipation of signals and entropic invariance the observers will infer that the four-dimensional world is dynamic. Such observers will be inclined towards the axiomatic method and construct a space-time representation on the basis of 'optical' facts.

The fact that the axiomatic method implies a different view of space-time – dynamic rather than static – shows that a more inclusive consideration of the history of space-time relations leads to an opposite but equally consistent view of four-dimensional space-time. In fact from the axiomatic point of view the block theorist's inference to a static universe from the relativity of simultaneity and time dilation appears to be premature. The space-time facts of the STR seem to be compatible with two incompatible interpretations of space-time. It is a clear case of underdetermination. If this suggestion is correct, the majority view can no longer claim that the passage of time is a human illusion and the only possible inference from the experimental evidence. From a purely geometric point of view of space-time, the inference seems reasonable but not from the axiomatic, relationist point of view. The latter is based on the view that temporal relations between events (in space-time) are grounded in the order of succession of events. Whilst Leibniz remained unspecific about the precise physical relations, which could serve as a basis of physical time, the axiomatic approach suggests that purely kinematic relations, based on time-reversal mechanical laws, are insufficient to establish physical time in Minkowski space-time. A space-time relationist will find the axiomatic method more amenable for it suggests that certain thermodynamic processes, like signal propagation, are both invariant and regular. They allow the space-time relationist to infer a dynamic view of four-dimensional space-time.

The following scenario presents itself: if the observers in Minkowski space-time concentrate on the flow of energy and the propagation of signals they will infer that 'local' time has a uniform direction and that space-time is dynamic. The relationist view entitles them to select such energy flows as examples of the invariant order of succession of events in space-time. They will disagree with the block theorists who derive their view from purely geometric and kinematic relations, like null-like related events. For the geometric approach light signals are timeless but for the axiomatic view they propagate.

Both the block theorist and the space-time relationist can only make inferences from measurable or observable phenomena to the nature of space-time. Are there ways to solve this underdetermination? The opponents would have to show that some relativistic effects are better indicators of the nature of space-time than others. The other strategy is patience: it is possible that some future measurable effect will be able to resolve the stalemate between the block theorist and the space-time relationist. For instance, Saunders (1996) holds that physics can decide between metaphysical views. The writer's own view is that it is unreasonable to suspect that science can be a judge in matters metaphysical. However, it is altogether reasonable to expect that some future observation will show that one metaphysical view is more compatible with the results of relativity than its opponent.

References

Aiello, M., M. Castagnino, O. Lombardi (2008): 'The Arrow of Time: From Universe Time-Asymmetry to Local Irreversible Processes'. *Foundations of Physics* **38**, 257–292

Albert, D. Z. (2000): *Time and Chance*. Cambridge (Mass.)/London: Harvard UP

Carathéodory, C. (1924): 'Zur Axiomatik der Relativitätstheorie'. *Sitzungsberichte der Preussischen Akademie der Wissenschaften*, Physikalisch-Mathematische Klasse **5**, 12–27

Cubero, D. *et al.* (2007): 'Thermal Equilibrium and Statistical Thermometers in Special Relativity'. *Physical Review Letters* **99**, 170601–1/4

Cunningham, E. (1915): *Relativity and the Electron Theory*. New York: Longmans, Green and Co.

Davies, P. C. W. (1974): *The Physics of Asymmetry*. London: Surrey University Press

Denbigh, K. G. (1981): *Three Concepts of Time*. Heidelberg: Springer

Dieks, D. (1988): 'Special Relativity and the Flow of Time.' *Philosophy of Science* **55**, 456–460

DiSalle, R. (2006): *Understanding Space-time*. Cambridge: CUP

Dorato, M. (2006): 'Absolute becoming, relational becoming and the arrow of time: Some non-conventional remarks on the relationship between physics and metaphysics.' *Studies in History and Philosophy of Modern Physics* **37**, 559–576

Drory, A. (2008): 'Is there a reversibility paradox'. *Studies in History and Philosophy of Modern Physics* **39**, 82–101

Duncan, T.L., J. S. Semura (2007): 'Information Loss as a Foundational Principle for the Second Law of Thermodynamics'. *Foundations of Physics* **37**, 1767–1773

Earman, J. (1967): 'Irreversibility and Temporal Asymmetry'. *Journal of Philosophy* **64**, 543–549

Earman, J. (1974): 'An Attempt to Add a Little Direction to "The Problem of the Direction of Time"'. *Philosophy of Science* **41**, 15–47

Earman, J. (2006): 'The "Past Hypothesis": Not even false'. *Studies in History and Philosophy of Modern Physics* **37**, 399–430

Einstein, A. (1907): 'Über das Relativitätsprinzip und die aus demselben gezogenen Folgerungen'. *Jahrbuch der Radioaktivität und Elektronik* **4**, 411–462

Einstein, A. (1920/1954, 15th edition): *Relativity*. The Special and the General Theory, London: Methuen

Einstein, A. (1949): 'Reply to Criticisms', P. A. Schilpp *ed.*: *Albert Einstein: Philosopher-Scientist*, 2 Volumes. La Salle (Ill.): Open Court, Vol. II, 665–688

Friedman, M. (1983): *Foundations of Space-Time Theories*. Princeton: PUP

Gold, T. (1966): 'Cosmic Processes and the Nature of Time', in R. G. Colodny *ed.*: *Mind and Cosmos*. Pittsburgh: University of Pittsburgh Press, 311–329

Grünbaum, A. (1955): 'Time and Entropy'. *American Scientist* **43**, 550–572

Grünbaum, A. (1967): 'The Anisotropy of Time', in: T. Gold *ed.*: *The Nature of Time*. Ithaca: Cornell University Press, 149–177

Harrington, J. (2008): 'Special relativity and the future: A defense of the point present'. *Studies in History and Philosophy of Modern Physics* **39**, 82–101

Hentschel, K. (1990): *Interpretationen und Fehlinterpretationen der Relatitivitätstheorie*. Basel: Birkhäuser

Huggett, N. (2006): 'The Regularity Account of Relational Spacetime.' *Mind* **115**, 41–73

Kittel, Ch., H. Kroemer (1980, 2nd edition): *Thermal Physics*. New York: W. H. Freeman & Company

Landsberg, P. T. (1982): 'Introduction', in P. T. Landsberg *ed.*: *The Enigma of Time*. Bristol: Adam Hilger

Leff, H. S. (2007): 'Entropy, Its Language and Interpretation'. *Foundations of Physics* **37**, 1744–1766

Lockwood, M. (2005): *The Labyrinth of Time*. Oxford: OUP

Minkowski, H. (1907): 'Das Relativitätsprinzip'. *Annalen der Physik* 1915

Pauli, W. (1981): *Theory of Relativity*. New York: Dover

Penrose, O., I.C. Percival (1962): 'The Direction of Time'. *Proc. Phys. Soc.* **79**, 605–616

Petkov, V. (2005): *Relativity and the Nature of Spacetime*. Berlin: Springer
Popper, K. (1956): 'The Arrow of Time.' *Nature* **177**, 538; *Nature*, **18**, 382
Popper, K. (1957): 'The Arrow of Time.' *Nature* **19**, 1297
Popper, K. (1965): 'Time's Arrow and Entropy.' *Nature* **207**, 233–234
Price, H. (1996): *Time's Arrow and Archimedes' Point.* Oxford: OUP
Reichenbach, H. (1924/1969): *Axiomatization of the Theory of Relativity*. Berkeley/Los Angeles: University of California Press
Reichenbach, H. (1956): *The Direction of Time*. Berkeley: University of California Press
Robb, A. A. (1914): *A Theory of Space and Time. Cambridge*: Cambridge UP
Saunders, S. (1996): 'Time, Quantum Mechanics and Tense'. *Synthese* **107**, 19–53
Schlick, M. (1917): 'Raum und Zeit in der gegenwärtigen Physik.' *Die Naturwissenschaft* **5**, 162–67, 177–186
Shallis, M. (1983): *On Time*. New York: Schocken Books
Stein, H. (1968): 'On Einstein-Minkowski Space-Time.' *Journal of Philosophy* **65**, 5–23
Stein, H. (1991): 'On Relativity Theory and the Openness of the Future.' *Philosophy of Science* **58**, 147–167
Stöckler, H. (2000, 4th edition): *Taschenbuch der Physik*. Frankfurt: Harri Deutsch
Teller, P. (1991): 'Substance, Relations and the Arguments about the Nature of Space-Time.' *The Philosophical Review,* **C/2**, 363–397
Torretti, R. (2007): 'The problem of time's arrow historico-critically reexamined.' *Studies in History and Philosophy of Modern Physics* **38**, 732–756
Uffink, J. (2001): 'Bluff Your Way in the Second Law of Thermodynamics'. *Studies in History and Philosophy of Modern Physics* **32**, 305–394
Wald, R. M. (2006): 'The arrow of time and the initial conditions of the universe.' *Studies in History and Philosophy of Modern Physics* **37**, 394–398
Weinert, F. (2006): in VIII. International Leibniz Congress, in H. Breger, J. Herbst, S. Erdner *eds.*: *Einheit in der Vielfalt*, (Hannover 2006)

No Presentism in Quantum Gravity

Christian Wüthrich

Abstract This essay offers a reaction to the recent resurgence of presentism in the philosophy of time. What is of particular interest in this renaissance is that a number of recent arguments supporting presentism are crafted in an untypically naturalistic vein, breathing new life into a metaphysics of time with a bad track record of co-habitation with modern physics. Against this trend, the present essay argues that the pressure on presentism exerted by special relativity and its core lesson of Lorentz symmetry cannot easily be shirked. A categorization of presentist responses to this pressure is offered. As a case in point, I analyze a recent argument by Monton (Presentism and quantum gravity, 263–280, 2006) presenting a case for the compatibility of presentism with quantum gravity. Monton claims that this compatibility arises because there are quantum theories of gravity that use fixed foliations of spacetime and that such fixed foliations provide a natural home for a metaphysically robust notion of the present. A careful analysis leaves Monton's argument wanting. In sum, the prospects of presentism to be alleviated from the stress applied by fundamental physics are faint.

1 Introduction

Presentism is the position in the philosophy of time that maintains that nothing exists that is not present. In other words, only present events and objects exist, but no past or future events or objects do. Furthermore, it usually assumes that there is a succession of presents, i.e. a moving *Now*. Although logically independent from the thesis that defines the position, most presentists thus take change, or becoming, to be a fundamental aspect of reality. Bradley Monton (2006, 264) has appropriately dubbed the package of presentism-cum-becoming "Heraclitean presentism". In logical space, as he rightly notes, there could also be a presentist metaphysics which holds that the spatially extended sum total of existence is completely static in that fundamentally, it does not involve change at all. Such a "Parmenidean" version

V. Petkov (ed.), *Space, Time, and Spacetime*, Fundamental Theories of Physics 167,
DOI 10.1007/978-3-642-13538-5_12, © Springer-Verlag Berlin Heidelberg 2010

of presentism, however, has rarely, if ever, been entertained.[1] What is of relevance to my present purposes is simply the core thesis of presentism according to which only present events and objects exist, and not whether this core is adorned with Heraclitean or Parmenidean plumes.[2]

There are a number of metaphysical objections against presentism in the literature, and they shall not be surveyed here. Moreover, some authors have denied that it presents the only, or even best, way to account for our intuitions about the phenomenology of temporality–traditionally considered the strong suit of presentism. But a much more powerful, and potentially devastating, challenge arises from modern physics: Einstein's special relativity (SR) provides strong, and perhaps conclusive, reason to view space and time not as two separable and quite distinct animals, but much rather as entangled aspects of the same underlying four-dimensional manifold that fuses the two into a "spacetime". It was Hermann Minkowski's great achievement to recognize the inseparability of space and time resulting from Einstein's theory when he solemnly declared at the Assembly of German Natural Scientists and Physicians in Cologne in September 1908: "The views of space and time which I wish to lay before you have sprung from the soil of experimental physics, and therein lies their strength. They are radical. Henceforth space by itself, and time by itself, are doomed to fade away into mere shadows, and only a kind of union of the two will preserve an independent reality." Minkowski was also the first to correctly describe the geometrical properties of this fused "space-time" structure that today we call *Minkowski spacetime*. Section 2 explicates how SR and its attendant Minkowski spacetime exert significant pressure on presentist positions and thus revisits the issue of compatibility of SR and presentism.

Although SR does not apodictically rule out presentism, it constrains it in a way that renders whatever presentism survives the relativistic revolution a metaphysically rather unattractive cripple. One might have expected that this would do it. But presentism dies hard, very hard. In fact, after a period of relative tranquility, it enjoys something of a renaissance in the philosophy of time. What is striking about this renaissance is that many of the hold-out (or born-again) presentists attempt to support their position by arguments of the kind that have traditionally been the weapon of choice for many of their opponents: arguments drawing on results from the physical sciences. Section 3 analyzes in some detail a particularly interesting case recently offered by Monton (op. cit.). His proposal is important in that it promises to breathe new, scientifically sophisticated life into the otherwise moribund

[1] Barbour (1999) can be read as offering a Parmenidean presentist view. Of course, there is lots more logical space available, e.g. containing a presentist position which subscribes to a moving Now without there being any change whatever. Furthermore, the basic presentist claim can be read as obtaining by necessity or merely contingently, which opens logical space for necessitarian and Humean brands of presentism. All these further varieties and distinctions, however, do not affect the present argument. I shall thus ignore them here.

[2] I understand that there is real worry about whether the debate between presentism and eternalism is well-formed and metaphysically substantive, cf. Callender (2000), Dorato (2006), and Savitt (2006a). As I argue in an unpublished essay, however, I believe that these worries can ultimately be dispelled. I wish to thank Steve Savitt for taking me to task on this issue.

idea of presentism. Section 4 then investigates the prospects of presentism in the so-called *constant-mean-curvature (CMC) foliation* approach to quantizing gravity, which Monton finds particularly amenable to his presentist inclinations. It will illustrate the many ways in which the CMC approach fails to vindicate presentism, despite its initial allure to the presentist. SR, while strictly speaking false of the actual world, at least in an unqualified sense, imposes an important constraint on feasible physical theories, or at least on all physically acceptable interactions. In this sense, it can also be considered a "second-order theory". This section, it should be warned, will be somewhat technical due to the nature of the material covered in it. Finally, Sect. 5 offers some conclusions.

2 Minkowski Spacetime and the Pressure from Special Relativity

The eternalist considers the four-dimensional "block universe" with all of spacetime and everything it contains to make up the sum total of existence. By contrast, the presentist maintains that the sum total of existence can be understood as consisting of a three-dimensional manifold of spatially distinct but temporally equally present, and thus simultaneous, events or objects. Presentism thus seems to require an objective "foliation" of Minkowski's spacetime into hyperspaces of three-dimensional "space" ordered by a one-dimensional "time" parameter.[3] In that it claims a different ontological status for those things present from those non-present, it (usually) presupposes that the distinction between the present and the non-present can be drawn in a principled, objective way. In other words, it requires a metaphysically robust, objectively valid concept of a spatially extended present.[4] Alas, SR provides a strong reason to believe that *that* can't be had.[5]

[3] A *foliation* slices up the four-dimensional spacetime into space and time via an equivalence relation interpreted as "simultaneity". A binary relation Rxy is an *equivalence relation* on a set S iff it is reflexive (for all $x \in S$, Rxx), symmetrical (for all $x, y \in S$, if Rxy, then Ryx), and transitive (for all $x, y, z \in S$, if Rxy and Ryz, then Rxz). *Space* at a time is then given by the corresponding three-dimensional "folium" and *time* is the one-dimensional linearly ordered quotient set induced by the equivalence relation, "lining up" the moments of simultaneity.

[4] At least standardly; Harrington (2008) has defended a "point present", a radically solipsistic version of presentism according to which not only temporally present events exist, but also only spatially present ones. For the point presentist, not even all of my present brain exists. Harrington's position evades the objection raised in this section–but at what price!

[5] While this paper focuses on presentism, a possibilist metaphysics defending a growing block or branching tree structure faces analogous challenges from SR. For instance, McCall (2000) maintains the reality of the past and the present, with the future as a branching set of four-dimensional alternatives. The "present" is the first branch surface, which is defined as a maximal set of pairwise spatially separated events. In order to uphold Lorentz invariance, the branch attrition along these surfaces is relativized to inertial frames. In this sense, McCall's view is the possibilist analogue of Fine's presentism, presented below.

In pre-relativistic physics, the notion of simultaneity of spatially distant events was unproblematic. In SR, however, it turned out that the requisite four-dimensional spacetime had a radically different structure: whether or not two spatially distant events are simultaneous was no longer an objectively and universally determinable fact of the matter. Two inertial observers at some relative velocity with respect to one another do not agree whether two events are simultaneous or not. The relation of simultaneity is thus relativized to reference frames. In a technical language, this means that there is no preferred foliation of spacetime into slices of three-dimensional spaces representing classes of simultaneous events. If we define "the present" as consisting of all those events which occur simultaneous with the point in spacetime representing the here and now, then the relativity of simultaneity seems to imply that the presentist is committed to relativize existence analogously: if we are two inertial observers moving at some relative speed, we take different distant events to be real!

Let's back up a little and have a closer look at how (classical and relativistic) physics conceives of time. Classical Newtonian mechanics does not postulate a *Now*, but is blatantly compatible with a metaphysically robust and objectively valid concept of a spatially extended present. In fact, a (non-relativistic) time-reparametrization-invariant theory, i.e. a theory in which the action remains invariant under redefinitions of time $t' = f(t)$, generally allows for the possibility of an objective spatially extended present, and even for temporal flux or becoming. In such a theory, two situations differing only in their parametrizations of time are really descriptions of one and the same physical situation. Consequently, time does not exist as an objectively measurable independent degree of freedom; more precisely, time is not a magnitude with an objectively privileged metric. In a theory like this, however, there exists an objective total ordering of events in time.[6]

Special-relativistic theories admit only a partial temporal ordering of events. The loss of absolute simultaneity leads to a loss of comparability: with an interpretation of the binary ordering relation as "being earlier than or simultaneous to"–it is a *temporal* ordering that we are seeking after all–, pairs of spacelike related events do not stand in this relation. There is simply no frame-independent fact of the matter as to whether event a is earlier than event b or the other way around for two spacelike related events a and b. In *general*-relativistic theories, where the topology of a spacetime may fail to even permit a non-unique foliation of spacetime into space and time, the possibility of causal loops entails that the temporal ordering is in general not even weakly asymmetric, i.e. there no longer is a partial temporal order of events. In fact, there is no global time deserving this title in general relativity (GR), a fact that finds a particularly vivid expression in the so-called "problem of time" arising in the Hamiltonian formulation of GR. Sic transit gloria temporis.

[6] A *total order* on a set S is given by a binary relation R that is reflexive (Raa for all a in S), weakly antisymmetrical (for all $a, b \in S$, Rab and Rba entails $a = b$), transitive (for all $a, b, c \in S$, Rab and Rbc entails Rac), and comparable (for any $a, b \in S$, either Rab or Rba). A *partial order* on a set is a binary relation with the first three properties, but not the last one. Thus, in a partially ordered set, there exist pairs of elements in the set which do not exemplify the relation.

Let's see what all of this implies for the prospects of presentism. Suppose one upholds the following basic commitments:

Naturalism: Our metaphysical positions must be compatible with physics, at least to the extent to which the latter is taken to be true of the world.
SR-Realism: Special relativity (SR) is taken to be (approximately) true of the world.
Presentism: There exists an objective spatially extended present and only events or objects *in* this present exist.

Of course, **Presentism** implicitly asserts that it is a coherent, non-trivial, substantive metaphysical position. **Naturalism** and **SR-Realism** jointly imply

Compatibilism: Whatever metaphysical view of the world we advance must be compatible with the fact that SR is (approximately) true.

My purpose here is not to defend any of these theses but only to ask whether a commitment to **Compatibilism** is consistent with maintaining **Presentism**. It is, as we shall see. The question, however, is whether **Compatibilism** leaves the presentist with an interesting position at all. The lesson gleaned from an argument independently advanced by Wim Rietdijk (1966) and Hilary Putnam (1967) suggests that it does not. Since their argument is well known, let me only briefly remind the reader how it essentially goes.[7]

The Rietdijk-Putnam argument assumes that the task is to figure out which of the spatially distant events in the four-dimensional spacetime are co-present with the *here-now*. To identify the objective, spatially extended present strikes me as an unavoidable task if presentism is characterized as I did above. Next, introduce an equivalence relation R interpreted as "being simultaneous with". Then, use R to construct the spatially extended present, starting out from the *here-now*. The problem essentially is, as mentioned above, that in SR simultaneity relations become frame-relative. This was the content of the relativity of simultaneity. If in Fig. 1, e designates the *here-now*, then the event denoted by a is simultaneous to e as far as the primed frame is concerned, but in the future of e according to the unprimed frame. In other words, in the primed frame, Rae, but in the unprimed frame, $\neg Rae$. Thus, there is no objective fact of the matter which spatially distant events are co-present with the *here-now*. It gets worse. Since simultaneity is a transitive relation, one would expect that what is co-present with a spatially distant event co-present with the *here-now* is also co-present with the *here-now*. Consider the situation as shown in Fig. 2. In the unprimed frame, e' is certainly simultaneous with e and because e represents the *here-now*, e' is also present (and thus exists). However, in the primed frame b is certainly simultaneous with e' and because e' is present, b is also present (and thus exists). Moving from one frame of reference to another in the course of the argument ought to be acceptable if simultaneity were objective, i.e. frame-independent. Of course in SR, it isn't. But that's the point. Consequently, a

[7] For a more basic and detailed rendering, see Savitt (2006b).

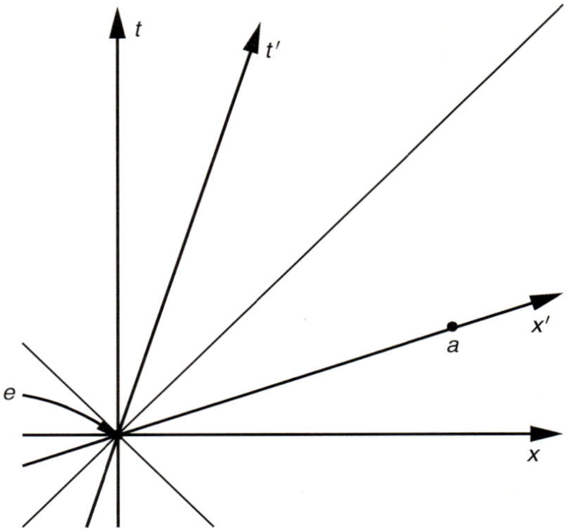

Fig. 1 The Rietdijk-Putnam argument illustrated

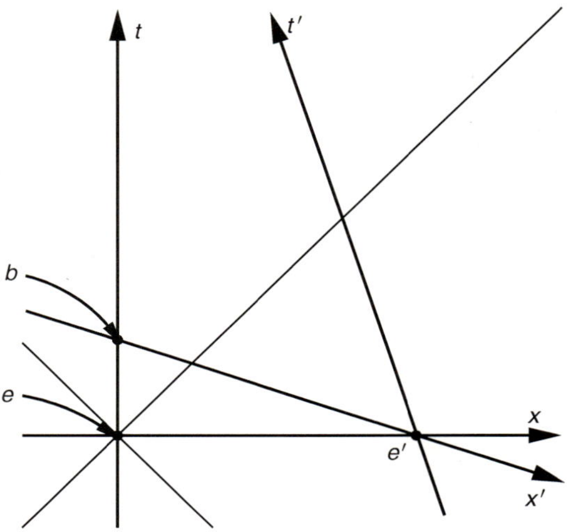

Fig. 2 An event to the future of the *here-now* (in any frame) is co-present with it

presentist is committed to the existence of event b which is in the future of e with respect to *all* frames of reference. But this is surely a reductio of the position.

Presentists have responded in a variety of ways to the pressure exerted by the Rietdijk-Putnam argument and I shall not list them in any detail, but just highlight the basic strategic options. Here are some *incompatibilist* strategies, i.e. responses rejecting **Compatibilism** in one form or other. First, a presentist could deny **Naturalism**. Such denial could take different forms. One could, as does Jonathan

Lowe,[8] claim that SR is not a theory *about time* but about something else instead. Alternatively, one could retort by accepting that SR speaks to the geometry of spacetime but reject that this has any ontological import, as does Dean Zimmerman (2008).[9] Second, a presentist might reject **SR-Realism**, simply asserting that SR is not approximately true of the world. This could occur simply on a priori grounds, an option I will not comment on. In fact, the remainder of this essay after this section will be dedicated to explore a posteriori exit strategies denying **SR-Realism**. Relevantly, Monton (op. cit.) can be read as a representative of this strategy, as will become clear below. Also, considerations from quantum mechanics can be invoked in an attempt to establish that SR is false or incomplete insofar as it lacks an absolute, privileged frame of reference. This response comes in different flavours: (a) (non-relativistic) collapse dynamics require a preferred frame in which the collapse occurs; (b) Bohmian interpretations are incompatible with SR; and (c) invoke Bell's theorem to argue that some tenets of SR must be given up. I concur with Craig Callender (2008) that these strategies don't succeed, but will not elaborate here.

What are the basic *compatibilist* responses at the presentist's disposal? First, the set-up of the Rietdijk-Putnam argument could be rejected as doing violence to a genuinely presentist metaphysics. What is more or less tacitly presupposed in the argument, it could be insisted, viz. that there is a four-dimensional manifold of spacetime events such as Minkowski spacetime of which it is then our task to determine which of these events are "determinate" as of the *here-now* or are objectively present, ought to be discarded by the presentist. While I think that it is still a perfectly justifiable task to ask of the presentist to describe the sum total of existence, to somehow tell a story as to how her position can be reconciled with SR, the Rietdijk-Putnam argument certainly still has force against an ersatzist version of presentism, which, as I have argued elsewhere (unpublished), we are forced into in order to save presentism from the threat of trivialization. On the other hand, a presentist might simply bite the bullet and consequently relativize existence, an option chosen by Kit Fine (2005; particularly Sect. 10, pp. 298–307): since what is present is relative to an inertial frame, what exists becomes fragmented in that it depends on the choice of frame. There is an intermediate strategy, somewhere between accepting the full consequences of the argument and rejecting the way it sets up the presentist commitments: define the objectively existing present purely in terms of the

---

[8] In a paper entitled "Experience of change and change of experience", delivered at the University of Geneva on 19 December 2008.

[9] Zimmerman, together with a number of present-day presentists, is hard to classify as either compatibilist or incompatibilist as he accepts SR, but not in the role a naturalist usually would. He thinks that SR leaves room for an additional relation of simultaneity not to be found in physics. This relation would only clash with physics if the latter were committed to a principle prohibiting extra relations of this sort, but such a principle, he thinks, would not be warranted. Of course, this relation would still effectively foliate spacetime. Such a foliation could either be observed, or it couldn't. If the former, **Compatibilism** would be denied; if the latter, we run into similar problems as the defense championed by Tooley and Craig, which is essentially of that type and shall be discussed below. I thank Jonathan Cohen for having reminded me of this connection.

Lorentz-invariant structure available in Minkowski spacetime. The solipsistic version mentioned earlier (in footnote 4) and defended by James Harrington (2008) trivially makes only use of the Lorentz-invariant structure, viz. a single spacetime point as representing the spatiotemporal location of the sum total of existence. But does this capture the true spirit of presentism? It can be doubted, as neither existence nor "becoming" remain universal on this proposal. Fine (2005, 304) puts it succinctly: presentists tend to be impressed by the distinction between space and time which they take to be metaphysically deep in that they think that there exists an objective "now", although there does not exist an equally objective "here". Of course, this intuition is lost in solipsistic presentism. Accordingly, it violates **Presentism** as I defined it above.

An alternative way of make exclusive use of the Lorentz-invariant structure has been proposed by Howard Stein (1991) and could be termed *past-light cone presentism*. The main idea is to identify the spatially extended present as the set of events on the past light cone of the *here-now*. Yes, you haven't misread: the idea is to define the *present* as the set of events on the *past* light cone. This proposal is Lorentz-invariant and can be motivated by an appreciation of epistemic accessibility, as causal signals reaching us *now* emanate from the events on the past light cone and thus appear to us as being co-present. While on the solipsistic version, the simultaneity relation remains, trivially, an equivalence relation, it is no longer symmetrical and transitive in past-light cone presentism. Symmetry, but not transitivity, can be restored by extending existence to events on the *future* light cone. But in what sense would this still be the present? Points on Andromeda some four million years apart in time, but at no distance in space according to some joint frame of reference for a generic observer on earth and one on Andromeda, would both be co-present with the *here-now*.

A final compatibilist strategy that ought to be mentioned is to accept that SR offers a perfectly *empirically adequate* theory, but to insist that absolute simultaneity still exists. It is just that we cannot possibly detect the privileged frame of reference which determines the present. In other words, absolute simultaneity is not empirically accessible. This strategy is, arguably, compatibilist only in letter, but not in spirit. Its motivations may be metaphysical or physical. A variant of the former is found in Michael Tooley (1997), one of the latter in neo-Lorentzian interpretations of SR, such as the one attempted by William Craig (2001).[10] In both cases, the metaphysics fully relies on postulated extra-structure that can't even in principle be observed. The extra-structure needed is not motivated by more than specific metaphysical agendas or a refusnik attitude toward SR. It violates Ockham's razor so crassly that the move cannot be justified by putting some post-verificationist philosophy of science on one's flag. An argument to the effect that since it is only because

[10] Craig also seems to think that SR is a kinematic theory that only underwrites electrodynamics, and not all or even most of physics. This is simply false. Physicists are working hard to make sure that all theories are Lorentz-invariant. If they fail in doing so, it is generally accepted that their theory faces a major problem.

of some ill-advised verificationist commitment that SR prohibits a privileged frame, and since we know that verificationism is false, we can infer that there is absolute simultaneity, does obviously not succeed. But note that even if we permitted the stipulation of this unobservable extra-structure, such as a simultaneity relation, it appears that it cannot do the work asked of it. If one's goal is to produce a metaphysics that vindicates our pre-theoretical (non-)ascriptions of simultaneity, then a postulated simultaneity relation will not help such vindication so long as it is epistemically inaccessible. And if it is epistemically accessible, **Compatibilism** is violated even in letter.

In sum, the prospects of compatibilist strategies appear bleak. Those of incompatibilist responses hardly seem brighter, at least not for those of us who accept **Naturalism**–except if we came to offer a strong a posteriori argument as to why SR does not approximately hold of the actual world. There is plenty of physics that such an argument could turn on. It could be that Lorentz symmetry only holds approximately and at large scales, e.g. if the underlying spacetime structure is discrete. Depending on how approximately it would hold, this may still lead to a compatibilist strategy. It could be that if gravity is turned on, or if we take quantum effects into considerations, or both, it will be seen that SR is invalid. To discuss, or even list, all the physics that such an argument could make use of is the task for another day. It is an interesting task that will lead the investigator into a thick, and almost impenetrable, forest of foundational issues in fundamental physics. Today, I will confine myself to an analysis of the suggestion in this vein recently made by Monton (op. cit.).

3 Monton's Incompatibilist Defence of Presentism

Monton sums up the Rietdijk-Putnam argument as follows (op. cit., 264):

(1) "Presentism is incompatible with [special] relativity [...]"
(2) SR "is our most fundamental theory of physics."
(3) "Presentism is incompatible with our most fundamental theory of physics (from (1) to (2))."
(4) "Presentism is false (from (3))."

While Monton recognizes that the step from (3) to (4) is non-trivial, he finds it preferable if the presentist wouldn't have to rely on blocking that step. In other words, at least for the sake of the present argument, he accepts **Naturalism**. Consequently, rejecting the argument will require denying either one of the first two premises or the inference from them to (3). But this inference is obviously valid. Offering an incompatibilist stance, Monton accepts premise (1). Remains premise (2): Monton finds it "relatively uncontroversial" that thesis (2) is false, i.e. that SR is not our most fundamental theory of physics. There is certainly a sense in which he is right: once gravity is taken into account, SR must be replaced by GR which is arguably more fundamental; GR is incompatible with quantum physics and both must be superceded by a quantum theory of gravity, which in turn may ultimately be supplanted by a "theory of everything". Thus, (2) is false. Of course, (3) could still be true, viz. exactly in those cases where it turns out that the final fundamental theory of physics is still incompatible with presentism, perhaps for reasons unrelated to the relativity of simultaneity. But, injects Monton, there are quantum theories of gravity which are compatible with presentism. What he has in mind here are approaches in so-called fixed-foliation quantum gravity (QG), such as QG relying on foliations of spacetime into hypersurfaces of constant mean (extrinsic) curvature or "CMC" for short. From the existence of such theories in QG, he infers that "(3) is false, and presentism is unrefuted" (ibid., 265). This inference is of course only valid if it is the case that one of those quantum theories of gravity compatible with presentism is in fact the most fundamental theory of physics. I will overlook, at least for now, this overly excited inferential step, but we will have to revisit it.

Monton's argument can be thought of as consisting of two steps: first, SR is marginalized as an irrelevant, and false, theory; second, the CMC approach to QG is then presented to add credence to the claim that fundamental physics is hospitable to presentism. The remainder of this section discusses the first part of the argument, the next section analyzes the second part.

Let me give three preliminary comments. First, I find it rather curious that Monton formulates the argument in terms of which theories are *fundamental*. Whether or not a theory–any theory–with which presentism's compatibility is tested is fundamental or not seems entirely beside the point. What matters is *truth*. Incompatibility with a theory which is true of the actual world seems a sufficient condition to rule out a metaphysical proposal. Presumably, fundamentality entails truth; no theory could reasonably be considered fundamental if it were not true. But of course fundamentality is not *necessary* for truth. There are many theories about higher-level phenomena, and presumably some of them are true without being fundamental. But that's the crux: by denying that SR is fundamental, Monton means to imply that it is false. Since incompatibility with a false theory is not problematic, presentism would be saved. In general, however, non-fundamentality does not entail falsehood. The situation at stake is more subtle, as it turns out. Strictly speaking, and if no qualifications about its domain of applicability are added, SR *is* a false theory: it is not *in toto* true of the actual world. However, it is still believed to impose a very rigid constraint on any candidate fundamental theory. Just exactly what this constraint is will ultimately be decisive in adjudicating whether presentism is compatible with the best physical theories true of our actual world. I will return to this below.

In a sense, it's even worse than this. Arguably, fundamentality imposes a partial ordering on theories. But this means that there may fail to be a fact of the matter as to which one of two particular theories is more fundamental. Furthermore, fundamentality may not be well-defined or philosophically justifiable as an important, or relevant, criterion. Thus, fundamentality appears to be a requirement which may be inapplicable in, as well as irrelevant to, the case at hand.

Second, let me illustrate the dialectical landscape as I see it. We have seen above that Monton's argument can only offer respite for presentism if it cannot only claim that SR is not a fundamental theory, but if it can also be made credible that presentism has good chances of being compatible with our most fundamental theory of physics, and that (3) is thus false. But in order to establish *that*, Monton must navigate between the Scylla of triviality and the Charybdis of falsehood. On the one hand, his argument could be interpreted as primarily expressing general scepticism about the current state of physics. Since we don't have the physics in the ideal limit of scientific enquiry, would be the thought, a presentist can maintain hope that she will ultimately be vindicated. But such a hope would be pious indeed. Thus, if the intended conclusion is simply that in principle it could be that presentism will eventually be compatible with fundamental physics, then it is disappointingly trivial.

On the other hand, Monton's argument may be read as offering a crystal ball from which the future of QG can be gleaned. Here, the idea would be to reach a prediction that, at least with reasonable probability, the final theory will be hospitable to presentism. But such a prediction would be audacious indeed. In fact, if the claim is that it is reasonably likely that presentism will eventually be compatible with fundamental physics, then the argument is unacceptably false.

It might be protested that I am striking Monton with an unfair dilemma. I am not: I don't claim that his conlusions are either trivial or false. What I *am* saying, however, is that he must strike a fine balance in order to end up with a substantive *and* true conclusion. What I will attempt to show in much of the remainder of this essay is that the room to manoeuvre between said Scylla and Charybdis is uncomfortably tight.

Third, Monton treats the classical and the corresponding quantum version of a theory curiously disanalogous. Such a disparity may sometimes be justified, but arguably not here. Let me explain. Monton seems to think that choosing a particular (CMC) foliation is inadmissible at the classical level, but entirely unproblematic once we go to the quantum theory. He asserts that presentism is incompatible with SR and GR because Minkowski and general-relativistic "spacetimes do not have a foliation into spacelike hypersurfaces as part of their structure." (ibid., 267) Such foliation, he admits, can sometimes be picked out, but "the foliation is not part of the spacetime structure as given, and thus imposing such a foliation amounts to changing the theory." (ibid., 268) It is somewhat mysterious why he has such qualms about changing the theory, particularly since at the end of his essay, he has no hesitation to proclaim that a committed presentist ought to demand that since string theory and loop quantum gravity do not account for presentist intuitions, they ought to be modified accordingly. Furthermore, as will become clear in Sect. 4, almost all of the work on the CMC approach has been done at the classical, not at the quantum, level. For canonical approaches, the classical and the quantum levels are not interpretationally independent: canonical quantization necessitates an interpretation of the classical theory to be quantized which will then be carried over into the corresponding quantum theory. Thus, for the fixed-foliation approach to QG that

Monton advocates, a CMC interpretation of the *classical* theory is presupposed and the disparity assumed by Monton seems ill-justified.

Leaving the preliminaries, Monton starts out stating this argumentative goal:

> [Because special and general relativity are not our most fundamental theories of physics], the compatibility of presentism with special and general relativity is *prima facie* irrelevant to the issue of presentism[5]. I will argue that this *prima facie* appearance is in fact correct. (ibid., 267)

Footnote 5 takes no prisoners: against Mark Hinchliff who asserted that SR is "one of our best-confirmed scientific theories of the nature of time" (1996, 131), Monton declares that

> [t]his claim is false: the special theory is a decisively refuted theory of the nature of time. Special relativity is incompatible with such phenomena as the gravitational redshift and gravitational lensing, phenomena that provide evidence for general relativity. (ibid.)

As Monton acknowledges, scientists do not reject *all* old ideas in a scientific revolution. Thus, one might require that the incompatibility of presentism with SR be carried over to any legitimate candidate fundamental theory. However, he quickly dismisses this answer on the basis that since there are many potentially viable approaches to QG, some of which frustrate the demanded incompatibility, there are no compelling grounds on which a presentist must concede an eventual incompatibility.

> Because of the lack of data to back up the claim that a good theory is incompatible with presentism,

Monton concludes that

> all the literature on the issue of whether presentism is compatible with [...] relativity is [...] irrelevant to the issue of whether presentism is true. (ibid., 269)

While it is certainly true that there is no empirical data *directly* suggesting an incompatibility with presentism, this conclusion can't be had that easily. The *Principle of Relativity*, i.e. the demand that the physics is the same in all inertial frames, is encoded in a theory as the Lorentz covariance of its dynamical equations, which means that there can't be any dynamical phenomena that would allow us to pick a privileged frame and thus an absolute simultaneity. In SR, this dynamical symmetry is carried over into the spacetime structure, leading to the geometry of Minkowski spacetime, which of course is invariant under Lorentz transformations. In GR, the *Principle of Equivalence* ascertains that at each point of spacetime, the spacetime structure exhibits the same symmetry. Quantum field theory (QFT) assumes the Minkowski spacetime as Lorentz-invariant background structure, and QFT on curved spacetime makes the same symmetry assumption for each point of the (curved) spacetime background. In fact, most physicists would agree that dynamical equations ought to be Lorentz-covariant and that the background spacetime at least of semi-classical theories must have the relevant symmetry at least in some local sense.

The fixed-foliation quantum theories of gravity to be discussed in the next section violate the Principle of Relativity in that they require a preferred frame of reference.[11] There is to date, of course, no empirical indication whatsoever that such a preferred frame of reference exists. In fact, Lorentz (or, more precisely, Poincaré) symmetry is fantastically well confirmed.[12] Thus, to require that Lorentz symmetry be valid is well justified. Now, this in itself does not entail an incompatibility of presentism with empirical data. As we have seen in Sect. 2, there are perfectly Lorentz-invariant ways of formulating a presentist position, although there is considerable doubt whether they succeed in fully capturing the spirit of presentism. Be this as it may, the fact that Lorentz symmetry is so well confirmed puts serious pressure on any approach that requires a preferred reference frame.[13]

Let me frame this in more general terms. SR can be thought of as a "first-order theory", i.e. a theory which makes claims about the world and as such can be true or false of the actual world. As it completely ignores gravity, a strong case can be made that it is, in fact, false. However, it might also be regarded as a "second-order theory", i.e. a theory that places certain constraints on other theories. More specifically, it requires that all possible physical interactions be governed by Lorentz-covariant dynamics. Second-order theories that provide constraints in the form of *necessary* conditions may be considered *true* if they correctly rule out false first-order theories and *false* in that they incorrectly rule out true first-order theories.

In sum, I submit that Monton is grossly underestimating the argumentative work that would be necessary to brush SR to the side. Thus, he has failed, in my view, to sufficiently establish the first part of his argument, viz. to marginalize SR as an irrelevant and false theory. It turns out that in exactly those aspects which are relevant to a presentist, SR is too pertinacious to be so easily blown away by the simple need of a quantum theory of gravity. Let us turn to the second step of the argument.

---

[11] Monton agrees: "the proponent of fixed foliation quantum gravity will agree that there is a preferred frame of reference, and can admit that [...] the theory makes sense only in one reference frame." (ibid., 271)

[12] For an authoritative review of experimental tests of Lorentz symmetry, cf. Will (2005a, b); for a recent review on phenomenological indications that Lorentz symmetry may be broken at the Planck scale, cf. Amelino-Camelia (2008).

[13] Monton addresses remarks by Gordon Belot and John Earman (2001) that could be framed as an objection to his view. They argue that fixed-foliation approaches to QG have few adherents because "[t]o forsake the conventional reading of general covariance as ruling out the existence of preferred co-ordinate systems is to abandon one of the central tenets of modern physics" (241). Monton disagrees vehemently: He flatly denies that fixed-foliation approaches require a preferred coordinate system. He bases this denial on Kretschmann's objection to general covariance as a physically contentful constraint on theories. While it is perhaps true that fixed-foliation theories can all be formulated in a generally covariant manner, the objection becomes impotent if general covariance is interpreted in the correct, substantive way, i.e. as a gauge symmetry of GR. Although the particular formulation chosen by Belot and Earman may be unfortunate, their point essentially stands: fixed-foliation theories break the symmetry for which we have excellent reason to believe that every viable theory must respect it.

4 The CMC Foliation Approach: A New Home for Presentism?

The constant mean curvature (CMC) foliation approach is a fixed-foliation theory as discussed in the previous section.[14] In fact–and this ends up undermining Monton's case–, it is not really an approach to QG in its own right, but merely a technique that is explored on the road to QG. It starts out, like other canonical approaches to gravity, from a formulation of GR as a Hamiltonian system with constraints, dealing with spacetimes of topology $\Sigma \times \mathbb{R}$–in itself a limitation. The canonical variables are the 3-metric induced on the spacelike hypersurface Σ, which describes the geometry of Σ, and its extrinsic curvature, which specifies the embedding of Σ in the four-dimensional manifold. The content of Einstein's field equations–the dynamical equations of the standard formulation of GR–is re-expressed in the constraint equations. These constraints define a subspace of the phase space Γ, the so-called *constraint surface* $\bar{\Gamma}$. In the CMC approach, only the subset $\Gamma_\tau \subset \bar{\Gamma}$ defined by the condition that the mean (i.e., the trace) of the extrinsic curvature is constant is considered. This mean (extrinsic) curvature is denoted by τ. A spacelike hypersurface Σ has constant mean curvature just in case τ is constant across Σ. Why does this condition deserve to be called a "time gauge", indicating that the spacetime is foliated into sets of "simultaneous" events? It just so turns out that a reasonably large open subset of the space of models of GR consist of spacetimes admitting a unique foliation into hypersurfaces parametrized by constant mean curvature. If a general-relativistic spacetime is sliceable into hypersurfaces of constant mean curvature–call these spacetimes *CMC-sliceable*–, then τ varies monotonically within a constant mean curvature foliation.

Starting out from the subset $\Gamma_\tau \subset \bar{\Gamma}$ of CMC-sliceable spacetimes, a particular foliation is chosen for every model in that subset: the CMC foliation. This move significantly reduces the technical difficulty of solving the constraint equations in that it effectively eliminates three of the four usual constraint equations, and three of the four functions to be solved for. Essentially, reducing $\bar{\Gamma}$ to Γ_τ amounts to fixing the gauge, hence "time gauge". The only gauge freedom left are reparametrizations of τ. Thus, general covariance is broken down to time-reparametrization invariance, which effectively brings the situation back to a time-reparametrization-invariant theory as characterized early in Sect. 2. Also, this step simplifies the remaining constraint equation to an equation linear in the momentum conjugate to τ. Given a particular τ-parametrization then, one can construct a Hamiltonian. The resulting time-dependent Hamiltonian $H(\tau)$ effectively measures the spatial volume of the universe. More precisely, it provides a measure for the volume of the Cauchy surface of mean extrinsic curvature τ. Thus, as Beig (1994, 77) concludes, by selecting

[14] This section is inevitably more technical than the rest of this essay, although an effort is made to provide a self-contained characterization of the approach. For more extensive and rigorous presentations of the approach, consult Beig (1994, 74–77), Fischer and Moncrief (1997), Isenberg (1995), and Rendall (1996). Cf. also Belot and Earman (2001, particularly 239f).

a distinguished parametrization, a time-dependent Hamiltonian system with the Hamiltonian given by the volume function can be constructed to mimic a cousin of GR. Once the classical Hamiltonian theory is in place, then, an attempt can be made at cooking it up into a quantum theory using the canonical recipe. It turns out that a canonical quantization of such a Hamiltonian system can successfully be completed for the $(2 + 1)$-dimensional cousin of GR, but not for the much more pertinent $(3 + 1)$-dimensional case of full GR.

The CMC approach has additional serious limitations, both at the classical and the quantum level. First, it is well-understood only for the vacuum case and for spatially closed spacetimes, i.e. for spacetimes with manifolds such that Σ is compact and without boundary. There are good reasons to believe that the actual universe exemplifies neither of these properties. Second, not all globally hyperbolic, spatially closed vacuum spacetimes admit a foliation into hypersurfaces of constant mean curvature.[15] Apart from the limitations noted above, this means that the CMC approach cannot deal with some general-relativistic spacetimes, *even if we restrict those to be globally hyperbolic*. There is no consensus as to how severe the restriction to globally hyperbolic spacetimes is. On the one hand, there are important classes of non-globally hyperbolic spacetimes.[16] On the other hand, important approaches to QG such as loop quantum gravity are confined to the same class of spacetimes. Also, the initial value problem can only meaningfully be addressed in the context of globally hyperbolic spacetimes. I will leave this question aside and instead turn to a brief discussion of the reach of the spacetimes amenable to a CMC-slicing.

Such a discussion starts out from the conformal reformulation of the standard constraint equations of Hamiltonian GR as proposed and developed by André Lichnerowicz and Yvonne Choquet-Bruhat and James York (1980). The conformal method has proved to be a potent means to approach the Cauchy problem and has important applications in numerical GR. The question that is being asked is not which part of a given spacetime can be covered by a CMC foliation. Rather, the idea is to simultaneously construct or recover a full four-dimensional spacetime as the solution of a Cauchy problem as well as to obtain a CMC foliation of it, using the mean curvature τ to parametrize the foliation and thus to provide a global time function. Naturally, this approach cannot hope to result in anything other than globally hyperbolic spacetimes.

The approach starts out from initial data on a spacelike hypersurface Σ, the induced metric λ_{ab} on Σ and a symmetric tensor field σ^{ab}, which is trace-free ($\lambda_{ab}\sigma^{ab} = 0$) and divergence-free (${}^{\lambda}\nabla_a\sigma^{ab} = 0$ where ${}^{\lambda}\nabla$ is the covariant derivative compatible with λ_{ab}) with respect to λ_{ab}. The tensor field σ^{ab} is the second fundamental form on Σ. Roughly, λ corresponds to the spatial components of the metric and σ to their time derivatives. To these, the scalar field τ is added. The triple $(\lambda_{ab}, \sigma^{ab}, \tau)$ on Σ, usually called the *conformal data*, then acts as initial data for

[15] Cf. Bartnik (1988) and Rendall (1996).

[16] Cf. Smeenk and Wüthrich (2010) for more on non-globally hyperbolic spacetimes.

the Hamiltonian equivalent of Einstein's field equations. This is to be understood in the sense that the usual initial data for the standard Hamiltonian field decomposition into induced 3-metrics and extrinsic curvature satisfy the standard constraint equations if and only if the conformal data satisfy the corresponding conformal constraint equations. These equations, which I am not going to reproduce here, constitute a coupled quasilinear elliptic system of partial differential equations that do not afford a solution for all choices of conformal data (and hence not for the corresponding standard situation). These equations pose formidable technical obstacles and do consequently not surrender to general solution. The reason why the CMC approach is pursued is because, as mentioned above, assuming that τ is constant–which is exactly what CMC does–offers a significant technical simplification at this point: It eliminates three of the four conformal constraint equations, as well as three of the four unknown functions to be solved for. The remaining constraint equation, often termed *Lichnerowicz equation*, although still not solved in the general case, permits the proving of theorems pertaining to the existence and uniqueness of solutions.

An important problem that arises in this context, the so-called *Yamabe problem*, is the issue of conformally rescaling a metric to obtain a metric of constant scalar curvature. It turns out that there is a solution to this problem for metrics on spatially compact manifolds. This is what the following theorem establishes (Isenberg 1995, 2252):

Theorem 1 (Yamabe). *Let λ_{ab} be a C^∞ Riemannian metric on a closed three-dimensional manifold Σ. Then there exists a C^∞ positive-definite function θ on Σ such that the scalar curvature of the metric $\theta^4\lambda_{ab}$ is constant.*

Yamabe's Theorem can be shown to imply, together with some propositions that require little extra work (ibid., 2253), that the set of all C^∞ Riemannian metrics on Σ can be partitioned into three *Yamabe classes*: Since each of these metrics will be conformal to a metric with constant scalar curvature $1, 0$ or -1, they fall exactly into one of the corresponding Yamabe classes denoted $\mathcal{Y}^+(\Sigma), \mathcal{Y}^0(\Sigma)$ or $\mathcal{Y}^-(\Sigma)$, respectively. For each λ_{ab}, its Yamabe class is thus a conformal invariant. It turns out that for some closed manifolds Σ, $\mathcal{Y}^+(\Sigma)$ and $\mathcal{Y}^0(\Sigma)$ are both empty, while $\mathcal{Y}^-(\Sigma)$ is never empty. Furthermore, $\mathcal{Y}^0(\Sigma)$ can only be empty if $\mathcal{Y}^+(\Sigma)$ is also empty, but the converse is not true.

James Isenberg (1995) systematically investigates for which sets of conformal data $(\lambda_{ab}, \sigma^{ab}, \tau)$ the Lichnerowicz equation can be solved and thus be mapped to a solution of the standard constraint equations, and for which sets it can't. As he shows, the solvability depends on three criteria. First, it depends on which Yamabe class λ_{ab} belongs to. Second, it relies on whether $\sigma^2 = \sigma_{ab}\sigma^{ab}$ is identically zero on Σ or not. Finally, it matters whether the constant τ is zero or not. These criteria are all conformal invariants. Of the resulting twelve classes of conformal data, six map to solutions of the Lichnerowicz equation and six don't. More precisely, Isenberg (1995, 2259) shows the following theorem:

Theorem 2 (Isenberg). *Let λ_{ab} be a (sufficiently smooth) Riemannian metric on Σ, σ_{ab} a symmetric tensor field on Σ which is trace-free and divergence-free with* *respect to* λ_{ab}, *and* τ *a constant. Then the following table indicates for which conformal data* $(\lambda_{ab}, \sigma^{ab}, \tau)$ *the Lichnerowicz equation does ("Yes") or does not ("No") admit a solution:*

	$(\sigma^2 \equiv 0, \tau = 0)$	$(\sigma^2 \equiv 0, \tau \neq 0)$	$(\sigma^2 \not\equiv 0, \tau = 0)$	$(\sigma^2 \not\equiv 0, \tau \neq 0)$
$\lambda_{ab} \in \mathcal{Y}^+$	*No*	*No*	*Yes*	*Yes*
$\lambda_{ab} \in \mathcal{Y}^0$	*Yes*	*No*	*No*	*Yes*
$\lambda_{ab} \in \mathcal{Y}^-$	*No*	*Yes*	*No*	*Yes*

For conformal data in the class $(\lambda_{ab} \in \mathcal{Y}^0, \sigma^2 \equiv 0, \tau = 0)$, *the solution is non-unique; for all others the solution is unique if it exists.*

For any given closed three-manifold Σ, Isenberg's Theorem offers a "complete function space parametrization" (ibid.) of the set of CMC solutions of the standard constraints. In fact, the set of CMC solutions of the standard constraints stand in a one-to-one correspondence with what is essentially the direct sum of the six classes of conformal data as given in the table in Theorem 2 (ibid.).[17]

Before we press on to more pertinent matters, let me note the fact that for conformal data of the class $(\lambda_{ab} \in \mathcal{Y}^0, \sigma^2 \equiv 0, \tau = 0)$ (i.e., in case the metric is conformal to another one with vanishing scalar curvature everywhere on Σ, the square of the tensor field essentially giving its temporal derivative is identically zero on Σ, and the constant mean extrinsic curvature vanishes on Σ), we are confronted with a kind of indeterminism. Given conformal data of this category on Σ, there exist multiple solutions to the dynamical equations. In other words, for this class of field values on Σ, the initial state of the physical system does not, in tandem with the dynamical equations, uniquely determine the state of the physical system for all times. The construction of the conformal method does not yield a unique four-dimensional spacetime. It is appropriate to speak of indeterminism since Σ can be considered a time slice on which the system's state is specified by the conformal data. From the fact that for a given folium with its constant mean curvature and initial data the dynamical development, and thus the construction of the full spacetime, is sometimes non-unique, it does not follow, as Earman (2008, 148) seems to suggest,[18] that for a given four-dimensional spacetime, its global foliation into hypersurfaces of constant mean curvature is sometimes non-unique, if it exists. The reason for this is that the different solutions will not correspond to different foliations of the same spacetime, but rather to different spacetimes altogether. Conversely, this in itself

[17] "Essentially" because the space of conformal data must be quotiented out by the action of the group of conformal transformations, as well as by the action of the spatial diffeomorphism group in order for the correspondence to be one-to-one.

[18] When he writes that "*[t]ypically such a foliation is unique when it exists*, but existence is guaranteed only for a limited class of solutions to Einstein's field equations, a class that does not exhaust solutions with causally nice features" (emphasis added). While I agree with every other part of the statement, I take issue with the first clause's suggestion that there may be cases where such foliation is not unique, for which I see no warrant in the literature.

does not imply that CMC-slicings will be unique for a given spacetime, where they exist.

Be this as it may, the main problem of the CMC approach is, already at the classical level, that only a limited, although arguably important, class of spacetime models of GR comply in that they are CMC-sliceable. Unfortunately, as mentioned above, this class does not even exhaust the globally hyperbolic spacetimes of GR. Furthermore, also as stated above, it is only tractable for spatially closed vacuum spacetimes. But there is an additional difficulty, as pointed out by Isham (1991, 200): time-dependent Hamiltonians, as we find them here, have odd consequences. First, they are typically interpreted to mean, at least at the quantum level, that energy can enter or leave the quantum system, i.e., that the system is not closed. But this is odd indeed, as the system at stake is supposed to be the entire universe. Second, as Isham continues, for systems with time-dependent Hamiltonians one cannot get the Wheeler-DeWitt equation for the reduced system from the relevant Schrödinger equation, which shows the inequivalence of different canonical approaches to QG. This may not ultimately amount to a strike against the CMC approach, but its advocate must find a way to accommodate this inequivalence.

One might dissent to using the CMC approach for presentist purposes with an analogue to Kurt Gödel's (1949, 562) objection to James Jeans's proposal to rest a robust notion of absolute time on the cosmological time of highly symmetrical spacetimes whose foliation into space and time sensitively depends on these symmetries. As Gödel insisted, whether or not absolute time existed should not depend on contingent matters of fact concerning the distribution of matter and energy in the actual universe. Similarly, a potential resuscitation of presentism by the CMC approach fails, the objection goes, on the grounds that the CMC foliation also depends on the same kinds of contingent facts. In defense of the CMC-inspired presentist, it should be noted, however, that the Gödel move is significantly weaker here than it was in the original case. The reason for this disanalogy is that CMC-sliceable spacetimes form, to repeat, open subsets in the space of solutions–unlike the highly symmetrical spacetimes relied on by Jeans. It is true that moving around the matter and energy content of the universe will in general deform the CMC foliation, but this will often not change the fact that there *is* a CMC foliation for the spacetime at stake.

Let us, for the sake of Monton's argument, ignore these limitations of the CMC approach and ask whether it would, if borne out, vindicate presentism, as Monton asserts. No, it would not; or at least not as easily as Monton seems to think. Apart from those limitations of the CMC approach already listed, it is far from clear whether the CMC approach can be exploited to underwrite a presentist metaphysics. In particular, it is far from obvious how the mean extrinsic curvature τ relates to physical time, despite the fact that it can be used as a global time parameter. What the presentist needs is an account of how τ gives raise to not just physical time, but a time that underwrites our presentist intuitions. The fact that the folia are Cauchy surfaces might help the presentist here, as this will permit to establish a direct connection to the initial value problem and issues of determinism, which, if anything, seem to be directly linked to the role of physical time.[19] In the absence of such an account, a presentist such as Monton may rightly claim that the CMC approach, to the extent to which it is to be taken seriously as a fundamental, or at least true, physical theory, relieves the pressure that presentism has felt since the advent of SR. He has not yet, however, produced a positive argument in favour of presentism. For this, an account relating the CMC approach to our allegedly presentist phenomenology is essential.

Finally, lest the presentist gets overly enamoured of the CMC approach, it ought to be noted that no one takes it seriously as a physically plausible full theory of classical or quantum gravity. The real interest in the approach is fueled by the fact that it so significantly simplifies the systems of constraint equations that the Hamiltonian approach to GR is usually confronted with. Thus, the sole reason the CMC ansatz is explored is because it offers a technically tractable toy theory of canonical gravity.[20] Overall, it is incapable of accommodating the full plethora of gravitational phenomena that a theory of gravity is expected to address. Finally, as a reminder, the irony that published work in the CMC approach has almost exclusively dealt with the classical level while Monton was really concerned with a fundamental quantum theory of gravity should not be lost on the reader.

5 Conclusion

Since there are no complete quantum theories of gravity available at present– let alone "theories of everything"–, the question of whether presentism is ultimately compatible with fundamental physics remains open. The most promising approaches to QG to date, string theory and loop quantum gravity, offer no respite for presentism. As far as I understand it, string theory is a fully Lorentz-invariant theory. Similarly, loop quantum gravity does not permit the introduction of preferred frames of reference and thus does not contain the resources to support a privileged foliation. As a matter of fact, there is a foreboding sense in which time evaporates completely as a fundamental physical magnitude in loop quantum gravity. Presumably, such physics could not underwrite Monton's project of reading a presentist metaphysics of time into the fundamental physics.[21] Even on its own limited terms, I have argued that those approaches to QG that rely on fixed-foliations such as the CMC proposal are not as hospitable to presentism as Monton seems to think.

---

[19] Although the potential non-uniqueness of CMC foliations would surely undermine such a connection if borne out.

[20] Cf. also Belot and Earman (2001, 241).

[21] Monton (op. cit., 277) thinks that the presentist can evade the problem of time by simply maintaining that the position does not speak to fundamental reality, but only to time. Thus, if time is emergent rather than fundamental, presentism would be true as long as the emergent time fits the presentist metaphysics. While I acknowledge this possibility, it doesn't offer an appealing option to the presentist.

Let me conclude with Callender (2000) who warns against permitting presentism (or, more generally, any tensed theory of time) to "push us away from the traditional understanding of relativity" (S596), a role to be reserved for developments in physics. Monton shrugs this charge off by explaining that no non-traditional interpretation of relativity is required, since the presentist can simply deny that SR or GR are true theories just *because* they are incompatible with presentism. One man's modus ponens is truly another man's modus tollens. But if a re-interpretation of relativity against the backdrop of presentism is not warranted by evidence or argument, then the whole-sale *rejection* of it will hardly be more acceptable! Monton seems to think that, at least as viewed from a point of view of a committed presentist, *since* presentism is true, science should not, for its own good, turn out to be incompatible with it. Since alternative approaches to QG are incompatible with becoming, and since the existence of becoming is a philosophical, not a scientific, issue for the presentist, "we should expect the correct theory of quantum gravity to be a fixed foliation theory" (op. cit., 274). But that's exactly the point: if we base the scientific decision among competing theories on metaphysical predilections, we better have good reasons to do so. A failure to appreciate this would mislead us into abandoning **Naturalism**, or anyway naturalism.[22]

In this vein, Callender continues by asking, quite pertinently in my view, "if science cannot find the 'becoming frame', what extra-scientific reason is there for positing it?" (S597) Monton (op. cit., 272) replies to this charge by insisting that he can't discern a reason why the presentist ought to be committed to the antecedent. The grounds for denying the antecedent of Callender's conditional statement, Monton believes, can be found in that the CMC foliation approach yields what can be interpreted as the becoming frame. To be sure, we would need some sort of account of how exactly the CMC foliation of a spacetime underwrites "becoming" for that move to be successful. Monton recognizes that he cannot offer any positive account from our experiences to the necessity of the becoming frame, or of how the becoming frame is coupled to a CMC foliation, but he defends himself by retorting that "just because we do not have a good argument for the presentist doctrine... does not mean that the doctrine is false" (ibid., 273n). True, but in the absence of such argument, there is little or no reason to take the CMC foliation approach seriously as a full-fledged physical theory potent enough to supplant GR. As we have seen, this approach is highly limited in its applicability, remains almost exclusively at the classical level, and does not offer a viable road to a resuscitation of presentism. More seriously still, if what I said above is true, then we do have reason to accept the antecedent of Callender's pronouncement.

If we accept the antecedent, however, then the presentist must give sound arguments that are sufficiently forceful to overturn time-honoured Lorentz invariance as a constraint on a future theory of QG. That does not seem to be forthcoming. On balance, I submit, the prospects of presentism look rather dim.

[22] Monton recognizes this possibility when he offers an alternative move: the presentist could decide to give up scientific, but not metaphysical, realism.

Acknowledgements I am indebted to Craig Callender, Jonathan Cohen, John Earman, Storrs McCall, Bradley Monton, Thomas Müller, Vesselin Petkov, and Steve Savitt for discussions and comments, and audiences in Montreal and Geneva for their engagement with this paper. I also thank Vesselin Petkov for his almost infinite patience with my procrastination. This project has been funded in part by the Swiss National Science Foundation ("Properties and Relations", 100011-113688), by the University of California, San Diego, and by the Hellman Family Foundation.

References

Amelino-Camelia, Giovanni, "Quantum gravity phenomenology", available online at http://arxiv.org/abs/0806.0339 (2008).

Barbour, Julian, *The End of Time: The Next Revolution in Physics*, Oxford: Oxford University Press (1999).

Bartnik, Robert, "Remarks on cosmological spacetimes and constant mean curvature surface", *Communications in Mathematical Physics* **117** (1988): 615–624.

Beig, Robert, "The classical theory of canonical general relativity", *Lecture Notes in Physics* **434** (1994): 59–80.

Belot, Gordon and John Earman, "Pre-Socratic quantum gravity", in Craig Callender and Nick Huggett (eds.), *Physics Meets Philosophy at the Planck Scale*, Cambridge: Cambridge University Press (2001), 213–255.

Callender, Craig, "Shedding light on time", *Philosophy of Science* **67** (2000): S587–S599.

Callender, Craig, "Finding 'real' time in quantum mechanics", in William L Craig and Quentin Smith (eds.), *Einstein, Relativity, and Absolute Simultaneity*, London: Routledge (2008), 50–72.

Choquet-Bruhat, Yvonne and James W York, "The Cauchy problem", in Alan Held (ed.), *General Relativity and Gravitation: One Hundred Years After the Birth of Albert Einstein*, Volume 1, New York: Plenum (1980), 99–172.

Craig, William L, *Time and the Metaphysics of Relativity*, Dordrecht: Kluwer Academic Publishers (2001).

Dieks, Dennis (ed.), *The Ontology of Spacetime*, Amsterdam: Elsevier (2006).

Dorato, Mauro, "The irrelevance of the presentist/eternalist debate for the ontology of Minkowski spacetime", in Dieks (2006), 93–109.

Earman, John, "Reassessing the prospects for a growing block model of the universe", *International Studies in the Philosophy of Science* **22** (2008): 135–164.

Fine, Kit, "Tense and reality", in his *Modality and Tense: Philosophical Papers*, Oxford: Oxford University Press (2005), 261–320.

Fischer, Arthur E and Vincent Moncrief, "Hamiltonian reduction of Einstein's equations of general relativity", *Nuclear Physics* B (Proc. Suppl.) **57** (1997): 142–161.

Gödel, Kurt, "A remark about the relationship between relativity theory and idealistic philosophy", in Paul A Schilpp (ed.), *Albert Einstein: Philosopher-Scientist*, New York: Tudor (1949), 557–562.

Harrington, James, "Special relativity and the future: a defense of the point present", *Studies in History and Philosophy of Modern Physics* **39** (2008): 82–101.

Hinchliff, Mark, "The puzzle of change", *Philosophical Perspectives* **10** (1996): 119–136.

Isenberg, James, "Constant mean curvature solutions of the Einstein constraint equations on closed manifolds", *Classical and Quantum Gravity* **12** (1995): 2249–2274.

Isham, Christopher J, "Conceptual and geometrical problems in quantum gravity", *Lecture Notes in Physics* **396** (1991): 123–229.

McCall, Storrs, "QM and STR", *Philosophy of Science* **67** (2000): S535–S548.

Monton, Bradley, "Presentism and quantum gravity", in Dieks (2006), 263–280.

Putnam, Hilary, "Time and physical geometry", *Journal of Philosophy* **64** (1967): 240–247.

Rendall, Alan D, "Constant mean curvature foliations in cosmological spacetimes", *Helvetica Physica Acta* **69** (1996): 490–500.
Rietdijk, C Wim, "A rigorous proof of determinism derived from the special theory of relativity", *Philosophy of Science* **33** (1966): 341–344.
Savitt, Steven F, "Presentism and eternalism in perspective", in Dieks (2006a), 111–127.
Savitt, Steven, "Being and becoming in modern physics", in Edward N Zalta (ed.), *Stanford Encyclopedia of Philosophy*, available online at http://plato.stanford.edu/entries/spacetime-bebecome/ (2006b).
Smeenk, Christopher and Christian Wüthrich, "Time travel and time machines", in Craig Callender (ed.), *The Oxford Handbook of Time*, Oxford: Oxford University Press (2010).
Stein, Howard, "On relativity theory and openness of the future", *Philosophy of Science* **58** (1991): 147–167.
Tooley, Michael, *Time, Tense, & Causation*, Oxford: Oxford University Press (1997).
Will, Clifford M, "Was Einstein right? Testing relativity at the centenary", in Abhay Ashtekar (ed.), *100 Years of Relativity; Space-Time Structure: Einstein and Beyond*, Singapore: World Scientific (2005a), 205–227.
Will, Clifford M, "Special relativity: A centenary perspective", *Séminaire Poincaré* **1** (2005b): 79–98.
Wüthrich, Christian, "Demarcating presentism", manuscript (unpublished).
Zimmerman, Dean, "The privileged present: defending an 'A-theory' of time", in Theodore Sider, John Hawthorne, and Dean Zimmerman (eds.), *Contemporary Debates in Metaphysics*, Malden, MA: Blackwell (2008), 211–225.

Part III
The Impact of Minkowski's Ideas Beyond the Philosophy of Space and Time

Space-Time, Phenomenology, and the Picture Theory of Language

Hans Herlof Grelland

Abstract To estimate Minkowski's introduction of space-time in relativity, the case is made for the view that abstract language and mathematics carries meaning not only by its connections with observation but as pictures of facts. This view is contrasted to the more traditional intuitionism of Hume, Mach, and Husserl. Einstein's attempt at a conceptual reconstruction of space and time as well as Husserl's analysis of the loss of meaning in science through increasing abstraction is analysed. Wittgenstein's picture theory of language is used to explain how meaning is conveyed by abstract expressions, with the Minkowski space as a case.

Hermann Minkowski predicted in his famous 1908 statement that "space by itself, and time by itself, are doomed to fade away into mere shadows, and only a kind of union of the two will preserve an independent reality". But for many philosophers and scientists, reality means what we experience, and what we experience is plainly space by itself and time by itself. In this article I will make the case for Minkowski's statement through a philosophical analysis of language, including the language of mathematics, and relate it to the views of empiricism and phenomenology.

I will do this by distinguishing two views of physics, or, to be more precise, two views on how meaning is constituted in a physical theory, *intuitionism* and *linguism*. As a science whose validity is completely dependent on observation and experiment, physics is linked to the experience of human beings. Every observation must, even with the extensive application of technology in modern science, eventually be lead back to human perception. But physics also depends on our ability to make abstractions, on language and mathematics. And an abstract language or a transparent mathematical structure may clarify matters that are not so easily seen on a level closer to experience. Here we can identify an epistemological duality which is worth a closer study.

Physics, which is created by human consciousness through the acts of perception and symbolic abstraction, cannot be isolated from the structure and function of the human mind, which gives meaning to the physical theories. Here we find the relevance of phenomenology, but, I will argue, a phenomenology which goes beyond Husserls one-sided emphasise on "the thing itself", as it is supposed to appear in the act of perception. But let me start by going further back in the history of philosophy.

V. Petkov (ed.), *Space, Time, and Spacetime*, Fundamental Theories of Physics 167, DOI 10.1007/978-3-642-13538-5_13,

Modern science is closely related to the empiricist tradition of philosophy, which has its roots back to Aristotle through thinkers like Roger Bacon, Thomas Aquinas, and Francis Bacon. Aquinas is known for his Peripatetic Axiom that "Nothing is in the intellect that was not first in the senses". However, the beginning of modern philosophical empiricism is usually associated with John Locke, who was a friend of Isaac Newton and Robert Boyle, and who, after comparing himself with these great scientists, presented himself as an "under-labourer in clearing the ground a little, and removing some rubbish that lies in the way of knowledge." (This may also be stated as the aim of the present article.) In addition to his emphasis on experience, it is interesting to observe the importance Locke attaches to language, and he reserves a substantial part of *Essay Concerning Humane Knowledge* (4th edition, 1700. Locke 2003) to questions relating precisely to this topic. An even stronger emphasis on language is put forth by the lesser known French empiricist Etienne de Condillac (*Essay on the Origin of Human Knowledge*, 1746. Condillac 2001), who was a great admirer of Locke. However, after the age of Condillac, interest in language seems to fade away in the philosophy of physics, becoming more or less a neglected subject in later empiricist philosophy and in the philosophy of science until the "linguistic turn" at the end of the twentieth century. A representative of this development and the most influential successor of Locke is David Hume, who inspired both Kant, Husserl, and – which is of particular interest in our context – Einstein. In some sense also Edmund Husserl's phenomenology may be said to be a developed and extended form of empiricism merged by the rationalist tradition of Descartes. Another development of empiricism is the sensualism or positivism of Ernst Mach, who we also know was of importance to Einstein in his attempt to rethink time and space upon formulating his theory of relativity.

Let us first consider one aspect of Husserl: Through his phenomenology, including the idea of intentionality as the constituting feature of consciousness and his analysis of consciousness in terms of conscious acts like perception, imagination, the signitive act, etc., he laid a philosophical foundation for the study of science. However, in addition to being a phenomenologist, Husserl was also an intuitionist. By "intuitionism" in physics, I imply the meaning-theoretical position that meaning of abstract symbols is provided by the concrete experiences to which it is related, be it observations made in a laboratory with the aid of sophisticated instruments, or experiences of ordinary life. The word "intuition" in this context is the English (and also the French) rendering of the German *Anschauung*. While traces of intuitionism can be found in Husserl's early works, it is taken to the forefront in his later work *The Crisis of European Sciences and Transcendental Phenomenology* (Husserl 1970) and in the small but famous Appendix VI to this work, *The Origin of Geometry* (Derrida/Husserl 1989). Here Husserl claims to be observing a loss of meaning in science through its historical development as a consequence of an increasing degree of abstraction. According to Husserl, this development was initiated by Galilei, who was inspired by the geometry of the ancient Greeks. I want to distinguish between Husserl's phenomenology and his intuitionism, claiming that the one does not implicate the other.

To clarify my line of argument, I will briefly describe the two traditions or lines of thought in the history of science which can be associated with intuitionism and its counterpart linguism, respectively. Of these, I judge linguism to be the correct approach, while intuitionism is an epistemological mistake, which unfortunately has become more or less a part of the common view among scientists. The intuitionist tradition may be associated with Hume, Mach, and the late Husserl, as well as the early Einstein, and Niels Bohr. The linguist tradition is represented by, among others, Heinrich Hertz, Ludwig Wittgenstein (in the *Tractatus Logico-Philosophicus*, commonly referred to as Wittgenstein I. Wittgenstein 2001), and, as I will suggest, Paul A.M. Dirac. In this tradition I would also like to include Locke, although with certain reservations. What is not open to discussion is the basic dependence of science on observation and experiment for its validation. Thus I would prefer to use the term *empiricism* for both traditions mentioned, distinguishing them by the further qualification *intuitionistic empiricism* and *linguistic empiricism*. It is the role of observation beyond validation which can be viewed differently, in particular how determining it is to the meaning-giving acts which transform a physical theory from a structure of signs into a description or picture of reality.

In short, intuitionism assumes that the meaning of the concepts and equations of physics is drawn exclusively from the sense perceptions and the related imagery of the reality which they are supposed to represent. For instance, the equations of hydrodynamics become meaningful through our experience and images of liquids in everyday life or in scientific laboratories. The system of mathematical signs and equations is looked upon as a "formalism", a way of structuring, communicating, and making numerical predictions from a set of observations or perceptions with a pre-mathematical and pre-linguistic meaning. The imagery based on perception is what *animates* the signs, giving them meaning and thus turning them into symbols (sign + meaning). Epistemological priority is given to the physical *intuition*, not the symbolic "formalisation". According to this point of view, there is a danger attached to the increasing abstraction of physics, as it moves the "formalism" further away from intuition. This was an issue when the abstract formalism of analytical mechanics by people like Lagrange and Hamilton replaced the original Newtonian mechanics. An acute problem appears when the mathematical structure of the theory may no longer be associated with any pictorial imagery, as in quantum theory, not to mention abstract conceptions like string theory. This is the problem of "interpretation" of those theories, a problem which is still hotly discussed. On the other hand, the existence of a theory like quantum mechanics may in itself indicate that the intuitionistic philosophy may be a blind alley.

Perhaps the strongest version of intuitionism is Mach's sensualism, in which the meaning of a scientific concept or symbol is nothing but the "sense impressions" it describes or codifies. A modern representative of this view is Stephen W. Hawking: "Any sound scientific theory, whether of time or of any other concept, should in my opinion be based on the most workable philosophy of science: the positivist approach ... According to this way of thinking, a scientific theory is a mathematical model that describes and codifies the observations we make." (Hawking 1988, 31).

A typical intuitionistic approach, inspired by Mach and Hume, is attempted by Einstein in *The Meaning of Relativity* where he first tries to define the concept of space in terms of perceptions. He states that "The natural sciences, and in particular, the most fundamental of them, physics, deal with ... sense perceptions" (Einstein 1974, 2). Thus, perception is not only a means to ensure that a physical theory is correct; it is what physics is all about. From this starting point, Einstein attempts to derive the concept of space from sense perceptions through the mediating notion of a physical body: "The conception of physical bodies, in particular of rigid bodies, is a relatively constant complex of such sense perceptions" (Einstein 1974, 2). The idea of defining space in terms of extended bodies and their (spatial) relations comes from Hume. Having given meaning to the notion of a rigid body, Einstein further proceeds through defining a *continuation* of a body A as putting another body up to it, and then he defines space (the space of body A, thus a space is always associated with a body) as the set ("ensemble") of all continuations of A. Einstein is then able to define length in terms of bodies acting as "measuring sticks", and so on. The whole theoretical development is based on the view that physics is about sense perceptions. (There are good reasons to believe, however, that Einstein, like many physicists, is not really consistent on this point, and that he in his scientific work more or less forgets about strong sensualistic statements like these).

The sensualistic approach soon leads to difficulties. The treatment of the concept of *time* in the text differs fundamentally from the concept of space. Although Einstein's intention is to define all the quantities in terms of sense impressions, this turns out to be difficult in the case of time, which can only be observed indirectly by a measuring instrument: the clock. Hence, Einstein's exposure depends of the notion of a clock. This move is obviously a break with the original simplicity of his method. While a physical body can be considered as an elementary and simple object, available for immediate perception, a clock is a complicated technological construction, and two clocks may in fact be built on very different physical principles. Even worse, how can one have any notion of a clock *before* the concept of time is given meaning? Einstein's solution in *The Meaning of Relativity* is simple: he does not try. Instead, he disturbs the logical simplicity of his exposure even more by *ad hoc* introducing another complicated physical phenomenon: light. Thus the basic concepts of space and time are defined in terms of mechanical clocks and electromagnetic light, thereby assuming both Newton's mechanics (applied locally) and Maxwell's electrodynamics. This is a demonstration of some of the internal difficulties in the intuitionistic project as such. The project of trying to give meaning to the abstract concepts and logical structures in physics in terms of "complexes of sense impressions" does not work.

In the same spirit, Husserl worries about the loss of meaning in modern science. This is a main subject in *Crisis* as well as in *Origin*. In *Crisis*, Husserl draws a broad picture of the development of science since the renaissance. The "crisis" of this science is the result of a gradual loss of meaning due to an increasing abstraction and, supporting this, mathematisation, a development similar to what had happened earlier in the particular science of geometry. Among the scientists who started this development Husserl points to Galileo Galilei with his mathematisation program expressed in his tenet that "the universe is a book written in the language of mathematics". According to Husserl, Descartes tried to reintroduce the human subject in European scientific thinking, but this attempt failed. According to *Crisis,* the aim of phenomenology is to lead the way out of this historically persistent crisis, taking its cue from Descartes' original point of departure in order to place science on its proper foundation.

This does not mean that Husserl the mathematician rejects mathematics as a tool in science. He calls it the decisive accomplishment which makes predictions possible, thus "going beyond the sphere of immediately experiencing intuitions and the possible experiential knowledge of the prescientific life-world." (Husserl 1970, 43).

Here I want to make a brief remark on the notion of the "life-world", which later came to be so important to Heidegger. Husserl (and later, Heidegger) talk about the "life-world" as being something simply given, independently of and prior to any scientific knowledge, and fundamentally different from the abstract objects of science – as if the world we experience does not have to be seen through the spectacles of language, with its built-in idealisations and abstractions; and, as if this world is something intersubjectively given, common to all inhabitants of the assumed human society. Husserl and Heidegger seem to overlook the fact that although there is a sharing of a language and tradition in a society, there is still the necessity for each individual to learn and adapt this language and these traditional views, creating his own interpretation of (and relation to) the world surrounding him. In this experienced world there are no sharp division between what is intersubjectively given and what is subject to individual interpretation and relation. Thus, paralleling my rejection of intuitionism, I also reject the notion of a "life-world" in the sense of Husserl and Heidegger. While it is still meaningful, in a loose sense, to talk about an individual life-world, to a scientifically educated and intellectually integrated person this life-world includes scientific knowledge in all its abstractions.

After accepting the accomplishments of mathematics, Husserl goes on to criticise the scientists who, according to him, confused the mathematical formulae with their "formula-meaning" with the "true being of nature itself" (Husserl 1970, 44). The technical or instrumental approach implied by these formulae lead to a "superficialisation" and, eventually, to an emptying of the meaning-content of the theory itself. "One operates with letters and with signs for connections and relations, according to the *rules of the game* for arranging them together in a way essentially not different from a game of cards or chess. Here the *original* thinking that genuinely gives meaning to this technical process and truths to the correct results are excluded." (Husserl 1970, 46).

A brief comment on this point: even if I later will argue that the abstract mathematical structure of a physical theory in itself contributes additional meaning of the theory, I do not deny that it is possible to use the formulae of the theory instrumentally as a means for producing numbers and completely miss their physical meaning. However, to Husserl this kind of instrumental activity is essential in highly mathematised physics.

To sum up Husserl's view in his own words: "But now we must note something of the highest importance that occurred even as early as Galileo: the surreptitious substitution of the mathematically substructed world of idealities for the only real world, the one that is actually given through perception, that is ever experienced and experiencable – our everyday life-world. This substitution was promptly passed on to his successors, the physicists of all the succeeding centuries." (Husserl 1970, 49) The mathematical idealities are substituted for the real world, which is our "everyday life-world" – and which, supposedly do *not* consist of "idealities". Husserl seems to be completely unaware of the idealisation implied by the use of language itself, and the implicit effect of language on our perception of the world surrounding us.

Husserl's judgement on Galileo himself, the originator of this development, is that he, the discoverer of physics, is at once "a discovering and a concealing genius."

This point of view is repeated again and again in *Crisis*, with small variations. The disease is loss of meaning through increased abstraction. The remedy is phenomenology.

Since the overall development of European science since Galilei in many ways parallels the earlier development of geometry, Husserl wrote *The Origin of Geometry* as a separate article, which by the editor of his works has been appended to *Crisis*.

In *Origin*, Husserl analyses the steps leading from ordinary human experience to abstract, mathematical science by taking classical geometry as a case. One should keep in mind that Husserl considers "physical" geometry, the science of spatial relations in physical space, not modern, axiomatic geometry based on logical constructions from set theory. Thus, his case may be considered as one on physics. *Origin* is written at a time when the general theory of relativity is well established, but this is not mentioned by Husserl. In any case, his analysis is not specific to the details of classical geometry, and he might as well have used curved space-time as his case. Husserl seems to be motivated by the role played by geometry in the science of Galilei.

The main problem addressed by Husserl is the establishment of objective, timeless truths concerning "ideal objects" in geometry, like the circle or the straight line. He imagines a development in five steps: (1) The immediate understanding by the "pre-geometer" of spatial relations based on practical experience with spatial objects. (2) The remembrance of such insight and the recognition of a problem at hand as being the *same* as the one encountered on an earlier occasion. This implies a certain degree of abstraction and idealisation by seeing the concrete event as an instance of something repeatable and thus general, an *idea* or an *ideal object* (*ideale Gegenständlichkeit*). (3) The establishment of *intersubjective* ideas through language, by the geometer's participation in a linguistic community. Thus, the idea has to have a linguistic expression. (4) By repeated use of linguistic communication, the geometric truth changes its appearance in the mind of the geometer. Instead of being thought of as a repetition of the same, it will seem to be something belonging to an objectively existing structure of ideas and truths: geometry. (5) The establishment of timelessness (and thus *eternal* truths) by the invention of written notation. Communication by speech is still something happening in time, while writing has a more lasting appearance, confirming the appearance of geometric truths as timeless and eternal.

According to Husserl, this development is accompanied by a loss of the original fullness of meaning which was present in the mind of the original pre-geometer handling spatial problems. Husserl describes the development of science as a continuous sedimentation of layers of abstraction, hiding the original meaning from later scientists.

Perhaps surprising to some scientists, the poststructuralist Jacques Derrida seems to be the one who identified the weak spot of Husserl's analysis. Derrida's comments on *Origin* appear in his famous introduction to its French translation. Husserl, in his emphasis on the immediate perception as the sole source of meaning, seems to turn a blind eye to the interdependence between meaning and ideality. The flow of sense impressions passing through human consciousness can hardly be ascribed any meaning without any (in a literal sense) *re-cognition*, seeing a particular phenomenon as *the same* as something we can think of, for instance an idea, or at least as something that may be repeated in our mind and perhaps in reality, something that may become, to the experiencing subject, an idea. If we then take the step (made in fact by Husserl himself, and as pointed out by Derrida) of associating ideas with words (or, more generally, with signs), we see that the word does more than *express* an idea, and thus meaning. The word is a necessary condition for establishing an idea. Earlier, already Locke had pointed out the necessity of words for "preserving essences" (i.e. ideas or ideal objects) and "giving them their lasting duration" (Locke 2003, IV.V.10). Reminding ourselves that stability is essential to an idea, we may conclude that ideas and language cannot be separated. In Husserl's words, "... ideal objects (*Gegenständlichkeit*) do exist objectively in the world, but only ... by virtue of sensible embodying repetitions (i.e. signs)". (Derrida/Husserl 1989, 161). This is the meaning of Derrida's statement that the word *constitutes* the ideal object. Moreover, by "words" in this context we also mean mathematical formulae and expressions as well as idealised drawings in geometry of objects such as lines and circles.

After this brief analysis of some representatives of intuitionism, I now turn to the other line of thought, which I have called linguism. It can be identified with thinkers like Heinrich Herz, Ludwig Boltzmann, and Ludwig Wittgenstein, Hermann Minkowski and Paul A. M. Dirac.

In his philosophically valuable introduction to *The Principles of Mechanics Presented in a New Form* (1891. Hertz 2003), Hertz refers to the mathematical models of physics as "Bilder" (pictures, or images). He lists four conditions which must be satisfied by such a mathematical "Bild". The first one is that it must be logically consistent or "permissible". The second is that it must be empirically verified or "correct". The third criterion is the more interesting one for our analysis, stating that out of two possible (logically consistent and empirically verified) models or "Bilder", one should choose the one which is most appropriate ("zweckmässig") "which includes in it more of the essential relation of the object (to be pictured)". How does one decide which is has the more essential relation? One has to choose the one which is the most distinct or clear ("deutlich"). This is an "inner" quality of the picture itself. The fourth criterion is that if one still has a choice between two pictures, one should choose the simplest. This too is quality of the picture itself.

The pictures are not just mappings of “sense impressions”. Rather, they are “models produced by our mind and necessarily affected by the characteristics of its mode of modelling them”.

Hertz compares the structure of the picture to a grammar, considering his own presentation of mechanics as a *systematic* grammar, in contrast to e.g. a grammar devised for the purpose of making a language easy to learn.

Even the limits of the picture (in Hertz’s case the limits of his model of classical mechanics) is, according to Hertz, shown by the picture itself, which clarifies all possible qualities and processes which belong to mechanics. (As an example of such a limit, Hertz mentions that mechanics is too simple and narrow to account for life processes).

Clarity and simplicity is often in physics simply called beauty. Thus, there are deep epistemological reasons for seeking out the most beautiful form possible for a theory. One of the advocates for the importance of beauty in theoretical physics is Dirac, who, by the way also is the person who brought the Minkowski space and quantum theory together in the Dirac equation when written in its manifestly covariant form. Dirac’s own texts are themselves examples of beautiful expositions, including his inventions in the mathematical notation. When the mathematician John von Neumann introduced a new standard of mathematical rigour to quantum mechanics in his book *Mathematical Foundations of Quantum Mechanics*, he still admitted, in the Foreword, that the Dirac bra- and ket-notation was “scarcely to be surpassed in brevity and elegance”. Today the Dirac notation is more widely used than ever, and I think it illustrates the point that clarity and simplicity conveys meaning. I believe that the reason for Dirac’s refusal to participate in the ongoing discussion on the interpretation of quantum mechanics was his implicit attitude that the theory explained itself better in a clear exposure rather than by being explained by an additional “interpretation”. Thus Dirac obviously stands in the tradition of Hertz.

Hertz’s point of view was further developed by Ludwig Boltzmann into the language of phase space or state space, showing all possible states and thus defining the boundary of the set of possible processes in the object pictured by this space. This has become a key concept for the development of general theories of physics, and also for mathematical models in other fields. We can note that Dirac was the one that introduced the concept of a state space in quantum mechanics.

There is a relation between Boltzmann’s thinking in terms of state space in physics to Wittgenstein’s view of language in general in *Tractatus*. In this book, after having made the initial statement that the world consists of facts, not of things, Wittgenstein defines facts as state-of-affairs that are the case, defining “logical space” (his more general version of Boltzmann’s state space) as consisting of all possible “state-of-affairs”. In later works, Wittgenstein admitted that language is more than just stating facts, and he invested the second part of his philosophical career in the development of a more extensive notion of language. But as long as we restrict ourselves to the declarative language of science, the restricted notion of language in *Tractatus* is still applicable, in particular his picture theory of language which, properly understood, clarifies many issues in modern, abstract, mathematical physics.

Mathematics is an extension of ordinary language, thus it is a language itself, although self-insufficient. In a sentence or a mathematical equation (or a complete scientific work), we find not only a collection of signs, but a collection which is structured in a certain way, having a grammatical or logical structure. However, according to Wittgenstein this structure, which carries meaning, is not itself explicitly stated *in* the sentence or equation; nonetheless, we can *see* it when we read the sentence or equation. It is *shown* by itself. Through its words and symbols, and through its logical and grammatical structure, the sentence or equation is a picture. A picture of what? It is not (or rarely) a picture of something visual that can be imagined, like a physical object. It is a picture of a *fact*.

Consider the following example: imagine two physical bodies, A and B. Imagine that A is visibly bigger than B. The picture you imagine contains, however, necessarily much more content than the simple fact that A is bigger than B, which is the only specification I have given. To imagine two bodies, you have to give them a shape (e.g. rectangular) and a colour (even if it is grey) and you even imagine them at some specific distance from you. It is not possible to imagine the size difference exclusively. But the imagined picture also conveys the fact that A is bigger than B. However, the written sentence "A is bigger than B" is, according to Wittgenstein, also a picture, but only of the fact. All properties of A and B irrelevant to this fact are deleted from this picture, it has been cleaned up or, as we say, abstracted. In this picture, A and B are represented (pictured) by their respective signs ('A' and 'B'), and there are signs symbolising the relation of one being bigger than the other, but the picture would not function as such without the symbols being placed in a certain *order*. This order is necessary for the meaning, but it is not explicitly expressed (stated), it must be seen by looking at the signs in their order, and in this way we can see the fact. Out of everything we can see in a mental image (things, like A and B, and additional facts like "A is rectangular" or "B is grey") in the sentence we have abstracted a pure picture, showing *only* the fact that A is bigger than B and nothing else about A and B.

I chose this example because the picture made by the sentence can be compared to a visual image in my imagination. But many facts cannot be imagined visually, for instance the statement that "Love does not last". This written sentence is a picture of a fact that cannot be pictured by a photographer or a painter, or in someone's imagination. Such facts are usually called abstract. Now you may doubt that the sentence "Love does not last" is true. If it is not, the sentence still retains its meaning. If the truth value is undecided, Wittgenstein calls it *Sachverhalt* (in English: state of affairs). A state of affair that is the case is a fact. So, to be precise, "A is bigger than B" and "Love does not last" are pictures of states of affairs.

But if a sentence, or a mathematical equation (or an extended text consisting of possibly both) is a picture, we can take into account that pictures differ in quality. Some pictures are more beautiful than others, some are clearer than others. The same applies to language and mathematics. And now we come to an important point, which is contrary to the assumptions of intuitionism: sometimes an abstract linguistic or mathematical expression is clearer than one which is more concrete, i.e. one closer to our visual (or perceptual) imagination. Thus *sometimes* abstraction does not imply a loss of meaning, but rather a *gain* of meaning.

From this point of view, where the presentation is assumed to be a picture of the facts it represents, the form of the presentation itself becomes crucially important. We can, as Hertz has reminded us, have a more or less *clear* picture and a more or less *simple* picture, i.e. a more or less beautiful picture. A beautiful picture communicates the beauty of the physical world. Thus, there are deep epistemological reasons for seeking out the most beautiful form possible for a theory.

One example of this is using the language and mathematics of the four-dimensional Minkowski space-time as a replacement for a language based on three-dimensional space and one-dimensional time. Although the Minkowski space is further removed from the elementary perception of spatial events happening in time, it conveys additional insight, and even more deeply, an insight which is hidden and obscured in the more muddled language of space and time. It fulfils Herts' criteria of being the clearest and simplest mathematical language for relativity theory. This is the reason for letting space by itself and time by itself "fade away" and giving place for a four-dimensional space-time conceptualisation as presented by Minkowski in 1908.

References

Condillac, Etienne Bonnot de: *Essay on the Origin of Human Knowledge*. Cambridge: Cambridge University Press 2001.

Derrida, Jacques/Husserl, Edmund: *Edmund Husserl's Origin of Geometry. An Introduction By Jacques Derrida*. Lincoln: University of Nebraska Press 1989.

Einstein, Albert: *The Meaning of Relativity*. Princeton: Princeton University Press 1974.

Hawking, Stephen W.: *A Brief History of Time*. London: Guild Publishing 1988.

Hertz, Heinrich: *The Principles of Mechanics Presented in a New Form*. New York: Dover 2003.

Husserl, Edmund: *The Crisis of European Sciences and Transcendental Phenomenology*. Evanston: Northwestern University Press 1970.

Locke, John: *An Essay Conserning Human Understanding*. New York: Dover 2003.

Wittgenstein, Ludwig: *Tractatus Logico-Philosophicus*. London: Routlegde 2001.

The Fate of Mathematical Place: Objectivity and the Theory of Lived-Space from Husserl to Casey

Edward Slowik

Abstract This essay explores continental/postmodern theories of place, or lived-space, as regards the role of mathematics, objectivity, and the relativist dilemma that afflicts the lived-space movement. By employing a geometric approach, such as Minkowski pioneered, it is argued that the lived-space theorists can gain a better insight into objectivity of spatial relationships.

1 Introduction

This essay explores space in contemporary continental philosophy and the philosophy of the social sciences, a popular movement often dubbed the study of "place", or "lived-space", due to its emphasis on the human experience of space, both personal and social. Among analytic philosophers of science, it is not widely recognized that there have been many contributions to the debate on the ontology and epistemology of space from this diverse field, which includes: contemporary philosophers of place (e.g., Edward Casey), prominent continental philosophers from the second half of the twentieth century (e.g., Deleuze, Derrida), and many renowned phenomenological investigations in the first half of the twentieth century (e.g., Husserl, Heidegger, and Merleau-Ponty). Many of these studies have sanctioned, often inadvertently, a form of relativism or social constructivism (Casey), or even metric conventionalism (Merleau-Ponty) as regards the ontology/epistemology of space. Accordingly, this essay will explores these highly popular works in order to determine both the general content of their claims and the overall philosophy of space either implicitly or explicitly advanced in their philosophies. As will be demonstrated, the theories of lived-space put forward by these philosophers, from the later Husserl to Casey, bare a number of uncanny similarities with work in the analytic study of space and spacetime, such as an emphasis on objectivity and an interest in structuralist forms of explanation.

Much of the examination will focus, however, on the role of mathematics within the lived-space approach to space, since a misunderstanding or mistrust of mathematics, which can be traced in part to the influence of the early phenomenologists,

V. Petkov (ed.), *Space, Time, and Spacetime*, Fundamental Theories of Physics 167,
DOI 10.1007/978-3-642-13538-5_14,

has been a major factor in the relativist dilemma that afflicts the lived-space movement (Sects. 2 and 3). By incorporating various geometrical concepts within the analysis of place, it will be argued that the lived-space theorists can better grasp the nature of objective spatial relationships–and, more importantly, that this appeal to mathematical content need not be construed as undermining the basic tenants of the lived-space approach (Sect. 4). In the final section, Deleuze's unconventional foray into differential geometry will serve as a means of demonstrating the inherent limitations of the lived-space conception of mathematics. Overall, the geometric approach to spacetime, as exemplified in Minkowski's interpretation of Special Relativity, is ideally suited to capture the objectivity of the spatial component of physical systems, unlike the contemporary lived-space school. Indeed, it will be argued that Minkowski's utilization of the group concept set the stage for the numerous philosophical investigations that later explored the subjectivity-objectivity issue (and which are based on these geometric techniques). Finally, it should be noted that one of the additional goals of this examination is to open up a largely unexplored field for researchers interested in the ontology and epistemology of space and spacetime, especially given the fact that this field, i.e., lived-space, has exhibited such a broad and popular appeal among present-day philosophers.

2 The Place Theory and the Subjective/Objective Dichotomy

In the more practice-oriented disciplines and philosophical schools of the late-twentieth century, considerable attention has been devoted to the concept of "place", or lived-space; which, put roughly, denotes the study of space (spatiality) as manifest within a human, usually social, order or practice (as in, dwelling, abode, local). The place theory of space relies on the insights gathered from a host of twentieth century philosophers and philosophical movements traditionally categorized as continental: for the philosophers, e.g., Husserl, Heidegger, Deleuze, and for the philosophical movements, e.g., phenomenology, environmental studies, literary theory, social geography. In particular, many researchers of place attempt to shed light on the relationship between our subjective, i.e., human, social, and practical experience of space, and the epistemological/ontological notion of objectivity.

2.1 Radical Spatial Subjectivism

Nevertheless, these studies have largely failed to address two important, and somewhat obvious, interrelated problems associated with the objectivity of space:

Problem (1): Can a theory of place successfully counter any radically subjectivist interpretation of the epistemology and ontology of space? As employed in this context, a "radical spatial subjectivist" rejects any objective or invariant spatial structure, and thus the geometric structure of space is *entirely relative* to different persons, cultures, or practices–example: space is Euclidean relative to geometer A, and space is non-Euclidean relative to geometer B, although both inhabit the same world.

Problem (2): How does the subjective or social aspect of the experience of space connect or interface with the underlying ontology of the physical world?

The relevance of space to the vexed subjective/objective problem assumes an obvious importance in the lived-space field, moreover: a subjective space is "a space that is tied to some feature of the creature's own awareness or experience . . . –the space of awareness within which it acts and with respect to which its actions are oriented and located" (Malpas 1999, 50). Objective space, conversely, "is a grasp of space that, while it requires a grasp of one's own perspective and location, is a grasp of space that is not centered on any particular such perspective or on any particular location . . . " (66). Yet, apart from a few instances, such as Malpas' forthright analysis, most of the investigations of place do not address adequately the exact structure or relationship between place and its objective and subjective components. If, indeed, any trend can de detected in this field, it would seem that many authors favor an interpretation of place that posits subjective space as primary, with objective space being derived, or "stitched together", from subjective experience. In a popular text, Edward Casey seems to endorse this reduction, viewing the objective, infinite, mathematical "space" of the Modern era as derived from the earlier, human-centered concept of bodily and social "place": "In a dramatic reversal of previous priorities, space is being reassimilated into place, . . . as a result of this reversal, spacing not only eventuates in placing but is seen to *require it to begin with*" (Casey 1997, 340; original emphasis); and, commenting on the social theorist Nancy, he proclaims, "spaces comes from places, not the other way around" (341). Among other examples, one can cite various difficult passages in Tuan's environmental study, where a person's experience "constructs a reality" (1977, 8), and Entrikin's appeal to "narrative" to connect the subjective and objective aspects of place (1991, 132–134), since "narrative" has strong subjectivist overtones.

However, without some set of *constraints* on the acceptable methods of explicating or constructing the global, objective structure of space (place) via the local, subjective spaces, the inevitable and unfortunate outcome is a radically divergent set of competing objective spatial structures. An obvious example is the spatial beliefs common to ancient Middle-Eastern civilizations, who interpreted the world as both flat and centered upon their home civilization–two "hypotheses" apparently confirmed through simple *bodily experience* and *common social practice.* Consequently, the subjectivist interpretation of space would seem to lack the conceptual resources needed to defuse problem (1): Was the earth "really" flat for ancient Middle-Eastern civilizations (e.g., the Genesis creation story in the Bible), but "really" spherical for modern Western societies? It is tempting to claim that modern science provides the "true" explanation of space; but, of course, modern science is just another social practice or narrative.[1]

[1] The place theorists might appeal to a deeper, non-metrical invariant to counter radical spatial subjectivism, such as a topological invariant, or a common geometric axiom (defined set-theoretically for all geometries). Yet, as will be explained, the lived-space school's antipathy to mathematics would almost certainly preclude this option. Throughout this essay, moreover, we will refer to objectivity as a joint epistemological and ontological notion, since lived-space theorists tend to blur the distinction between the two; see Rescher (1997), on the many forms of objectivity, including

2.2 Internal and External Constraints

Some place theorists strive to avoid the impending conflict of spatial schemes by means of constraints imposed either internally or externally to all potential subjectivist theories: either by openly endorsing the "irreducibility" of the objective aspect of spatial experience, or by acknowledging the intervention of an underlying physical space in the subjective act of spatial construction. One of the more notable efforts to address issues related to our problem (1) appears in Malpas (1999), which also draws upon both of the above methods for undermining a radical subjectivism. First, Malpas persistently rejects the view that objective space can be derived "from a mere concatenation of subjective spaces" (61; as does Campbell 1994, 5–37). By claiming that the two are "correlative concepts" (Malpas 1999, 36), or have a "complex interconnection" (70), the suggestion would seem to be that the irreducibly objective facet of spatial experience sets up barriers, an internal constraint, concerning different subjective formulations of space (i.e., the proposed objective-subjective irreducibility prevents, say, the flat and spherical models of earth's geometry from being equally successful constructions). Yet, this strategy seems consonant with a full-blown objectivism regarding space–and, of course, the intention was to develop a theory that shuns a strong objectivism through the utilization of an irreducibly subjective aspect of spatial experience. Since it is the objective aspect of space that, returning to our example, rules out the flat-earth case, the subjective element would appear to be idle. Moreover, this form of response does not explain *how* the interrelated objectivity-subjectivity of place prevents conflicting constructions of place.

Possibly in response to such worries, Malpas invokes an external constraint on the subjective constructions of space by means of a supervenience relationship between place and the underlying physical space: "In some sense place must 'supervene' upon physical space, and upon the physical world in general, such that the structure of a particular place will reflect, in part, the structure of the physical region in relation to which that place emerges" (34). Yet, since no further details are offered on this quite mysterious form of supervenience, which must be an ontological relationship of some sort, this attempt to resolve problem (1) comes at the considerable expense of inflaming problem (2): i.e., what does it mean to say that the subjective, socially-oriented conception of place supervenes on physical space?

2.3 Physical Space and Merleau-Ponty's Metric Conventionalism

Since few in the lived-space tradition address problems (1) and (2), it is possible that, given the continental/postmodern leanings of this movement (see Sect. 3 below),

ontological. Furthermore, "subjective", as used in this essay will refer to either a personal ("egocentric" in some texts) or social conception of space; i.e., as a non-objective conception. Finally, references to "place", or "lived-space", theory and theorists signify the contemporary, largely continental or continental-influenced, approach to space (e.g., Casey, Malpas, Lefebvre, etc.), and not the early phenomenologists.

a radical spatial subjectivism is, in fact, acceptable to many place theorists. As a means of undermining the primacy of objective space, a radical subjectivist might, for instance, appeal to Quine's "indeterminacy of translation" in order to claim that each separate subjective space is incommensurable, i.e., not communicable, with other subjective spaces. Most analytic philosophers of physics will no doubt find this type of argument inadequate, since it turns on an uncritical acceptance of a (controversial) theory of reference that may not hold in the more quantitative, mathematical domain of physics and, hence, physical geometry (as opposed to common language). While the role of mathematics will be taken up in more detail below, an interesting question does arise in the context of this hypothetical response to a "spatial incommensurability": Have the place theorists employed any philosophical arguments utilizing mathematical/physical evidence or premises in support of radical spatial subjectivism?

In short, the answer is apparently negative, at least among the contemporary advocates of the lived-space school. Merleau-Ponty, who foreshadows the place theorists, did invoke a Poincaré-style metric conventionalist argument to undermine the "reality" of physical geometry, if not its objectivity. Metric conventionalist arguments attempt to reveal the underdetermination that plagues the ascription of spatial geometry, in conjunction with the physical hypotheses, for any would-be geometer: e.g., Poincaré's disc-world (1905), where the measurements conducted by the hypothetical inhabitants disclose a non-Euclidean metric structure. On Poincaré's disc, two theories are consistent with the evidence: (a) that the geometry is Euclidean but "universal forces" distort the measuring apparatus, or (b), the geometry is non-Euclidean and there are no such universal forces. Does this outcome support a radical spatial subjectivism, as some place theorists might contend?–No, because not all aspects of the choice between (a) and (b) are conventional. If one chooses to preserve a flat space, then one *must* postulate forces that distort the measuring instruments. Alternatively, if one accepts the non-Euclidean measurements, then one *must* conclude that the space is curved. Given a strong form of spatial subjectivism, however, any geometry, and any stipulation on spatial measuring instruments, should apply equally: for instance, option (c), where the geometry is Euclidean but no universal forces alter the measuring device.[2] Consequently, option (c), which is available to the radical spatial subjectivist, but not the metric conventionalist, thereby demonstrates that the latter cannot serve as an argument to support radical spatial subjectivism.

[2] That is, since radical spatial subjectivism is based on a strong form of relativism (either to personal experience, society, practice, etc.), it follows that the physical assumptions implicit in the construction and application of the measuring apparatus are also relative–if the spatial subjectivist were to deny this, and claim instead that the measurement devices are somehow non-relative, it would open up the spatial subjectivist to the charge of inconsistency. Moreover, as argued by, among others, Einstein (1949), since our understanding of the physical constitution of these measuring devices also relies on geometrical assumptions (via the force laws that govern their behavior), the spatial subjectivist can always claim that these lower-level applications of geometry are likewise subjective (relative), thus opening the way for option (c).

Turning to Merleau-Ponty's analysis, he comments: "'Real', i.e., perceived, triangles, do not necessarily have, for all eternity, angles the sum of which equals two right angles, if it is true that the space in which we live is no less amenable to non-Euclidean than to Euclidean geometry" (1962, 391); and, in more detail:

> It is impossible to relate this or that proposition concerning space to the structure of space, and some other [proposition] to a physical influence. . . . The same physico-geometrical ensemble is capable of covering both flat space and curved space. . . . If we take relativistic science seriously, we must say that Riemannian space is not real, but objective to the extent that it allows for Einstein: it allows for better integrating the results of modern physics than does Euclidean space. We can thus speak of a closed space, such that in pursuing it we return to the same place. The experimental verification is relative to it. If space is closed, it is clear that there can be a double image of the same star, the whole difficulty being only to identify them. . . . In this sense, the idea of closed space must not be considered . . . as an overcoming of Kantian relativism, but on the contrary, as its accomplishment . . . . (2003, 103)

Briefly, Merleau-Ponty's estimate is misleading in that he seems to imply that any geometry is straightforwardly consistent ("no less amenable", 1962, 391) with the totality of physical evidence. Rather, as described above, specific physical assumptions are required in order to render a particular geometry consistent with empirical data, and these assumptions can be challenged in numerous ways: e.g., the peculiar universal forces needed to retain Euclidean geometry may be inconsistent with our best, well-confirmed physical theories. So, when all of the evidence is taken into consideration, many alternative theories of "geometry plus physics" may be excluded, such that only a handful, or just one, may be supported by the evidence.

Merleau-Ponty's claim that, "the same physico-geometrical ensemble is capable of covering both flat space and curved space", thus can only be maintained if one takes a rather impoverished view of the criteria for constructing and evaluating the "physico-geometrical ensemble". His own example of a Riemannian (spherical) space, where the experimental verification could be "a double image of the same star", practically makes this point: Is Merleau-Ponty suggesting that a flat space interpretation of the same evidence would be as equally plausible and consistent as the spherical depiction (since "the experimental verification is relative to it")? One can easily imagine evidence that would begin to unravel this assumption, such as the simultaneous super-nova explosion of both stars (i.e., the same star), or simply measuring the angles of a (very large) triangle, à la Gauss, in order to determine if the interior angles match the Euclidean prediction. So, unless wildly ad hoc and implausible physical hypotheses are invoked, the evidence hardly seems "relative" to the particular geometry used in the theory.

Ironically, Merleau-Ponty relies on a similar tactic–namely, the constraints imposed by the world/evidence–to dispel his own version of our problem (1), which he describes as a potential solipsism that may ensue from a subjective-based conception of space: "Since there are as many spaces as there are distinct spatial experiences, . . . are we not imprisoning each type of subjectivity, and ultimately each consciousness, in its own private life" (1962, 291–292)? Answer: invoke physical space as an external constraint, which, presumably, prevents spatial solipsism by connecting all of our separate spatial experiences to the same spatial world. He states that, "I never wholly live in varieties of human space, but am always ultimately rooted in a natural and non-human space" (293); and that "Human spaces present themselves as built on the basis of natural space, . . ." (294). Merleau-Ponty is quick to add, however, that "natural and primordial space is not geometrical space" (294), which accords with his other claims, cited earlier, that the objectivity of Riemannian space only "allows for better integrating the results of modern physics than does Euclidean space", and thereby does not overcome a "Kantian relativism". In essence, Merleau-Ponty's own phenomenological theory appeals to physical space in order to counter problem (1), radical spatial subjectivism–but this maneuver is no different than the scientist who appeals to the physical evidence, in conjunction with the consistency of our best physical theories, to counter the metric underdetermination brought about by a host of divergent "geometry plus physics" combinations. Put differently, how can Merleau-Ponty be so sure that subjective spatial experience is somehow constrained by the physical world, but that the determination of metric properties (in conjunction with the best physics) is not? Indeed, if metric conventionalism does hold true, such that the physical component is powerless to help (in the manner advocated by Merleau-Ponty), then falling back on the physical world cannot free a subjective spatial theory of the same underdetermination.

Finally, like Merleau-Ponty, some of the other phenomenological investigations that inspired the contemporary lived-space movement may have employed physical space as a form of external constraint. In Husserl's theory, since the phenomenal realm of the subject presupposes a physical body, a pre-existing "continuum of places" is postulated for the body's occupation (see Husserl 1981, 225). As for Heidegger, the complexities of the relationship between Dasein (roughly, human existence) and spatiality are enormous (see e.g., Vallega 2003, Malpas 2006), but a similar dependence on a pre-given world may be in evidence: "space is . . . 'in' the world in so far as space has been disclosed by that being-in-the-world which is constitutive of Dasein . . ." (1962, 146). Overall, these early phenomenological theories of the human and social construction of space–which are not modern lived-space theories, by the way–run afoul of problem (2); namely, the manner by which the underlying physical ontology interacts with, and thus *constrains,* subjective space constructions.

3 The Place Theory and The Mathematization of Space

Unlike the early phenomenologists, contemporary exponents of the place theory seldom appeal to either the underlying ontology or the objectivity of space to resolve problems (1) and (2), likely due to the perception that it situates the human/social element of space in a decidedly inferior and subordinate status with respect to the more quantitative and mathematical, and thus less qualitative and subjective, aspects of space and science. Ironically, the modern bias against the use of mathematics in attempting to meet the relativist challenge can be traced, at least in part, to these same early twentieth century phenomenologists, most notably, Heidegger and the later Husserl. Unlike recent treatments of place, which either ignore or quickly dismiss mathematics as relevant to the place theory, these early phenomenological tracts openly discussed the relationship between mathematics, especially geometry, and their new conception of a subjective, lived-space (much like Merleau-Ponty above). Husserl, in particular, will comprise a major part of the remainder of our investigation, for the difficulties associated with Husserl's theory of subjective space in his later work are identical to the problems just described for the contemporary practitioners of the lived-space theory, and hence Husserl's more forthright analysis of the interrelationship of objectivity and mathematics will serve as an ideal basis for diagnosing the viability of contemporary place theory. As will be disclosed, one of the more intriguing puzzles that emerges in the early phenomenological works concerns the status of mathematics, especially geometry, in its seemingly unavoidable mediating role between, on the one hand, physical space, and on the other, subjective lived-space.

3.1 Husserl and the Early Phenomenological Influence

Despite the presence of an a priori factor in spatial experience, which allows an immediate grasp of general geometric truths ("essential seeing"), objective space and geometry in Husserl's middle period (e.g., *Ideas* I (1982)) are ultimately constructions based on subjective experience, much like the earlier theories put forward by, among others, Helmholtz, Mach, Wundt, and Lotze.[3] The geometry of our subjective experience is Euclidean, furthermore, whether in a single intuited act of spatial perception/imagination (as just described), or as one goes beyond these single acts to construct the larger space that results from the accumulation of spatial experience through bodily motion (and spatial variations in imagination).[4] In his late period, a more subjectivist tone is supposedly struck in several of Husserl's works that cover space and geometry, foremost being, *The Crisis in European Sciences* (1970), along with its associated appendices ("The Origin of Geometry", in particular). These writings would prove a source of inspiration for the later place school, for they bring to the forefront several concepts central to the contemporary approach

[3] Husserl studied Lotze's theories of space and geometry in his early years; see Mohanty (1995, 51). On empiricist theories of space and geometry, see Torretti (1978).

[4] Husserl (1997) is his first extended treatment of these issues. It is also worth noting that Husserl's student, Oskar Becker, strived to remove the apparent contingency associated with the geometry of Husserl's theory. Influenced by Weyl's work, Becker relies on a group-theoretic argument to prove that our subjective experience of moving freely through space (via the Helmholtz-Lie theorem) singles out Euclidean geometry as the only candidate for objective space. Weyl's own theory also employs a group-theoretic approach, but only preserves a Euclidean structure infinitesimally for each point of the spacetime manifold, while repudiating a Euclidean global structure for physical space (as mandated by the variably curved spacetime of General Relativity; see Mancosu and Ryckman 2005, and Sect. 4). Finally, Becker's own student, Elizabeth Ströker, would develop a theory like Becker's in her (1987) text, a work that is often cited approvingly by contemporary place theorists.

to place: principally, the "life-world", and the "mathematization of nature". The life-world, as defined in the *Crisis,* is "the spatiotemporal world of things as we experience them in our pre- and extra-scientific life" (1970, 138). The emerging mathematization of the world, which takes the form of Euclidean geometry, cannot capture the life-world in its entirety, however, for mathematical idealizations and abstractions can only indirectly apply to the purely qualitative aspects of the life-world (32–37). Nevertheless, Husserl does not question the objectivity of physical geometry, for he repeatedly rejects any historicist, relativist conception that would regard space and geometry as merely contingent constructs of a particular society: "geometry, with all its truths, is valid with unconditioned generality for all men, all times, all peoples, and not merely for all historically factual ones but all conceivable ones" (377). In effect, Husserl grounds the unconditioned validity of all geometric practices on an invariant human feature common to all individuals and societies. This invariant feature, which we will explore further in Sect. 4, would thereby preclude our problem (1), since it acts as a form of internal constraint on the construction of geometric schemes.

Returning to the topic of Husserl's impact on the later place theory, in particular, for the prospects of a mathematical conception of lived-space, it was probably his methodology of "bracketing off" the objective sciences that would prove to be most influential. The process of bracketing, also termed the epochē in the *Crisis,* is designed to isolate the objective sciences in order to ascertain the unique or principle characteristics of the life-world, which "must have their own 'objectivity', even if it is in a manner different from our [objective] sciences . . ." (133). This theme, that the proposed objective principles of the life-world may be "different" than the developed sciences of the day, persists throughout the *Crisis:*

> A certain idealizing accomplishment is what brings about the higher-level meaning-formation and ontic validity of the mathematical and every other objective a priori on the basis of the life-world a priori. . . . What is needed, then, . . . [is] a division among the universal inquiries according to the way in which the "objective" a priori is grounded in the "subjective-relative" a priori of the life-world . . . . (140)

By separating the different "a prioris" of the objective sciences and the life-world, the implication is that mathematics and geometry must be, or should be, confined to the objective a priori (via the epochē) in order to ascertain the true nature of the life-world. Not surprisingly, ensuing generations of place theorists would almost certainly interpret Husserl's late research as advocating a complete and total ban on the use of mathematical techniques in their study of the "subjective-relative" sphere of human spatial practices. In fact, with respect to space, Husserl is quite clear that geometric content is not "internal" to the life-world: "Prescientifically, the world is already a spatiotemporal world: to be sure, in regard to this spatiotemporality there is no question of ideal mathematical points, of 'pure' straight lines or planes, no question at all of mathematically infinitesimal continuity or of the 'exactness' belonging to the sense of the geometrical a priori" (139–140). In other words, the life-world has its own kind of space, a space which is radically different from the space utilized in mathematical physics, i.e., physical geometry.

Finally, Heidegger's *Being and Time* also expounds a subjectivist-based hypothesis of space, yet his skeptical critique of the concept of objectivity arguably influenced the place school in a more profound and radical fashion. Despite a general similarity of content between Heidegger's and Husserl's theories–Heidegger's "Dasein" and the "Mathematical Projection of Nature" functioning somewhat analogously to Husserl's life-world and mathematization of nature–the type of a priori science of the life-world championed in Husserl's later work would seem quite incompatible with Heidegger's finite, historical understanding of human experience. In *Being and Time,* Heidegger refers to "the manifold questionableness of the phenomenon of 'validity', which since Lotze has been fondly passed off as a not further reducible 'basic phenomenon'", and he proceeds to outline various meanings of "validity": "as manner of being of the ideal, as objectivity, and as bindingness [for all people]" (1927, 155–156). Therefore, any attempt to locate an invariant structure underlying all human spatial practices would likely draw the Heideggerian charge of invoking timeless "essences"; or, Husserl's project errs by trying to explicate our social engagements in the world, the "ready-to-hand", by means of the "present-to-hand", which are the theoretical idealizations derived from those practices–but this turns Heidegger's philosophy exactly on its head, for the defining trait of Dasein is its "being-in-the-world" (existence).[5]

3.2 Contemporary Social Trends

There are a number of themes in these major phenomenological tracts that, directly or indirectly, shaped the course of the place theory's approach to space and mathematics: first and foremost is the primacy of subjective lived-space, which thus serves as the basis for deriving objective geometric space; second, that subjective space is essentially qualitative, and *not* quantitative, geometrical or mathematical; and third, as a direct result of the rise of mathematical physics in the Early Modern period, that objective geometrical space is Euclidean, infinite, and homogeneous.

To demonstrate the mathematical aversion that is prevalent among many place theorists, one need only consult Casey's influential history, *The Fate of Place* (1997), which is representative of much contemporary work on the topic of lived-space.

> "The ultimate reason for the apotheosis of space as sheerly extensional is that by the end of the seventeenth century place has been disempowered, deprived of its own dynamism. . . . The triumph of space over place is the triumph of space in its endless extensiveness, its coordinated and dimensional spread-outness, over the intensive magnitude and qualitative multiplicity of concrete places. . . . Space on the modernist conception ends by failing to locate things or events in any sense other than that of pinpointing positions on a planiform geometric or cartographic grid. Place, on the other hand, situates, and it does so richly and diversely. It locates things in regions whose most complete expression is neither geometric nor cartographic" (200–201).

---

[5] In addition, see Friedman (2000, 13–23), for the development of Heidegger's quite hostile attitude towards modern mathematical logic and physics.

Presumably, the motivation for this line of thought is derived from many sources, but Husserl's later work may have played a major role: prior to quoting from the *Crisis* (where Husserl declares that in the life-world "we find nothing of geometrical idealities, no geometrical space or mathematical time with all their shapes"; 1970, 50), Casey explains that "the organic body singled out by Husserl opens onto the 'primary world' that is not amenable to direct mathematization" (223). Furthermore, in Casey's chronological survey, Husserl is one of the first philosophers examined who supposedly favors a view, like Casey's, concerning the (alleged) non-mathematical essence of subjective place.

While these extracts help to corroborate the importance of Husserl, other passages make an explicit link with a Heideggerian brand of subjectivism, such that mathematics, logic and language are relative, at least in part, to culture or practice (i.e., place): "Treatments of logic and language", he cautions, "are . . . place-blind, as if speaking and thinking were wholly unaffected by the locality in which they occur" (xii). He also hints at a theme common among many in the lived-space movement, specifically, an attempt to link an objective, mathematical conception of space with various forms of social and political totalitarianism or exploitation: "Is it accidental that the obsession with space as something infinite and ubiquitous coincided with the spread of Christianity, a religion with universalist aspirations" (xii)? In Casey's defense, some exponents of subjective space go much further, as in the case of Henri Lefebvre, who categorizes "abstract" space, which is geometric, with a "phallic" attribute that "symbolizes force, male fertility, masculine violence" (1991, 287).

Leaving aside the gross implausibility of these last few allegations, what is equally troubling in these texts is the woeful treatment of the historical development of the concept of subjective space. Casey's treatise, which claims to be a history of place, discusses neither the rise of the empirical approach to geometry and space that began with Helmholtz and Mach (among many others), nor the *Lebensphilosophie* movement that drew encouragement from these nineteenth century mathematical developments (with the *Lebensphilosophie* school serving as the starting point for Heidegger). For many of the place theorists, there is a (postmodern/continental) tendency to interpret modern, or post-Kantian, philosophy as having began with the later Husserl and Heidegger, hence contributing to an impoverished conception of the significance of mathematics in the evolution of the subjective space idea. That a long "dry spell" came between the German Idealists and the phenomenologists is evident in Casey's book: "Starting with Kant and continuing in Husserl and Whitehead and Merleau-Ponty, place is considered with regard to living organisms and, in particular, the lived human body" (332)–which suggests that nothing of importance for the development of subjective spatial theories occurred between Kant and Husserl!

4 Towards a Mathematical Conception of Subjective Space

As outlined in Sects. 1 and 2, an interpretation of spatial objectivity that, in some fashion, includes a subjective component might possibly provide a means of combating the radical subjectivist dilemma, problem (1), while simultaneously upholding the subjective experience of the individual, culture, or practice. Despite being largely ignored by contemporary place theorists, late nineteenth century mathematicians actually developed many techniques that can be seen as offering just this kind of strategy, and it is discussed in the philosophical writings of Eddington, Weyl, and a host of others. These later philosophical explorations, moreover, were likely spurred by Minkowski's singular achievement in 1908.

4.1 Geometry and the Subjective

The lived-space theorists are fond of characterizing geometrical space as the flat, lifeless plain of Euclidean geometry (as the above quotes by Casey indicate), but the history of geometric theories and constructions undermines this simplistic assumption. For our purposes, two of the most important innovations concern the analytic method of geometric construction and the investigation of the intrinsic structure of manifolds (differential geometry), which originated in the pioneering work of Gauss and Riemann, in particular. A Euclidean understanding of geometrical objects, such as "point" or "line", was no longer necessary given the new analytic methods, since algebraic equations are essentially neutral and uninterpreted as regards their geometric meaning. The analytic approach allowed, in turn, the creation of differential geometry, which could furnish a characterization of surfaces in terms of their intrinsic, as opposed to extrinsic, curvature (where "intrinsic" curvature is determined from a perspective confined entirely to that surface, and "extrinsic" from outside). In short, curvature could now be characterized intrinsically for each point on a surface without requiring a larger, Euclidean space in which to embed the surface. The intrinsic geometry of a surface can be regarded, roughly, as its geometry as determined by geometers confined to that surface using (idealized mathematical) measuring procedures (e.g., comparing vectors between neighboring points, etc.). Consequently, by conceiving geometric structures from a local, surface-bound basis, there is a tacit affinity between the intrinsic methodology of differential geometry and the theory of lived-space.

A second point of comparison between the place theory and the geometric techniques invented in the nineteenth century can be found in the latter's utilization of coordinate frames, the transformations rules that link these frames, and the invariants preserved among these translations; a branch of differential geometry known as tensor analysis (and which is intimately connected with the intrinsic geometric method just described). The kinship with the place theory's idea of subjective space is immediately apparent in tensor analysis, since this branch of mathematics can be roughly characterized as the study of "what remains the same" (invariant) under different spatial perspectives (frames). On the Euclidean plane, distance is an invariant feature, such that the distance between any two given points is measured to be the same regardless from which position, or coordinate point, one measures it: if $u = (x_1, y_1)$ and $v = (x_2, y_2)$, then the distance between these points is $d(u, v) = \sqrt{(x_2 - x_1) + (y_2 - y_1)}$, which will be an identical numerical value from all perspectives. The transformations on the plane (the space $\Re^2$ of ordered pairs of real numbers) that leave distance invariant includes all rotations, U, and translations, a, such that: $t(x) = U(x) + a$, for the vector $x = (x, y)$. This distance (metric) function can be generalized to incorporate different coordinate systems and different geometries (Euclidean, Spherical non-Euclidean, etc.) as given in the well-known formula for the line element, $ds^2 = g_{ij}dx^i dx^j$ (for Riemannian and semi-Riemannian spaces). Overall, the group of allowable transformations on a space specify the type of geometry–consequently, the *same* space (say, $\Re^2$) can allow different groups of transformations, and thus different invariants, and thus different geometries. That is, some perspectives in $\Re^2$ will reveal an invariant quantity that other frames will not uphold. Only a limited number of transformations among frames will preserve the invariants of Euclidean geometry (length and angle), for example, whereas a wider class of transformations will preserve the ratios along parallel lines (affine geometry).[6] Unlike the monotonous, uniform geometry caricatured by the place theorists, the picture that differential geometry presents is quite complex and varied, with a host of different geometrical structures and invariants all residing in the very same space. More importantly, differential geometry constructs these invariants from the subjective perspective of diverse coordinate positions or frames (and the transformations among frames), thereby revealing an indispensable, or *non-reducible,* contributing role for a subjective (i.e., perspectival and non-global) component of space and geometry in securing the objective invariants.

Finally, this methodology resolves both problems (1) and (2) in a more consistent and plausible manner than the (non-mathematical) lived-space approach can supply. The relativism of subjective space, problem (1), is resolved since many subjective perspectives (frames) in a space are *not* incorporated within any particular group of transformations: that is, the long sought after "constraint" on possible spatial constructions is a direct consequence of the type of transformation group and its corresponding invariant, which thus accounts for the absence of incommensurable geometries (i.e., it explains why there is only a determinate number and order of interrelated, non-incommensurable geometries). Likewise, the fixed interrelationship between the invariants of the geometry and the group of transformations also resolves our problem (2). The geometrical invariants are often regarded

[6] More carefully, The Euclidean transformations are a subgroup of the affine transformations, hence Euclidean geometry is a subgeometry of affine geometry (and both, in turn, are subgroups and subgeometries of the larger projective transformations and projective geometry). Also, $\Re^2$ is the vector space of ordered pairs of real numbers allowing addition and scalar multiplication. Parts of this discussion are based on Brannan et al. (1999). See e.g., Nozick (2001) and Debs and Redhead (2007), for similar approaches to objectivity and invariance.

as representing the *objective* features of the underlying spatial ontology, although these features can only be accessed through the subjective-bound group of transformations. In short, the groups of transformations among frames secure the needed constraints, and constraints indicate, or correspond to, the world's "real" structure.

In the realm of spacetime theories, Minkowski (1964 [1908]) offered one of the first applications of these differential geometric techniques, providing a formulation of Special Relativity that emphasized the invariance of the spacetime interval, $c^2dt^2 - dx^2 - dy^2 - dz^2$, under a group of transformations (often dubbed, Lorentz transformations) required to preserve the laws of physics (i.e., the constancy of the speed of light, c, and the independence of the laws of classical physics over the choice of inertial system): "the existence of the invariance of natural laws for the relevant group *GC*" (301). Put roughly, Minkowski was able encapsulate or encode a specific domain of our experience of the material world, both actual and potential, via a geometric method that relates the content of this experience (that is, it "saves the phenomena" from multiple perspectives). His comments on the "pre-established harmony between mathematics and physics" (312) can therefore be seen, in retrospect, as a milestone in the application of these geometrical techniques to model actual phenomena. More importantly, Minkowski makes an explicit link between the spacetime group invariant and "independent reality", i.e., objectivity, in the famous opening of his 1908 paper: "only a kind of union of the two [space and time] will preserve an independent reality" (297). As a consequence, Minkowski's achievement, which is seldom examined from this perspective, can be viewed as a foreshadowing the goals of the lived-space theory, since a group of transformations naturally incorporates the objective and the subjective (as outlined above).

It is important to note that Minkowski himself never actually discussed the ramifications of his spacetime group of transformations as regards the metaphysics of the objectivity/subjectivity divide, but a host of others philosophers and physicists, inspired by his work, soon would, most notably, the brand of neo-Kantian inspired structuralism exhibited in the work of Weyl, Eddington, and Cassirer (albeit Weyl was more directly inspired by Husserlian phenomenology; see Ryckman 2005, and Sect. 5). With Minkowski's achievement as a guide, and given the tensor calculus framework of General Relativity as well, it naturally led to a new appraisal of the subjective/objective relationship, as the following comments by Weyl indicate: "[The] *objective* world is of necessity *relative* [subjective]; it can be represented by definite things (numbers or other symbols) only after a system of coordinates has been arbitrarily carried over into the world" (1949, 116). This objective-subjective interrelationship, moreover, "contains one of the most fundamental epistemological insights which can be gleaned from science", since "whoever desires the absolute must take the subjectivity and egocentricity into the bargain" (116). In a similar vein, Eddington emphasizes "the subjectivity of the universe described in physical science" (1958, 85). It is worth quoting Eddington's argument at length:

> Relativity theory allows us to remove (if we wish) the subjective effects of ... *personal* characteristics of the observer; but it does not remove the subjective effects of *generic* characteristics common to all "good" observers [the allowable transformations] .... [The mathematician] has invented a transformation process which enables us to pass very quickly

> from one [possible] observer's account to another's. The knowledge is expressed in terms of tensors which have a fixed system of interlocking assigned to them; so that when one tensor is altered all the other tensors are altered, each in a determinate way. . . . A tensor may be said to symbolize absolute knowledge; but that is because it stands for the subjective knowledge of all possible subjects at once. (85–87)

Eddington, like the lived-space theorists, also cautions against envisaging the universe from a subject-less, "view from nowhere": "There does not seem to be much difficulty in conceiving the universe as a three-dimensional structure viewed from no particular position", but he notes that "it is perhaps rather unfortunate that it is, or seems to be, so easy to conceive; because the conception is liable to be mischievous from the observational point of view" (86)–that is, it is mischievous as judged from the new subjective-based physics that utilizes tensor calculus.

4.2 *The Problem of Quantifying the Qualitative*

The lived-space theorists may nevertheless reject any application of mathematics, such as the one outlined above, on the grounds that the essentially qualitative nature of subjective space–the "intensive magnitude and qualitative multiplicity of concrete places", quoting Casey–is just not amenable to mathematical analysis. Eddington's conclusion draws upon a distinction between the "personal versus the generic" understanding of subjectivity, so it might be claimed that subjective space is really the personal (which Eddington denies as relevant for the new physics). Unfortunately, interpreting a practice-based (or praxis) theory of space as akin to personal experience is quite problematic, for the manner by which space, as a social practice, acquires this individual, "personal" trait is left unexplained–and, it may contradict the very idea of a *social practice,* which must rise above the experience of individual practitioners in some fashion. Put another way, the difficulty with associating these non-quantitative, non-mathematical "intensive magnitudes" with subjective spatial *experience,* either at the personal or social level, is that it fails to provide a rationale for criticizing the mathematization of *physical* space–unless, of course, they hold that physical space actually possesses irreducibly qualitative properties, like color or pain, which is patently absurd.

Moreover, the foundational role that the purely qualitative magnitudes play for many lived-space theorists inevitably yields anxieties over an impending radical subjectivism. Husserl's *Crisis* provides a clear example of this dilemma, for he denies that the life-world a priori is geometrical (see Sect. 3.1) while simultaneously rejecting an historicist, relativist interpretation of life-world schemes. He contends that the relativism worry disappears "as soon as we consider that the life-world does have, in all its relative features, a *general structure*", such that "this general structure, to which everything that exists relatively is bound, is not itself relative" (139). Finally, "as life-world the world has, even prior to science, the 'same' structure that the objective sciences presuppose" and "are the same structures that they presuppose as a priori structures and systematically unfold in a priori sciences"(139). In other words, there is an invariant structure that underlies both the life-world a priori and the a priori of the objective sciences.

Deprived of mathematics, however, it is not exactly clear what Husserl has in mind in declaring that the life-world and the objective sciences share the "same" structure. Given the epochē, mathematics has been bracketed away from the life-world a priori, so the similarity of structure cannot be mathematical/geometrical structure–yet, what "structure" remains? It is possible that Husserl has in mind a basic similarity between the scientific a priori's mathematical structures, on the one hand, and the relational structure of the mental content associated with kinesthetic awareness (of the life-world a priori), on the other; where "kinesthetic" refers to the experience of one's body in moving or resting, "each being an 'I move', 'I do' [etc.]" which "are bound together in a comprehensive unity" (106). Nevertheless, a relational similarity of this sort would seem to warrant an analysis employing some form of deeper mathematical structure (such as set theory, topology, or category theory?), since *one of the relata* is, in fact, the mathematics of the natural sciences. But, any non-life-world idealization, like set theory, is apparently ruled out by the epochē; that is, these more abstract structures are also idealizations ultimately derived from the life-world (43–48), so there can be no more basic structure that *underlies both* social practices and mathematics.[7]

Another tactic might be to simply assert that this similar structure is a metaphysical primitive or unanalyzable notion, a maneuver that Heidegger may have exploited in his later works.[8] Alas, recourse to metaphysical expedients of this type would seem incompatible with Husserl's claims for the scientific status of the life-world a priori, nor does a primitive metaphysical concept really explain how the impending relativism has been averted. Moreover, a purely metaphysical means of overcoming the radical subjectivism quandary leaves the relationship between the subjective aspect of space and the underlying ontology a mystery, our problem (2). Husserl's *Crisis,* a notable precursor to the modern place theories, thus demonstrates the inherent vulnerability of any position that seeks objective scientific status for a

[7] It is not being claimed, here, that psychological/social factors, such as a language or conceptual scheme, cannot provide a structural foundation from which mathematics emerges (see e.g., Lakoff and Nunez 2001). Yet, these attempts to derive mathematics from human practices then run into another version of problem (1), since it would appear that different human practices might then generate conflicting mathematical schemes. Husserl's long antipathy to "psychologism" was based on this very concern.

[8] In Joseph Kockelmans' excellent survey, the relativism problem for Heidegger is discussed with respect to the "aboriginal Event", the Ereignes, which is "ontologically prior to Being as well as to time, because it is that which grants to both what they properly are" (1992, 162–163; from Zur Sache Des Denkens 1988). Given the Ereignes, "one understands, or perhaps more accurately stated, experiences that the various epochs [different manifestations of Being's history] are no longer mysteries, but are the necessary consequence of the inherent finitude of an aboriginal Event which presents the Open [the bestowing of past, present, future] and grants Being" (167). Yet, it difficult to understand how Heidegger's appeal to this sort of quasi-mystical insight can constitute a serious resolution of the relativism problem. In fact, seeking divine revelation from Ereignis in this manner would seem to be just another way of introducing the "God of the philosophers", which is a strategy that he thoroughly repudiates.

subjective/practice oriented conception of space while simultaneously bracketing away mathematical methods.

Part of what may be driving Husserl's "bracketing away" of mathematics from the life-world a priori can be labeled the "circularity argument": since mathematics is a by-product of human practices, thus mathematics cannot be used to explain its own origin in human practices. Yet, this argument is fallacious, since the manner by which mathematics came about does not provide any information on the domain of mathematical application. Indeed, the fact that human practices can be given a fairly sophisticated mathematical description serves as direct counter-example to the circularity argument. Specifically, there is a long tradition of attempts to integrate modern mathematical techniques with the psychological and social aspects of space,[9] such as the child psychologist Jean Piaget, who used many of the geometrical structures explained above: e.g., topological, projective, and Euclidean (Piaget 1967). Later researchers have extended these geometrically-informed hypotheses to the larger social realm as well (see Sack 1980, Hillier and Hanson 1984, to name a few).[10] For example, the spatial constructions of different cultures could be tied to different geometrical invariants. Since these invariants, and their associated geometries, are nested within one another in a natural and determinate way, the diversity of geometrical practices does not entail, therefore, problem (1). Moreover, since the geometric invariants are normally construed as providing a link to, or a representation of, the underlying physical ontology– via the invariant relationships among frames and their associated constraints on possible frames–problem (2) is also resolved.[11] In conclusion, the utilization of mathematical techniques and concepts to capture the spatial (and temporal) aspects of experience might lead to

[9] A notable declaration of the need for mathematical investigations in the social studies of space is the following: "It is clear that environmental 'objects' and human 'subjects' are deeply entangled with each other . . . . Nor is it the case that the object side of the urban system can be dealt with mathematically and the subject side only qualitatively. The fact that the city is shaped by the human cognitive subject does not lessen its mathematical content . . . . [T]he cognitive processes by which the subject intervenes reflect mathematical laws. . . . The project for space syntax research must now be to engage with the problematics of both the mathematical and humanistic paradigms in the hope and expectation that by finding how each is present in the other we will progress towards synthesis" (Hillier 2003, 19).

[10] In one of Cassirer's last works, he also advocates the expansion of these geometrical concepts to other disciplines. Cassirer notes that psychologists are "not especially interested in mathematical speculation", and that mathematicians do "not care about psychological problems"; yet, he insists that this "separation is of questionable value" (1979, 285). He continues: "Of course we cannot mix up the two fields of investigation; we must make a sharp distinction between the mathematical and psychological problem of space. But that ought not to prevent us from looking for a connecting link between the two problems; and I think that the concept of a group [of transformations] may be regarded as such a connecting link" (285).

[11] A recent interpretation of Husserl's philosophy of space/geometry, which also utilizes transformation groups, is Tieszen (2005, Chap. 3), although the problematic issues pertaining to the life-world (as argued above) would seem to pose an obstacle to his reconstruction of Husserl's later philosophy. See also, Carr (1977), on the complexities of Husserl's life-world.

new breakthroughs, as Minkowski successfully demonstrated long before the later Husserl, Heidegger, or the modern lived-space movement.

5 A Concluding Case Study: Deleuze on Differential Geometry

The case presented thus far can be briefly summarized: by unfairly purging mathematical/geometrical concepts, the practice-oriented philosophers of place have unwittingly deprived their theory of a useful means of answering both the relativism problem (1), as well as the problematic relationship between the subjective experience of space and the underlying physical ontology, our problem (2).

Not all place theorists have ignored mathematics, however. Therefore, by way of conclusion, we will examine what is probably the most famous (or infamous) instance of the application of mathematics within the theory of lived-space. In *A Thousand Plateaus,* Giles Deleuze and Felix Guattari invoke a plethora of modern geometrical concepts in their exposition of "smooth" space and "striated" space, which, more or less, corresponds to the qualitative and quantitative aspects of space, respectively.[12] On the whole, Deleuze and Guattari make some interesting claims that are relevant to the potential utilization of mathematics within the place theory. Smooth space, in various passages, is described as topological, non-metrical, local, and "is therefore a vector, a direction and not a dimension or metric determination" (1987, 478); whereas striated space is characterized as metrical. Deleuze and Guattari cite Riemann's theory of manifolds as a model for viewing smooth space; and, although their analysis is rather incongruous mathematically, one of the goals seems to be a sketch of the relationship between the subjective, qualitative space of the individual (smooth space) and a formal quantitative method of objectifying that space (striated space, which they link with, not surprisingly, Euclidean space; 371). Their main contention is that smooth space, as the environs of the individual, need not be necessarily conceived as merely a part of a larger metrical, striated space. They introduce Riemann's theory to demonstrate that a point of the manifold (smooth space) can be connected to an adjacent point in a number of ways, such that a metrical connection need not be assumed, a process they call "accumulation" (485): put in the modern mathematical parlance, while the infinitesimal neighborhood (or tangent space) of each manifold point is Euclidean, a vector in this tangent space can be compared with another vector in a separate tangent space in such a way that their (locally defined) Euclidean properties are lost in the transfer. Consequently, Deleuze and Guattari have relied upon the intrinsic approach to geometry

[12] On Deleuze and Guattari on chaos theory, see Sokal and Bricmont (1998). While controversial in their own right, these types of critiques of postmodern thought do shed light, at least tangentially, on an apparent trend among some contemporary theorists of lived-space; namely, the appropriation of mathematical and scientific terms or ideas, such that they are no longer used in their strictly technical sense, but rather are exploited to present an array of different meanings or notions (many possibly literary in origin). Furthermore, this essay cannot explore all of the discussions of geometry in Deleuze's work, but merely examines a notable instance.

and its concept of a manifold (as outlined in Sect. 4) to correlate separate subjective spaces; albeit without really laying to rest the relativism issue addressed in this essay, since the many possible connections among infinitesimal neighborhoods raises problem (1) in a new guise.[13]

Finally, Deleuze and Guattari attribute to Riemann the (quite implausible) legacy of "the beginning of a typology and topology of multiplicities", i.e., an alternative method of conceiving quantities, broadly construed, which would ultimately come to fruition in Henri Bergson's qualitative concept of "duration" ("as a type of multiplicity opposed to metric multiplicity or the multiplicity of magnitude", 483). On these grounds, they conclude that "we consider Bergson to be of major importance (much more so than Husserl, or even Meinong or Russell) in the development of the theory of multiplicities" (483). Yet, in an ironic twist, the results in differential geometry that Deleuze and Guattari refer to, on the multiple connections among points on a manifold, are largely the product of Hermann Weyl–and Weyl's motivation in these mathematical investigations was, in part, to adapt Husserl's phenomenological work on subjective space perception to the new conception of physical space that followed in the wake of the General Theory of Relativity. Each point of the manifold, for Weyl, is linked to the infinitesimal Euclidean space of a hypothetical observer, such that it guarantees the kinesthetic experience of the free mobility (in three-dimensions) of objects in that infinitesimal neighborhood (via the Helmholtz-Lie theorem). Each point of the manifold is, accordingly, a separate Husserlian (subjective) space, as well as the remnant of Kant's synthetic a priori "form of intuition" of space. Yet, the mutual orientation of the Euclidean metrics located at separate points may differ, and thus the overall space (manifold) may not be Euclidean, a fact that can only be determined by experience (see Ryckman 2005, Chaps. 5 and 6, and endnote 4). Deleuze and Guattari may have been unaware of this bit of mathematical history, or of the direct relevance of Husserl's philosophy for the story. Yet, given their very peculiar efforts to proclaim Bergson's importance for the evolution of their quasi-mathematical concept of "multiplicity" (see 482–483), a more sinister take on this entire discussion is that Deleuze and Guattari favor Bergson because he is a more faithful exponent of the *Lebensphilosophie* movement, as

[13] That is, there is an underdetermination of the exact method of connecting the separate smooth (tangent) spaces. Another problem is that a non-metrical connection, such as an affine connection that preserves linearity but not length, does not guarantee a metrical connection, although the latter does contain the former. Deleuze and Guattari might think, erroneously, that the two concepts are necessarily and sufficiently conjoined, since they claim that "the two [non-metrical "accumulation" and "Euclidean conjunction"] are linked and give each other impetus" (486). But, an affine connection does not require a metrical connection at all. In trying to capture the alleged interdependence of smooth and striated spaces sought by Deleuze and Guattari, a better case from differential geometry might be found in the distinction between tangent vectors (or contravariant vectors) and 1-forms (or covariant vectors), which are inter-defined and equally necessary for the mathematical presentation (see e.g., Burke 1980, Chap. 2): the 1-forms, which provide the gradient for smooth functions, are often given the pictorial representation of a contour map, and thus its role in "numbering" the tangent vectors would nicely fit the category of striated space, while the tangent vectors obviously play the role of smooth space.

opposed to the more objectivist, scientifically-oriented project of Husserl (with its sharp repudiation of any relativism or historicism). If this interpretation is correct, it would reveal a deeper tension within the lived-space approach as a whole, namely, the continuing battle between its objectivist, scientifically inclined and subjectivist, non-scientifically inclined contingents–and the outcome of this contest will largely determine the role of mathematics for the practitioners of the theory of place. Interestingly, in a recent collection of articles entitled, *Deleuze and Space* (Buchanan and Lambert 2005), none of the mathematical themes in *A Thousand Plateaus* are taken up, which would indicate that the prospects for an integration of mathematical methods within the place theory are, at least for the foreseeable future, not very promising.

References

Buchanan, I. and Lambert, G., eds. (2005). *Deleuze and Space.* Edinburgh: Edinburgh University Press.

Burke, W. L. (1980). *Spacetime, Geometry, Cosmology.* Mill Valley, Cal.: University Science Books.

Brannan, D., Esplen, M., Gray, J. (1999). *Geometry.* Cambridge: Cambridge University Press.

Campbell, J. (1994). *Past, Space and Self.* Cambridge, Mass.: MIT.

Carr, D. (1977). "Husserl's Problematic Concept of the Life-World", in *Husserl: Expositions and Appraisals,* edited by F. Elliston and P. McCormick, Notre Dame, IN: University of Notre Dame Press, 202–212.

Casey, E. (1997). *The Fate of Place.* Berkeley: University of California Press.

Cassirer, E. (1979, [1945]). "Reflections on the Concept of Group and the Theory of Perception", in *Symbol, Myth, and Culture,* ed. by D. P. Verene. New Haven: Yale University Press.

Debs, T. A. and Redhead, M. (2007). *Objectivity, Invariance, and Convention: Symmetry in Physical Science.* Cambridge, MA.: Harvard University Press.

Deleuze, G. and Guattari, F. (1987). *A Thousand Plateaus,* trans. by B. Massumi. Minneapolis: University of Minnesota Press.

Eddington, A. (1958, [1939]). *The Philosophy of Physical Science.* Ann Arbor: University of Michigan Press.

Einstein, A. (1949). "Remarks on the Essays". In P. Schilpp (Ed.), *Albert Einstein, Philosopher-Scientist* (pp. 663–688). La Salle, Ill.: Open Court.

Entrinkin, J. N. (1991). *The Betweeness of Place.* Baltimore: Johns Hopkins University Press.

Friedman, M. (2000). *A Parting of the Ways: Carnap, Cassirer, and Heidegger.* Chicago: Open Court.

Heidegger, M. (1962, [1927]). *Being and Time,* trans. by J. Macquarrie and E. Robinson. New York: Harper and Row.

Heidegger, M. (1988). *Zur Sache des Denkens.* Tbingen: Verlag.

Hillier, B., and Hanson, J. (1984). *The Social Logic of Space.* Cambridge: Cambridge University Press.

Hillier, B. (2003). "The Knowledge that Shapes the City: The Human City beneath the Social City", Proceedings of the 4th International Space Syntax Symposium, London, 1–20.

Husserl, E. (1970, [1937]). *The Crisis of European Sciences and Transcendental Phenomenology,* trans. by D. Carr. Evanston: Northwestern University Press.

Husserl, E. (1981, [1934]). "Foundational Investigations of the Phenomenological Origin of the Spatiality of Nature", trans. by F. Kersten, in *Husserl: Shorter Writings,* ed. by P. McCormick and F. Elliston (South Bend: University of Notre Dame Press).

Husserl, E. (1982, [1913]). *Ideas Pertaining to a Pure Phenomenology and to a Phenomenological Philosophy, First Book,* trans. by F. Kersten. Dordrecht: Kluwer.
Husserl, E. (1997, [1907]). *Thing and Space,* trans. by R. Rojcewitz. Dordrecht: Kluwer.
Kockelmans, J. (1992). "Heidegger on Time and Being", in *Martin Heidegger: Critical Assessments,* ed. by C. Macann. London: Routledge.
Lakoff, G. and Nunez, R. (2001). *Where Mathematics Comes From: How the Embodied Mind Brings Mathematics into Being.* New York: Basics Books.
Lefebvre, H. (1991). *The Production of Space,* trans. by D. Nicholson-Smith. Oxford: Blackwell.
Mancosu, P. and Ryckman, T. A. (2005). "Geometry, Physics, and Phenomenology: Four Letters of O. Becker to H. Weyl", in *Oskar Becker und die Philosophie der Mathematik,* ed. by V. Peckhaus, Munich: Verlag, 153–227.
Malpas, J. E. (1999). *Place and Experience.* Cambridge: Cambridge University Press.
Malpas, J. E. (2006), *Heidegger's Topology: Being, Place, World.* Cambridge, MA: MIT.
Merleau-Ponty, M. (1962, [1945]). *Phenomenology of Perception,* trans. by C. Smith. London: Routledge.
Merleau-Ponty, M. (2003). *Nature,* edited by D. Sglard, translated by R. Vallier, Evanston, IL: Northwestern University Press.
Minkowski, H. (1964, [1908]). "Space and Time", trans. by W. Perrett and G. B Jeffery, in *Problems of Space and Time,* ed. by J. J. C. Smart. New York: Macmillan, 297–312.
Mohanty, J. N. (1995). "The Development of Husserl's Thought", in *The Cambridge Companion to Husserl,* ed. by B. Smith and D. W. Smith. Cambridge: Cambridge University Press.
Nozick, R. (2001). *Invariances: The Structure of the Objective World.* New York: Belknap
Piaget, J. (1967). *The Child's Conception of Space.* New York: W. W. Norton.
Rescher, N. (1997). *Objectivity: The Obligations of Impersonal Reason.* Notre Dame, In.: University of Notre Dame Press.
Ryckman, T. (2005). *The Reign of Relativity.* Oxford: Oxford University Press.
Sack, D. (1980). *Conceptions of Space in Social Thought.* Minneapolis: University of Minnesota Press.
Sokal, A., and Bricmont, J. (1998). *Fashionable Nonsense: Postmodern Intellectuals' Abuse of Science.* New York: Picador.
Ströker, E. (1987). *Investigations in Philosophy of Space,* trans. by A. Mickunas. Athens, OH: Ohio University Press.
Tieszen, R. (2005). *Phenomenology, Logic, and the Philosophy of Mathematics.* Cambridge: Cambridge University Press.
Torretti, R. (1978). *Philosophy of Geometry from Riemann to Poincaré.* Dordrecht: D. Reidel.
Tuan, Yi-Fu (1977). *Space and Place.* Minneapolis: University of Minnesota Press.
Vallega, A. (2003). *Heidegger and the Issue of Space.* University Park: Pennsylvania State University Press.
Weyl, H. (1949, [1927]). *Philosophy of Mathematical and Natural Science.* Princeton: Princeton University Press.

Index

Printed by Printforce, the Netherlands